Ecological Studies

Analysis and Synthesis

Volume 253

Ecological Studies is Springer's premier book series treating all aspects of ecology. These volumes, either authored or edited collections, appear several times each year. They are intended to analyze and synthesize our understanding of natural and managed ecosystems and their constituent organisms and resources at different scales from the biosphere to communities, populations, individual organisms and molecular interactions. Many volumes constitute case studies illustrating and synthesizing ecological principles for an intended audience of scientists, students, environmental managers and policy experts. Recent volumes address biodiversity, global change, landscape ecology, air pollution, ecosystem analysis, microbial ecology, ecophysiology and molecular ecology.

Qing Zhao

Bayesian Analysis of Spatially Structured Population Dynamics

Springer

Qing Zhao
Bird Conservancy of the Rockies
Fort Collins, CO, USA

ISSN 0070-8356 ISSN 2196-971X (electronic)
Ecological Studies
ISBN 978-3-031-64517-4 ISBN 978-3-031-64518-1 (eBook)
https://doi.org/10.1007/978-3-031-64518-1

This Springer imprint is published by the registered company Springer Nature Switzerland AG
The registered company address is: Gewerbestrasse 11, 6330 Cham, Switzerland

To my mom

Foreword[1]

I first met Qing Zhao when he was a post-doc working with the U.S. Fish and Wildlife Service on problems of climate change effects on waterfowl populations at Patuxent Wildlife Research Center in 2012–2017. He visited my office one day having an interest in N-mixture models and their use for modeling abundance in the presence of imperfect detection. This began our collaboration that led to a paper on modeling spatial dynamics using the N-mixture model (Zhao et al. 2017) and subsequent work that was motivated by his fascination for the modeling of spatial dynamics of animal populations and the use of models of species abundance to explain, describe, and predict population state and environment effects. Another persistent theme of Qing's work since his days as a post-doc at Patuxent is the use of hierarchical modeling, a topic that has grown at an astronomical rate in ecology over the last 20 years due to the conceptual clarity it provides in formulating models, and the computational ease it provides in fitting complex models by Markov chain Monte Carlo (MCMC). These themes are exemplified in Qing Zhao's book "Bayesian analysis of spatially structured population dynamics", which synthesizes and extends in new directions his own work on these subjects over the last 10 years.

Qing's curiosity and drive to improve the realism of statistical models has produced a unique and tremendously useful book on hierarchical modeling that fits one of the dominant themes of statistical ecology in recent years: the increasing spatial explicitness of statistical models, and their improving computational accessibility. In this book, Qing develops sophisticated spatially explicit models for the demography of animal populations. The scope of the book is discrete space, discrete time models, which is a reasonable focus of such a book because the resulting models are conceptually easier for non-statisticians to understand, and arguably simpler to fit to real data sets.

One reason for the growth of hierarchical modeling in recent years is the increasing availability of MCMC algorithms and software that simulate from posterior distributions of models given only pseudo-code representations of those models (e.g., the Nimble, JAGS, and Stan software packages; Kéry and Kellner 2024).

[1] Any use of trade, firm, or product names is for descriptive purposes only and does not imply endorsement by the U.S. Government.

These software platforms don't work effectively for all models that are interesting or useful for ecologists. Therefore, it's very valuable to be able to write code for your own MCMC algorithms. To do this, it is immensely helpful to have templates you can work with and expand. This is an approach that has been adopted by others including notable works in statistical ecology such as Clark (2007) and Hooten and Hefley (2019). In his book, Qing provides MCMC algorithms implemented entirely within the R programming language (R Core Team 2020). This will maximize their value to the readers who are not specialists in computer science or programming.

The introductory chapter of the book establishes the context of spatially structured population dynamics, namely, the BIDE concept (Conroy and Carroll 2011), which describes population dynamics in terms of Births, Immigration, Deaths, and Emigration. These are foundational concepts in spatial ecology and play a role in common approaches, like metapopulation models and species distribution models. Importantly, modern views of these models accommodate observation error, usually in the form of non-detection bias. This chapter also introduces the reader to Bayesian analysis and the use of Markov chain Monte Carlo to analyze the models described, as well as the reliance on simulated data to provide a demonstration for each class of models. Simulation and the use of simulated data to understand models and the performance of statistical procedures are still greatly under-appreciated and under-utilized by ecologists. Indeed, data simulation is helpful both to understand the models and to assess the performance of model-fitting algorithms.

Chapter 2 is the first subject matter chapter. Here, Qing considers basic occupancy models, dynamic models and spatially explicit models that allow for adjacency-based movement, distance-based movement, and environment-mediated movement. These models are conceptually and methodologically fairly simple, allowing the reader to understand some of the main structural elements that are reused in different contexts in all later chapters. A novel direction he takes the reader in this chapter is spatially dynamic occupancy models for interacting species.

The topic of Chap. 3 is N-mixture models for abundance, estimated from surveys that produce binomial counts. The main focus is on the model of Dail and Madsen (2011)—conceptually I think one of the most important models to come out of statistical ecology in the last few decades—and related models, and their extension to spatially organized populations in which individuals move from one population to another based on spatial proximity. The realism in which movement rates might depend on environmental conditions and be density dependent is also considered. Finally models of interacting species along the lines of Zhao et al. (2022) are developed. These models go a long way toward connecting observations to ecological theory and hence the models provide a powerful framework for testing such theory that I imagine can be useful in many contexts.

Chapter 4 on integrated population models (IPMs) is a timely chapter given the intense current interest in these types of models in ecology, as a recent book on IPMs (Schaub and Kery 2022) attests. The core structure addressed here is the famous Cormack-Jolly-Seber (CJS) model, which can be used both to model the survival processes of individuals or the movement processes of individuals among sites.

Considering such CJS data when you also have imperfect count data, such as described in the previous chapter, gives rise to a new class of IPMs—spatial integrated population models (Sect. 4.4)—which promise a full accounting for all elements of the BIDE conceptual model.

The final topical chapter is on spatial-capture-recapture (SCR) models. These are models for encounter history data in which the spatial organization of individuals within a population is an explicit component of the hierarchical model, in the form of a model for individual home range centers. The chapter includes dynamic models in which individuals survive and are recruited as well as generalizations to allow for movement of the home ranges across years. SCR models are essentially the individual-based analogs of the models covered in Chaps. 2–4, and they represent an important and fertile ground for much future development in statistical ecology.

In summary, I enjoyed reading this book, thinking about the dynamic models described therein and the types of problems to which they might be applied in practice. The book is valuable to all students of hierarchical modeling and quantitative population ecology and will foster an improved, clear thinking about more complex and realistic models and also serve as a "recipe book" for extending the complexity of models to fit their own problems.

USGS Eastern Ecological Science Center
Laurel, MD, USA
August 1, 2024

J. Andrew Royle

References

Clark JS (2007) Statistical computation for environmental sciences in R: an introduction. Princeton University Press

Dail D, Madsen L (2011) Models for estimating abundance from repeated counts of an open metapopulation. Biometrics 67(2):577–587

Hooten MB, Hefley T (2019) Bringing Bayesian models to life. CRC Press

Kéry M, Kellner KF (2024) Applied statistical modeling for ecologists: a practical guide to Bayesian and likelihood inference using R, JAGS, Nimble, Stan and TMB. Academic Press

R Core Team (2020) R: a language and environment for statistical computing. R Foundation for Statistical Computing, Vienna, Austria. URL https://www.R-project.org/

Schaub M, Kéry M (2022) Integrated population models: theory and ecological applications with R and JAGS. Academic Press

Zhao Q, Royle JA, Boomer GS (2017) Spatially explicit dynamic N-mixture models. Popul Ecol 59:293–300

Zhao Q, Fuller AK, Royle JA (2022) Spatial dynamic N-mixture models with interspecific interactions. Methods Ecol Evol 13(10):2209–2221

Preface

The writing of this book was motivated by my enthusiasm for modeling spatially structured population dynamics. When I tried to recall what inspired this enthusiasm, memories started to come back. The enthusiasm was first of all inspired by the works of Ilkka Hanski and his colleagues in metapopulation ecology. I first learned about the idea of metapopulations in graduate school in the early 2000s. The idea was fascinating, but to be honest, without any background in mathematics or statistics, I couldn't figure out how to study metapopulations at all. As time flies, I earned a master's degree in statistics in 2010. I then joined Patuxent in 2012, and later that year, at Patuxent, I attended a workshop for hierarchical modeling taught by Marc Kéry, Richard Chandler, and Andy Royle. At this workshop I learned about this truly amazing Dail-Madsen model. Ever since then, I have been trying to extend the Dail-Madsen model to some spatial versions (or integrate it with some other models) that might be useful for metapopulation studies. I had more failures than successes for sure, but the work only interested me more and more. At some point, I decided to write a book about what I have learned from my own and others' studies.

I wouldn't be able to implement the Dail-Madsen model, among many other models, without knowledge and coding skill in hierarchical Bayesian modeling. The work introduced in this book was undoubtedly inspired by the development of hierarchical Bayesian modeling framework, which allows us to construct statistical models according to our understanding (or assumption) about how ecological systems function and how data are obtained. The framework also allows people like me, who (still) don't know much about math but can do some coding, to implement quite complicated models in a relatively easy way. In addition to the workshop mentioned above, I have been learning and polishing my coding skill by reading and replicating code from books including but not limited to the Royle and Dorazio 2008 book, the Kéry and Schaub 2012 book (the one with a red butterfly on its cover), and the two-volume Kéry and Royle 2016 (volume 1 with a green dragonfly on its cover) and 2021 (volume 2 with a red dragonfly on its cover) "book of books."

Last but not the least, I have learned to write MCMC algorithm in R first from Mevin Hooten in a class he taught at Colorado State University in 2013 and then from the Hooten and Hefley 2019 (phoenix) book. It has been really fun to look into the algorithms behind the JAGS/Nimble code I have been using forever. I understand that there must be at least a handful of people who want to see some JAGS/Nimble code (or even an R package), which will allow them to implement the models introduced in this book more conveniently (than using home-grown R code). I nonetheless thought it would be more fun to write and more valuable to provide algorithms in R for these models. JAGS/Nimble code for some of the models introduced in this book could be found in the books mentioned above as well as separate papers for the specific models. The algorithms in R, on the other hand, are more difficult to find but will satisfy the curiosity of readers who want to gain a deeper understanding of these models. The code will hopefully also become a valuable source of information for researchers who want to write their own sampler in Nimble (which I believe is the future). I encourage readers to replicate the code carefully, which, according to my own experience, is the best way to learn.

That pretty much summarized why I wrote this book. Enjoy reading and coding!

Fort Collins, CO, USA Qing Zhao

Acknowledgments

I want to thank my employer, Bird Conservancy of the Rockies, for supporting the writing of this book.

I am grateful to (in an alphabetical order) Ben Augustine, Kayla Davis, Paige Howell, Marc Kéry, Christopher Kilner, Quresh Latif, Chris Latimer, Bryan Nuse, David Pavlacky Jr., Andy Royle, Yunyi Shen, Cat Sun, and Chris Wikle for providing valuable comments on drafts of the book.

A special shout-out to Andy Royle for long-term support, education, inspiration, and discussion about modeling ideas and stuff.

Mr. Stephen Chow's movies, in particular *King of Comedy*, have encouraged me a lot during tough times.

Finally, I want to thank my families: my mom, Xiaomin Ding, for raising me and selflessly encouraging me to pursue my dream; my wife, Shuang, for love, understanding, support, patience, cooking the best food in the world to keep me healthy, and taking care of business to allow me to work a lot; and my kids, Luna and David, for cheering me up.

Contents

Background 1

But as I said it, swift there passed me by
On noiseless wing a 'wildered butterfly
Seeking with memories grown dim o'er night
Some resting flower of yesterday's delight
And once I marked his flight go round and round
As where some flower lay withering on the ground
And then he flew as far as eye could see
And then on tremulous wing came back to me
I thought of questions that have no reply
And would have turned to toss the grass to dry
But he turned first, and led my eye to look
At a tall tuft of flowers beside a brook
—Robert Frost

1.1 Introduction

Populations, collections of interacting organisms of the same species, have a variety of interesting structures. For example, populations contain individuals of different ages (newborn, 1-, 2-year-old, and so on) and sexes (male and female), and thus have age-sex structures. Another universal and fascinating characteristic of populations, which will be the focus of this book, is that they have spatial structures, meaning that we can always find more or fewer individuals in some areas than in other areas, while sometimes we find none at all. On top of that, such spatial structures are dynamic, meaning that they change over time (e.g., an area with none of a species in a year may have a few individuals of that species in the next year). The processes underlying spatially structured population dynamics are called BIDE, which stands for birth, immigration, death, and emigration (Bartlett 1960; Pulliam 1988; Thomas and Kunin 1999). While birth and immigration make populations increase, death and

Supplementary Information The online version contains supplementary material available at https://doi.org/10.1007/978-3-031-64518-1_1.

Q. Zhao, *Bayesian Analysis of Spatially Structured Population Dynamics*, Analysis and Synthesis 253, https://doi.org/10.1007/978-3-031-64518-1_1

emigration make them decrease. Furthermore, immigration and emigration are linked, in the sense that immigrants in an area must be emigrants from some other areas. The BIDE process is also called demography, while immigration and emigration combined can be called dispersal movement, or simply movement. Spatially structured population dynamics and underlying BIDE processes are driven by heterogeneity and dynamics of abiotic environment, along with intra- and inter-specific interactions.

That sounds simple, right? However, in fact, understanding spatially structured population dynamics has been very challenging to ecologists, partly due to our inability to obtain "complete" observations of populations at relevant spatial and temporal scales. For example, we might mistakenly think that an area was not used by a skink species (or say the local population size of the skink was 0 for that area) simply because the two little skinks living in that area happened to be hiding under a rock when we were trying to find them during the survey of the area. Or, we could be unsure if a sifaka was dead or alive, and if it was alive, where it was, after we captured it for the last time. Without complete observations about populations, ecologists heavily rely on statistical models to estimate the whereabouts, numbers, birth, death, and movement of organisms from partial observations (i.e., data). Nowadays, these statistical models become more and more complex to satisfy our curiosity about natural populations and their complexity, making these models challenging to build and compute.

Thus, with fascinating ecological phenomena and inevitable statistical challenges in mind, this book is devoted to the introduction of some powerful Bayesian analytical tools for understanding spatially structured population dynamics. In this first chapter, we will start by reviewing, briefly and somewhat incompletely out of necessity, some of the key earlier efforts of modeling spatially structured populations. We then will learn some basics of Bayesian modeling and computation which makes the implementation of complex models feasible and convenient. After that, we will talk about the advantages of simulations in statistical methodology, and describe how to simulate environmental covariates to set the stage for later chapters. We will also briefly talk about the importance of dealing with missing data, model selection and model checking, and point the readers to the corresponding chapters and sections for detailed descriptions about these topics. Lastly, we will briefly talk about the connection between the models introduced in this book and another famous family of models, species distribution models (Elith and Leathwick 2009). We will end the first chapter with a brief summary.

In later chapters, we will talk about Bayesian models for the analysis of a variety of types of data that are commonly collected by ecologists. More specifically, in Chap. 2, we will introduce occupancy models (MacKenzie et al. 2002; Tyre et al. 2003) that analyze detection/non-detection data; in Chap. 3, we will introduce N-mixture models (Royle 2004) that analyze count data; in Chap. 4, we will introduce integrated population models (Besbeas et al. 2002; Brooks et al. 2004) that jointly analyze count and capture-recapture data; and in Chap. 5, we will introduce models for analyzing spatial capture-recapture data (Efford 2004; Royle and Young 2008). In the last chapter, Chap. 6, we will summarize what we have learned so far and talk about some future directions.

Models introduced in this book are all discrete-space, discrete-time models; even when they are applied to a continuous habitat, the data need to be collated into discrete spatial units (e.g., grids). There are other families of models that could be used to understand spatially structured population dynamics at continuous space and/or time (e.g., reaction-diffusion models; Turing 1952, Cantrell and Cosner 2003), but they are excluded from this book, mainly due to my limited knowledge. Models for species-rich communities are left out of the book due to the fact that these models have been systematically described in other books (e.g., Ovaskainen and Abrego 2020) and papers (e.g., Dorazio and Royle 2005; Ovaskainen et al. 2017). In that sense, this book does not mean to provide a complete analytical toolbox for readers. Even for the models introduced in this book, hardly any of them are in an "ultimate" form, meaning that there are always room to further extend them. Therefore, the goal of this book is to provide ideas and coding techniques for readers who are willing to develop their own models for theoretical, applied, or methodological purposes. The R programing language (R Development Core Team 2019) is used in this book due to its popularity. Customized R code for Markov chain Monte Carlo algorithm is provided so that readers can have an in-depth understanding of how these models work. Other programming languages (e.g., Nimble; de Valpine et al. 2017) can be used to implement these models, and readers are highly encouraged to investigate them.

1.2 History of Modeling Spatially Structured Population Dynamics

In the poem quoted at the beginning of this chapter, Robert Frost described his observation about a butterfly (of course it is a butterfly!) that first flew over some inhabitable area due to habitat destruction and eventually moved into a habitat remnant. Interestingly, ecologists and conservation practitioners have been fascinated by the same kind of movement in patchy habitats, particularly considering the enormous and widespread habitat destruction due to human activities in the modern era. Such a phenomenon and its potential implications in ecological research and conservation practice intrigued studies of metapopulations, while a metapopulation is defined as a set of connected local populations that persist in a balance between local extinctions and colonization (Levins 1969, 1970; Hanski and Gilpin 1997; Hanski 1998; Ovaskainen and Saastamoinen 2018).

While the concept of metapopulation tends to be constrained to local populations that inhabit patchy habitats and have a high rate of turnover (i.e., extinction-colonization; Hanski and Gilpin 1997, p. 9), later it has been generalized to any spatially structured populations that can inhabit patchy to continuous habitats (as long as the habitat is heterogeneous) and have any rate of turnover, even including the situation without any turnover at all. Another distinction between "classic" metapopulation models and "modern" statistical models is that, while early metapopulation models tend to assume perfect observation of presence and absence, modern models have been developed to separate observation from the true ecological processes (MacKenzie et al. 2002; Tyre et al. 2003). Therefore, modern statistical models can be considered the generalized forms of classic metapopulation models in terms of both their ecological realism and statistical elegance. Such generalization is undoubtedly thanks to the broad application of Geographic Information System (GIS) and synthesis of metapopulation ecology and landscape ecology (Howell et al. 2018; Royle et al. 2018). Next, we will talk about the evolution of mathematical and statistical models that have been developed and used to understand metapopulations and more generally spatially structured populations over the last half century.

1.2.1 Levins's Spatially Implicit Metapopulation Model

The earliest model that was used to understand metapopulation dynamics was introduced by Levins (1969, 1970). This mathematical model considers how the fraction of occupied habitat patches p change over time t given rates of extinction e and colonization c such that $dp/dt = c \times p \times (1 - p) - e \times p$. This model is spatially implicit in the sense that it ignores the spatial locations of habitat patches. In addition, it also ignores the specific condition of each habitat patch, and thus patch-specific extinction and colonization rates. Therefore, the model is probably not very useful for analyzing real-world data to understand complex ecological processes that ecologists are facing nowadays. Levins's model nonetheless forms the theoretical foundation of metapopulation ecology as it demonstrates the importance of extinction and colonization rates in driving metapopulation dynamics and viability.

1.2.2 Spatially Explicit Metapopulation Models

Later on, ecologists have put a lot of effort into extending metapopulation models by explicitly considering the spatial locations of habitat patches as well as variation in habitat patch characteristics that influence extinction and colonization rates (e.g., Hassell et al. 1991; Caswell and Etter 1993). Among these models, a particularly influential model is Hanski's incidence function model, which assumes decreased extinction rates for local populations in larger habitat patches and increased colonization rates for empty patches with higher connectivity from other patches (Hanski 1994). In the same paper, Hanski illustrated the practical value of this model by analyzing presence/absence data from three butterfly metapopulations. The ideas of this and other spatially explicit metapopulation models are still used and further extended in "modern" statistical models that we will talk about next.

1.2.3 Modern Models

While early metapopulation models tend to assume perfect observations, studies have shown that ignoring observation errors can lead to strong biases in parameter estimation (Moilanen 2002). Numerous statistical models have been developed to account for imperfect observations in a variety of data types (Williams et al. 2002; Royle and Dorazio 2008; Kéry and Royle 2016, 2021): occupancy models can be used to analyze detection/non-detection data while accounting for imperfect detection (MacKenzie et al. 2002; Tyre et al. 2003); N-mixture models can account for imperfect detection in count data to provide inference for true abundance (Royle 2004); capture probability has long been explicitly modeled and separated from survival process in capture-recapture models (Cormack 1964; Jolly 1965; Seber 1965); and varying survey effort and thus detection rate can be accounted for in spatial capture-recapture models (Fuller et al. 2016). The effort of separating observation from true ecological processes in these models has undoubtedly revolutionized ecological research.

Modern models also tend to extend their applications from metapopulations to any spatially structure populations, due to their flexibility in dealing with a wide range of patchiness in habitats. In particular, these modern models can be used to analyze data that are collated into any kind of spatial units, either natural habitat patches (e.g., ponds and forest fragments) or artificially defined polygons that cover a continuous landscape (e.g., grids). Modern models can also deal with any rate of turnover in population dynamics. That being said, static occupancy models are useful only when there are both 0's and 1's in the true occupancy status (which will be detailed in Chap. 2), and dynamic occupancy models (MacKenzie et al. 2003) are useful only when there is a high rate of turnover (i.e., changes from 0's to 1's or from 1's to 0's in true occupancy). N-mixture models are more flexible than occupancy models as they can deal with situations with or without 0's in true abundance. The latter can be considered as a situation with only 1's in a perspective of occupancy, but the variation in abundance is still meaningful for characterizing the spatial structure and dynamics of a population. Dynamic N-mixture models (Dail and Madsen 2011) can also provide some basic demographic information underlying the spatiotemporal variation of abundance, although more reliable inference of complex demographic processes is achieved only when capture-based data and count data are jointly analyzed under an integrated population modeling framework. While occupancy models, N-mixture models and integrated population models are applicable when spatial units are larger than the territory of the study species, spatial capture-recapture models operate at a spatial resolution that is smaller than the territory of the study species, allowing them to explain habitat selection and its demographic consequence.

Overall, many modern statistical models are generalized forms of classic metapopulation models; they cover a wide range of ecological and observation situations, and provide information for different aspects of populations at different spatial and temporal scales. These modern models form the body of this book and will be introduced in more detail in later chapters.

1.3 Bayesian Modeling and Markov Chain Monte Carlo

Modern statistical models that account for imperfect observations, including but not limited to the models introduced in this book, often are implemented under the hierarchical Bayesian framework due to the flexibility and convenience of doing so. The hierarchical Bayesian framework is a statistical approach that expresses models in a hierarchical form (Gelman et al. 2014). For ecological models that are implemented under the hierarchical Bayesian framework, they often contain three levels, an observation or data sub-model that describes how data are obtained given ecological process, a process sub-model that describes the ecological process given parameters, and a parameter sub-model that describe the prior knowledge represented by prior distributions of the parameters (Wikle 2003). However, before we dive into these complicated models, we first need to understand Bayes' theorem (Bayes 1763). Bayes' theorem specifies that the probability of an event (e.g., rainfall has a negative effect on the occurrence of silk-cotton tree in Ghana) is based on our observations (data) and prior knowledge about the event. More specifically, we have

$$Pr(\theta|y) = \frac{Pr(y|\theta) \times Pr(\theta)}{Pr(y)}, \tag{1.1}$$

in which $Pr(\theta|y)$ is our interest, the posterior distribution of parameter θ given data y, which represents, e.g., what we have learned about the relationship between rainfall and silk-cotton tree occurrence from our data, $Pr(y|\theta)$ is the likelihood of y given θ (e.g., how likely we have obtained our data given that rainfall negatively affects silk-cotton tree occurrence), $Pr(\theta)$ is the prior distribution of θ representing our prior knowledge (e.g., $\theta \sim Normal(-0.5, 0.1)$ represents a specific level of negative effect of rainfall on silk-cotton tree occurrence and our confidence about such knowledge or belief), and $Pr(y)$ is the marginal probability of obtaining the data y.

If we can integrate $Pr(y|\theta) \times Pr(\theta)$ numerically, we can easily calculate $Pr(y)$ because $Pr(y)$ is none other than a constant that makes the whole thing, $\frac{Pr(y|\theta) \times Pr(\theta)}{Pr(y)}$, integrate to 1. In other words, if $Pr(y|\theta) \times Pr(\theta)$ integrates to 0.37, $Pr(y)$ has to be 0.37. However, even though we do know $Pr(y|\theta)$ and $Pr(\theta)$, it is often very challenging, and sometimes even impossible, to integrate $Pr(y|\theta) \times Pr(\theta)$. Therefore, in many cases, we will have to rely on the Markov chain Monte Carlo (MCMC) approach to characterize $Pr(\theta|y)$ by simulation.

The idea of MCMC is that, while it may be difficult to integrate $Pr(y|\theta) \times Pr(\theta)$, we actually do not have to know its precise mathematical form. Instead, we can simply sample (i.e., randomly draw) a bunch of values for θ, and summarize these random values to learn about important features of $Pr(\theta|y)$ such as measures of location and spread. But of course, we cannot use any values of θ; we do need to know whether $Pr(\theta^{[candidate]}|y)$ is greater than $Pr(\theta^{[current]}|y)$, in which $\theta^{[candidate]}$ and $\theta^{[current]}$ are just different values of θ, to pick up some reasonable values for θ. We can compare $Pr(\theta^{[candidate]}|y)$ and $Pr(\theta^{[current]}|y)$ by having

$$\frac{Pr\left(\theta^{[candidate]}|y\right)}{Pr\left(\theta^{[current]}|y\right)} = \frac{Pr\left(y|\theta^{[candidate]}\right) \times Pr\left(\theta^{[candidate]}\right)/Pr(y)}{Pr\left(y|\theta^{[current]}\right) \times Pr\left(\theta^{[current]}\right)/Pr(y)}$$
$$= \frac{Pr\left(y|\theta^{[candidate]}\right) \times Pr\left(\theta^{[candidate]}\right)}{Pr\left(y|\theta^{[current]}\right) \times Pr\left(\theta^{[current]}\right)}, \tag{1.2}$$

in which $Pr(y)$, the denominator in Eq. (1.1), is cancelled and thus gone for good (remember that $Pr(y)$ is a constant that has nothing to do with the value of θ). On top of that, we need to find a rule to decide which value we want, $\theta^{[candidate]}$ or $\theta^{[current]}$. The general rule of selecting between $\theta^{[candidate]}$ and $\theta^{[current]}$ is called the Metropolis-Hastings algorithm, which we will illustrate below with an example of a logistic regression. For more details (and a book-load of awesome examples) of the Metropolis-Hastings algorithm, readers can refer to Hooten and Hefley (2019).

1.3.1 Metropolis-Hastings

Recall that a logistic regression assumes that a binary event (e.g., occurrence or presence/absence of silk-cotton tree) follows a Bernoulli distribution. Further, the "success" probability of the Bernoulli distribution could be a function of covariates (e.g., rainfall). Therefore, we have

$$\left.\begin{array}{c} y_i \sim Bernoulli(\psi_i) \\ \psi_i = logit^{-1}\left(\beta_0 + \beta_1 \times x_{i,1} + \ldots + \beta_{n_{covs}} \times x_{i,n_{covs}}\right) \end{array}\right\}, \tag{1.3}$$

in which y_i represents the presence ($y_i = 1$) or absence ($y_i = 0$) of the study species at the ith site among a total of n_{site} sites, ψ_i is occupancy probability, x's are n_{covs} covariates, and β's are the intercept and slope parameters representing the effects of covariates on occupancy. It is called a logistic regression because an inverse logit function is used to transform $\beta_0 + \beta_1 \times x_{i,1} + \ldots + \beta_{n_{covs}} \times x_{i,n_{covs}}$, which can take any real-numbered value, into ψ_i, which has to be between 0 and 1. Also, note that y_i's are called detection/non-detection data in later chapters, but are called presence/absence data here under the assumption of perfect detection (i.e., the species is present if it is detected, and vice versa).

Next, let us use R code provided below to simulate some data based on the abovementioned model specification (also see Appendix 1.1). All the R code used in this book can be found in the appendices or from the following GitHub repository https://github.com/QingZhaoQE/The-BOOK. Note that we set the random number seed to 1 here so that anyone can generate the same data by using the same code below. Let us consider 360 sites (i.e., $n_{site} = 360$) and three covariates (i.e., $n_{covs} = 3$) which have positive ($\beta_1 = 0.8$), negative ($\beta_2 = -0.4$) and positive ($\beta_3 = 0.2$) effects on logit occupancy probability, respectively. The intercept is set to 0 (i.e., $\beta_0 = 0$) so that occupancy probability equals 0.5 when all covariates equal 0 ($logit^{-1}(0) = 0.5$) as the values of these covariates, x's, are randomly generated from a standard Normal distribution.

Also note that the code uses something that is equivalent to Eq. (1.3), but is written in the manner of matrix calculation such that

$$\left.\begin{array}{l} \boldsymbol{y} \sim Bernoulli(\boldsymbol{\psi}) \\ \boldsymbol{\psi} = logit^{-1}(\boldsymbol{X\beta}) \end{array}\right\}, \tag{1.4}$$

in which $\boldsymbol{y}$ is a vector of $\{y_1, y_2, \ldots y_{n_{site}}\}$, $\boldsymbol{\psi}$ is a vector of occupancy probabilities $\{\psi_1, \psi_2, \ldots \psi_{n_{site}}\}$,, $\boldsymbol{X}$ is an n_{site} by n_{covs} + 1 matrix, the first column of which contains 1's and the second to the last columns contain the values of n_{covs} covariates (i.e., x's), and $\boldsymbol{\beta}$ is a vector of $\{\beta_0, \beta_1, \ldots, \beta_{n_{covs}}\}$, the intercept and slope parameters. Such matrix calculation is used throughout the book for improved efficiency in computing.

```
library(boot) # for using logit and inv.logit functions

set.seed(1)

# Basic values
nsite <- 360 # number of sites
ncovs <- 3  # number of environmental covariates
beta <- c(0, 0.8, -0.4, 0.2) # intercept and slopes for occurrence

# Define function for data simulation
data_sim <- function(nsite, ncovs, beta) {
 x <- matrix(rnorm(nsite*ncovs, 0, 1), nsite, ncovs) # environmental
covariates
 psi <- inv.logit(cbind(1, x) %*% beta) # occurrence probability
 y <- rbinom(nsite, 1, psi) # observations of occurrence

 list(x=x, y=y)
} # data_sim
```

Our interest is to estimate $\boldsymbol{\beta}$. To obtain posterior samples of $\boldsymbol{\beta}$, we need the likelihood of $\boldsymbol{y}$ given $\boldsymbol{\beta}$, $Pr(\boldsymbol{y}|\boldsymbol{\beta})$, as well as the prior distributions of $\boldsymbol{\beta}$, $Pr(\boldsymbol{\beta})$. We can easily calculate $Pr(\boldsymbol{y}|\boldsymbol{\beta})$ in R since we know that $\boldsymbol{y} \mid \boldsymbol{\beta}$ follows a Bernoulli distribution (Eq. (1.3)) and thus have $Pr\ (\boldsymbol{y}|\boldsymbol{\beta}) = Bernoulli(\boldsymbol{y}|\boldsymbol{\beta})$. Also, because we have no prior knowledge of $\boldsymbol{\beta}$ here, we use a so-called vague prior *Normal*($\mu = 0, \sigma = 2$) on each element in $\boldsymbol{\beta}$ and thus have $Pr(\boldsymbol{\beta}) = Normal(\boldsymbol{\beta}|\mu, \sigma)$. It is a very common practice in Bayesian statistics to use vague priors, i.e., distributions that are "flat" and thus provide almost no information to the posterior, but readers should not hesitate to use informative priors if prior knowledge is available.

Next, we will generate some random values and apply the Metropolis-Hastings rule of acceptance to obtain posterior samples of $\boldsymbol{\beta}$. In the first step, we set all elements in $\boldsymbol{\beta}$ to 0. In the following steps (which are also called iterations), each time we change the values of $\boldsymbol{\beta}$ a little bit randomly to obtain a candidate value of the parameter for potentially replacing the current value. This candidate value is represented by $\boldsymbol{\beta}^{[candidate]}$ and is randomly generated from a Normal distribution such that $\boldsymbol{\beta}^{[candidate]} \sim Normal(\boldsymbol{\beta}^{[current]}, \delta)$, in which $\boldsymbol{\beta}^{[current]}$ is the current value, and δ is the standard deviation (and is called tuning factor in the code below) that governs the magnitude of the change from $\boldsymbol{\beta}^{[current]}$ to $\boldsymbol{\beta}^{[candidate]}$. In the following code, we used the same δ value, 0.15, for all β's; we could extend δ to a vector to allow a different value for each β if needed.

Once a candidate value is generated, our task is to determine if we want to keep the current value $\boldsymbol{\beta}^{[current]}$ or update it to $\boldsymbol{\beta}^{[candidate]}$ using the Metropolis-Hastings algorithm. For this purpose, we first need to calculate the Metropolis-Hastings ratio such that $mh = \frac{Bernoulli\left(y|\boldsymbol{\beta}^{[candidate]}\right) \times Normal\left(\boldsymbol{\beta}^{[candidate]}|\mu, \sigma\right)}{Bernoulli\left(y|\boldsymbol{\beta}^{[current]}\right) \times Normal\left(\boldsymbol{\beta}^{[current]}|\mu, \sigma\right)}$. We then compare mh to a number that is randomly drawn from $Uniform(0, 1)$. If mh is greater than this number, we update $\boldsymbol{\beta}^{[current]}$ to $\boldsymbol{\beta}^{[candidate]}$, otherwise we keep $\boldsymbol{\beta}^{[current]}$, but as a new sample. This is equivalent to say, we will update $\boldsymbol{\beta}^{[current]}$ to $\boldsymbol{\beta}^{[candidate]}$ for sure if the latter is equally good or better than the former (i.e., $mh \geq 1$), but we will still have a chance to update even if the latter is worse than the former (i.e., $mh < 1$). We repeat this process of updating (or not updating) many times to get the posterior samples of $\boldsymbol{\beta}$. Updating the entire vector of $\boldsymbol{\beta}$ simultaneously is called block updating, which normally provides faster convergence than updating each $\boldsymbol{\beta}$ parameter at a time, and thus will be used throughout the book. Because these posterior samples are dependent ($\boldsymbol{\beta}^{[candidate]}$ is based on $\boldsymbol{\beta}^{[current]}$), they are also called a posterior chain, or simply a chain.

The R code for implementing this algorithm is provided below.

```
logistic_mcmc <- function(y, x, nmcmc) {

 # Setup variables
 nsite <- dim(x) [1]
 ncovs <- dim(x) [2]

 beta_save <- matrix(0, ncovs+1, nmcmc)

 # Priors
 beta_mean <- 0
 beta_sd <- 2

 # Starting values
 beta <- rep(0, ncovs + 1)

 # Tuning factor
 beta_tune <- 0.15

 # Begin MCMC loop
 for (k in 1:nmcmc) {
```

```
    ### Sample beta
    beta_star <- rnorm(ncovs+1, beta, beta_tune)
    mh1 <- sum(dbinom(y, 1, inv.logit(cbind(1,x) %*% beta_star), log=T)) +
        sum(dnorm(beta_star, beta_mean, beta_sd, log=T))
    mh2 <- sum(dbinom(y, 1, inv.logit(cbind(1,x) %*% beta    ), log=T)) +
        sum(dnorm(beta    , beta_mean, beta_sd, log=T))
    mh <- exp(mh1 - mh2)
    if (mh > runif(1)) {
     beta <- beta_star
    }

    beta_save[,k] <- beta
  } # k

  # Write output
  list(beta_save=beta_save)
} # logistic_mcmc
```

Now let us run the following code to implement the algorithm. A few lines are included to allow parallel computing, which will speed up the process. Computing time is also tracked: it turns out to only take 9 s to run 50,000 iterations in 3 parallel chains.

```
  # Check convergence
library(foreach)    # for parallel computing
library(doParallel) # for parallel computing

numCores <- round(detectCores() / 2) # use half of the cores for parallel
computing
registerDoParallel(numCores) # setup parallel computing

data <- data_sim(nsite=nsite, ncovs=ncovs, beta=beta) # simulate data
nmcmc <- 50000 # number of iterations
chain <- 3     # number of chains

start_time <- Sys.time() # start time of computing
out <- foreach (i=1:chain, .packages='boot') %dopar% {
 logistic_mcmc(y=data$y, x=data$x, nmcmc=nmcmc)
} # i
stopImplicitCluster() # clean up the cluster
end_time <- Sys.time() # end time of computing
time_taken <- end_time - start_time # computing time
time_taken
```

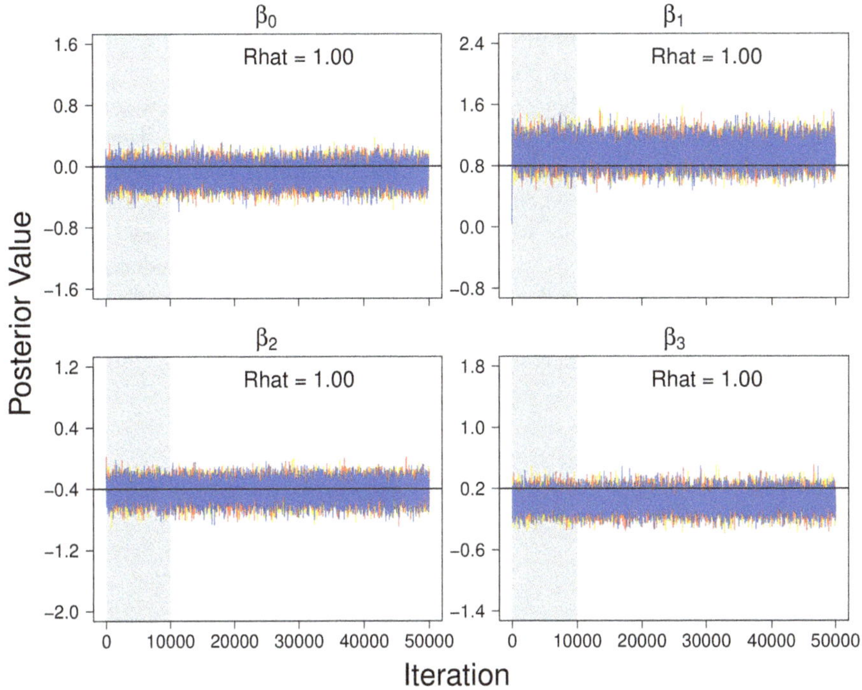

Fig. 1.1 Traceplots showing the posterior chains of β's in the logistic regression described in Eq. (1.3). Horizontal lines represent the true values used in data simulation. Rhat statistics are calculated after discarding the first 10,000 iterations. A Rhat statistics that is close to 1 indicates good convergence

After the algorithm is executed, we need to run some diagnostics to make sure that we have obtained appropriate and adequate information for $\boldsymbol{\beta}$. The first approach of diagnostics is to create some traceplots of the posterior chains (Fig. 1.1), using the code below. Because the code for creating traceplots is quite repetitive for other models, it is put in appendices for later chapters. We want to visualize whether the chains are stable, meaning that they are not shifting upwards or downwards. If they are still shifting, that means the chains are still exploring the parameter space of $\boldsymbol{\beta}$, and we should run more iterations. For our case, we can see that the chains are stable. We also want to visualize the mixing of the chains, which is highly driven by the magnitude of δ. When δ is too small, the chain explores the parameter space too slowly, in which case the posterior samples may not have covered the parameter space adequately when we stop computing. On the other hand, when δ is too big, the chain will be stuck at the same value for too long, which again is unlikely to cover the parameter space adequately. For our case, we can see that the chains are well mixed, indicating that the magnitude of δ, and thus the rate of updating, is about right. These features of the chains basically tell us that the posterior samples are likely to provide appropriate and adequate information for $\boldsymbol{\beta}$.

The code also contains lines to calculate Rhat statistics (Brooks and Gelman 1998) which can be used to further confirm convergence. An Rhat statistic that is close to 1 indicates satisfactory convergence, which is the case for all the parameters shown in Fig. 1.1. Since we have plotted the entire chains in Fig. 1.1, we can see that the model probably has converged within the first 5000 iterations. Therefore, we discarded the first 10,000 iterations of the chain, the burn-in, which contains values before the model converges when calculating Rhat statistics. There is another trick called "thinning," which means we can keep one value out of every, say, 100 values (we say the chain is thinned by 100 in this case) which sometimes reduces autocorrelation within the chain. We do not thin at all in this book because all the models converge well without thinning. In addition, some studies have found that thinning is not all that useful anyway (Link and Eaton 2012).

```
  library(rstan) # for calculating rhat statistics

cols <- c('gold', 'tomato', 'royalblue')
ylab_beta <- c(expression(beta[0]), expression(beta[1]), expression
(beta[2]), expression(beta[3]))

par(mfrow=c(2,2))
par(mar=c(1,3,3,1))
par(oma=c(4,3,0,0))

for (i in 1:(ncovs+1)) {
 tt <- matrix(, nmcmc, chain)
 for (j in 1:chain) {
  tt[,j] <- out[[j]]$beta_save[i,]
 } # j
 yint <- ceiling(sd(tt[(nmcmc*0.2+1):nmcmc,]) * 10) / 10
 yint <- ifelse (yint < 0.1, 0.1, yint)
 ymin <- beta[i] - yint * 8
 ymax <- beta[i] + yint * 8
 plot(1, xlim=c(0,nmcmc), ylim=c(ymin, ymax), type='n', xlab='',
ylab='', axes=F)
 rect(0, -100, nmcmc*0.2, 100, col = 'grey88', border = NA)
 if (i %in% 1:2) {
  axis(1, at=seq(0,nmcmc,length.out=6), labels=rep('',6))
 } else {
  axis(1, at=seq(0,nmcmc,length.out=6), cex.axis=1.5)
 }
 axis(2, at=round(seq(ymin, ymax, yint*4), digits=1), cex.axis=1.5,
las=2)
 box()
 for (j in 1:chain) {
  lines(tt[,j], col=cols[j])
 } # j
 abline(h=beta[i], col='grey16', lwd=1.5)
```

```
 title(main=ylab_beta[i], cex.main=2, line=1.2)
 text(x=nmcmc*0.6, y=ymax, pos=1, cex=1.8,
   labels=paste(c('Rhat = ', format(round(Rhat(tt[(nmcmc*0.2+1):
nmcmc]), digits=2), nsmall=2)), collapse=''))
} # i

title(xlab='Iteration', cex.lab=2.5, line=2.6, outer=T)
title(ylab='Posterior Value', cex.lab=2.5, line=1.2, outer=T)
```

Readers may have noticed that the posterior chains are not centered on the true values. This is just natural, as there is no guarantee that the posterior means obtained from analyzing one simulated data set are the same as the true values. Therefore, the result based on one simulation is not sufficient for evaluating bias in parameter estimates. However, the posterior means do look close enough to the true values, giving us a preliminary idea that the model seems to work well. A strict evaluation of bias would be to calculate $E\left(\widehat{\theta}\right)$, in which $\widehat{\theta}$ is the estimator of a certain parameter θ (e.g., β_1), and E means expectation. We say that an estimator is unbiased if we can mathematically prove that $E\left(\widehat{\theta}\right) = \theta$. However, we often cannot calculate $E\left(\widehat{\theta}\right)$ due to the difficulty in integrating $Pr(y|\theta) \times Pr(\theta)$, as mentioned earlier. An alternative (and often more practical) approach is to repeat the simulation multiple times (remember to remove the seed so a different set of data is generated each time), and calculate the grand mean for each parameter from all the posterior samples of all simulations. The grand mean should get very close to the true value when the simulations are repeated many times. The code for repeating the simulations 100 times is provided below.

```
# Evaluate bias
numCores <- round(detectCores() / 2) # only use half of the cores for
parallel computing
registerDoParallel(numCores) # setup parallel computing

nsims <- 100  # number of simulations
nmcmc <- 50000 # number of iterations
chain <- 3   # number of chains

set.seed(NULL) # remove seed to generate a different dataset each time
out_nsims <- list()
for (sim in 1:nsims) {
 data <- data_sim(nsite=nsite, ncovs=ncovs, beta=beta)
 out_nsims[[sim]] <- foreach (i=1:chain, .packages='boot') %dopar% {
  logistic_mcmc(y=data$y, x=data$x, nmcmc=nmcmc)
 } # i
} # sim
stopImplicitCluster() # clean up the cluster
```

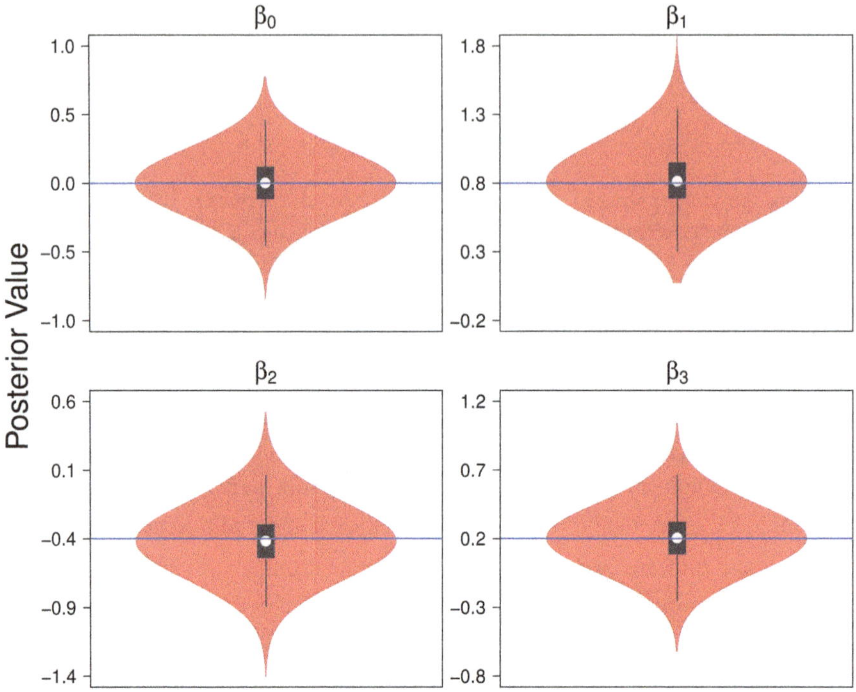

Fig. 1.2 Violin plots showing the posterior distributions (the "violin") as well as key statistics (mean: white dot, 50% Credible Interval: box, 95% Credible Interval: whiskers) of β's in the logistic regression described in Eq. (1.3). All the posterior samples from 100 simulations are pooled. Horizontal lines represent the true values used in data simulation

Then using the code below to create some violin plots (Fig. 1.2), we can see that the grand mean of each parameter is indeed very close to its true value, indicating that the model is likely to provide unbiased parameter estimates. In later chapters, we will only run one simulation for each model to save some time. Bias of estimators in many of the models introduced in this book has been evaluated in the corresponding publication of the model. In addition, readers are encouraged to develop their own code to evaluate potential bias and other operating characteristics of these models; as a matter of fact, experimenting with simulated data produced under a model of your interest can be insightful (see also Chap. 3 in Kéry and Royle 2016).

```
library(vioplot) # for making violin plots

par(mfrow=c(2,2))
par(mar=c(1,3,3,1))
par(oma=c(0,3,0,0))

for (i in 1:(ncovs+1)) {
 tt <- array(, dim=c(nmcmc*0.8, chain, nsims))
```

```
 for (s in 1:nsims) {
  for (j in 1:chain) {
   tt[,j,s] <- out_nsims[[s]][[j]]$beta_save[i,(nmcmc*0.2+1):nmcmc]
  } # j
 } # s
 plot(1, type='n', ylim=c(beta[i]-1,beta[i]+1), xlab='', ylab='',
axes=F)
 vioplot(tt, col='lightcoral', rectCol='grey36', lineCol='grey36',
border=NA, wex=0.8, cex=1.8, add=T)
 abline(h=beta[i], col='royalblue', lwd=1.5)
 axis(2, at=seq(beta[i]-1,beta[i]+1,length.out=5), cex.axis=1.5,
las=2)
 title(main=ylab_beta[i], cex.main=2, line=1.2)

} # i

title(ylab='Posterior Value', cex.lab=2.5, line=1.2, outer=T)
```

1.3.2 Gibbs Sampling

In rare cases, we can work out the distribution of $Pr(y|\theta) \times Pr(\theta)$, which allows us to directly sample posterior values for θ. This distribution can be the posterior distribution of θ, $Pr(\theta|y)$, if θ is the only parameter to estimate in the model, but more commonly the full conditional distribution of θ, $Pr(\theta|-\theta, y)$, if the value of θ depends on the (temporary) values of other parameters, denoted $-\theta$, in the model. Now, let's consider a simple case in which we want to estimate the mean of y_i for $i = 1, \ldots, n$ observations that follow independent and identical Normal distributions with an unknown mean μ (the parameter to estimate) and a known standard deviation σ. We also assume that the prior distribution of μ is a Normal distribution with a mean μ_0 and standard deviation σ_0. Thus, we have $Pr(y|\mu) = Normal(\mu, \sigma)$ as the likelihood of y given μ and $Pr(\mu) = Normal(\mu_0, \sigma_0)$ as the prior. After a series of cool mathematical derivations detailed in, e.g., Hooten and Hefley (2019, pp. 60–61), we know that the posterior distribution of μ is a Normal distribution, i.e., $Pr(y|\mu) \times Pr(\mu) \propto Normal(\mu_P, \sigma_P)$, in which $\mu_P = \sigma_P^2 \times \left(\frac{\sum_{i=1}^{n} y_i}{\sigma^2} + \frac{\mu_0}{\sigma_0^2}\right)$ and $\sigma_P = \sqrt{1/\left(\frac{n}{\sigma^2} + \frac{1}{\sigma_0^2}\right)}$. Now, we can simply draw random numbers from this distribution, an approach called the Gibbs sampling (Geman and Geman 1984).

Now, let us use the following code to simulate data. The same code is also provided in Appendix 1.2.

```
set.seed(1)

# True values
n <- 100 # sample size
```

```
mu <- 1 # mean of y, we are going to estimate this
sigma <- 1.5 # SD of y, assumed to be known

# Simulate data
y <- rnorm(n, mu, sigma) # data
```

We then can use the following code to implement the Gibbs sampling to draw posterior samples for μ:

```
# Prior
mu0 <- 0 # mean of prior of mu
sigma0 <- 20 # SD of prior of mu

# Posterior
mu_var <- 1 / (n / sigma ^ 2 + 1 / sigma0 ^ 2) # variance of posterior of mu
mu_mean <- mu_var * (sum(y) / sigma ^ 2 + mu0 / sigma0 ^ 2) # mean of
posterior of mu

# Sampling
library(foreach)   # for parallel computing
library(doParallel) # for parallel computing

numCores <- round(detectCores() / 2) # use half of the cores for parallel
computing
registerDoParallel(numCores) # setup parallel computing

nmcmc <- 50000 # number of iterations
chain <- 3    # number of chains

start_time <- Sys.time() # start time of computing
mu_gibbs <- foreach (i=1:chain) %dopar% {
 rnorm(nmcmc, mu_mean, sqrt(mu_var)) # posterior samples of mu based on
gibbs sampling
} # i
stopImplicitCluster() # clean up the cluster
end_time <- Sys.time() # end time of computing
time_taken <- end_time - start_time # computing time
time_taken
```

We also have lines of code to implement the Metropolis-Hastings sampling to draw posterior samples for the same parameter μ.

```
# Define algorithm
mu_mcmc <- function(y, nmcmc) {
```

```
 mu_save <- rep(0, nmcmc) # vector to save the posterior samples of mu
based on MH sampling
 mu_keep <- 0 # starting value of mu
 mu_tune <- 0.05 # tuning factor of mu

 for (k in 2:nmcmc) {
  mu_star <- rnorm(1, mu_keep, mu_tune) # proposed new value of mu
  mh1 <- sum(dnorm(y, mu_star, sigma, log=T)) + dnorm(mu_star, mu0,
sigma0, log=T)
  mh2 <- sum(dnorm(y, mu_keep, sigma, log=T)) + dnorm(mu_keep, mu0,
sigma0, log=T)
  mh <- exp(mh1 - mh2)
  if (mh > runif(1)) {
   mu_keep <- mu_star
  }

  mu_save[k] <- mu_keep # posterior samples of mu based on MH sampling
 } # k

 # Write output
 list(mu_save=mu_save)
} # logistic_mcmc

# Sampling
numCores <- round(detectCores() / 2) # use half of the cores for parallel
computing
registerDoParallel(numCores) # setup parallel computing

nmcmc <- 50000 # number of iterations
chain <- 3   # number of chains

start_time <- Sys.time() # start time of computing
mu_mh <- foreach (i=1:chain) %dopar% {
 mu_mcmc(y=y, nmcmc=nmcmc)
} # i
stopImplicitCluster() # clean up the cluster
end_time <- Sys.time() # end time of computing
time_taken <- end_time - start_time # computing time
time_taken
```

Finally, let us use the following code to plot some traceplots and compare the results from both sampling approaches. From Fig. 1.3, we can see that both approaches provide very similar results. However, Gibbs sampling only takes 0.1 s to run for 50,000 iterations, which is much faster than the 1.3 s the Metropolis-Hastings sampling takes. In addition, Gibbs sampling is more efficient because the acceptance rate is 1.0, whereas the acceptance rate is <1.0 for the Metropolis-

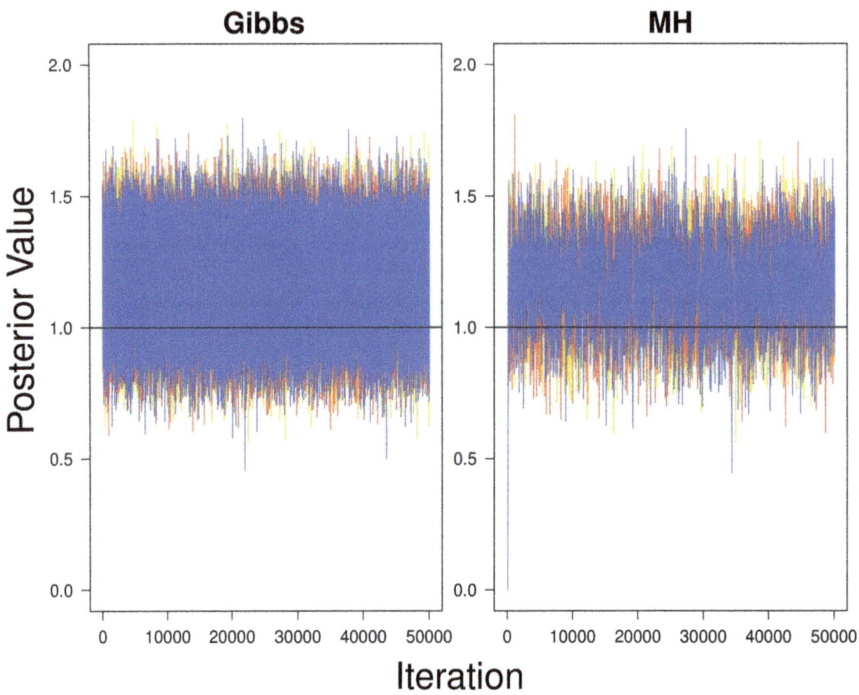

Fig. 1.3 Traceplots showing the posterior chains of the mean parameter μ of a Normal distribution with a known standard deviation obtained from Gibbs sampling (left panel) and Metropolis-Hastings sampling (right panel). Horizontal lines represent the true values used in data simulation

Hastings algorithm. Therefore, we should consider using Gibbs sampling whenever the full conditional distribution of the parameter of interest is known.

```
  cols <- c('gold', 'tomato', 'royalblue')

par(mfrow=c(1,2))
par(mar=c(1,2,3,1))
par(oma=c(4,3,0,0))

plot(1, xlim=c(0,nmcmc), ylim=c(0, 2), type='n', xlab='', ylab='',
axes=F)
axis(1, at=seq(0,nmcmc,length.out=6), cex.axis=1.2)
axis(2, at=round(seq(0,2,0.5), digits=1), cex.axis=1.2, las=2)
box()
for (j in 1:chain) {
 lines(mu_gibbs[[j]], col=cols[j])
} # j
abline(h=mu, col='grey16', lwd=1.5)
title(main='Gibbs', cex.main=2, line=0.6)
```

```
plot(1, xlim=c(0,nmcmc), ylim=c(0, 2), type='n', xlab='', ylab='',
axes=F)
axis(1, at=seq(0,nmcmc,length.out=6), cex.axis=1.2)
axis(2, at=round(seq(0,2,0.5), digits=1), cex.axis=1.2, las=2)
box()
for (j in 1:chain) {
 lines(mu_mh[[j]][[1]], col=cols[j])
} # j
abline(h=mu, col='grey16', lwd=1.5)
title(main='MH', cex.main=2, line=0.6)

title(xlab='Iteration', cex.lab=2.5, line=2.6, outer=T)
title(ylab='Posterior Value', cex.lab=2.5, line=1.1, outer=T)
```

1.4 Role of Simulation in Statistical Methodology and this Book

Now, we have gotten a taste of how we can write our own MCMC algorithms, let us set the stage for the more exciting models we will talk about in later chapters by simulating some environmental covariates, shall we?

But first, let us talk about the rationale (and beauty) of using simulations for this book. Using simulations is a key methodology in statistical modeling. In fact, simulations have been a standard part of methodological papers in statistical ecology in recent decade (e.g., Broms et al. 2016; Rossman et al. 2016; Hostetter et al. 2022; Rakhimberdiev et al. 2022), with some papers being entirely based on simulations (e.g., Bellier et al. 2016)! One big advantage of using simulations is that we can assess if the model is able to recover its parameters under the assumption that the true (or data-generating) model is known. While we all understand that such an assumption is way too strong for real-world studies, how could we trust a model to provide reliable knowledge about our study system if it cannot recover its parameters even when such a strong assumption is made? In that sense, simulations provide baseline information about the inferential performance of a model under an "ideal" situation. That being said, a model will not be too useful if it is built upon a super abstract, up-in-the-air concept. Therefore, next, we want to simulate environmental covariates that can represent some key realism in real-world studies.

First, let us generate 360 habitat patches (e.g., meadows) over a time period of 20 years. The patchy habitat seems to follow the tradition of metapopulation studies. However, it is worth pointing out that none of the models introduced in this book are limited to patchy habitats. In fact, these models can be applied to continuous habitats just fine; we only need to define spatial units as polygons (e.g., grids) that cover the entire landscape. The key point is not whether the habitat is patchy or not; the key point is about considering the spatial locations of spatial units and the spatiotemporal variation in their characteristics.

For the habitat patches considered in this book, let us randomly select their locations within a rectangular area. We can think of this rectangular area as some region representing our study area. We consider two characteristics of the habitat patches, their size, and temperature, which will be used as covariates in the models introduced later. Mean patch size is randomly drawn from a log-Normal distribution, but we do want to make sure these patches do not overlap with each other (at least in the plots). Then, let us add a small amount of temporal variation to the mean patch size so that they change a little bit over the years. Mean temperate is generated to have a moderate spatial gradient such that the northeast corner of the area is the warmest and the southwest corner is the coolest. Furthermore, let us add a strong oscillation by using a sine function to make temperature warmer in some years than others and there is some temporal autocorrelation in this covariate.

We use the following code to generate these covariates. The same code is also provided in Appendix 1.3. The random number seed is again set to 1 so that anyone can generate the same values for these covariates. Alternatively, readers can download these covariates at https://github.com/QingZhaoQE/The-BOOK

```
library(boot) # for using inv.logit functions

set.seed(1)

nsite <- 360 # number of sites
nyear <- 20 # number of years

lon <- runif(nsite, -10, 10) # longitude or easting
lat <- runif(nsite, - 8, 8) # latitude or northing
d <- matrix(, nsite, nsite) # distance between sites
for (i1 in 1:nsite) {
 for (i2 in 1:nsite) {
  d[i1,i2] <- sqrt((lon[i1] - lon[i2]) ^ 2 + (lat[i1] - lat[i2]) ^ 2)
 } # i2
} # i1
diag(d) <- NA

size <- matrix(, nsite, nyear) # habitat patch size
size_mean <- numeric(nsite)
for (i in 1:nsite) {
 size_mean[i] <- exp(rnorm(n=1, mean=log(min(d[i,],na.rm=T) / 2),
sd=0.05))
 for (t in 1:nyear) {
  size[i,t] <- exp(rnorm(n=1, mean=log(size_mean[i]), sd=0.05))
 } # t
} # i

temp <- matrix(, nsite, nyear) # temperature
temp_mean <- inv.logit(lat / 30 + lon / 20) * 10
```

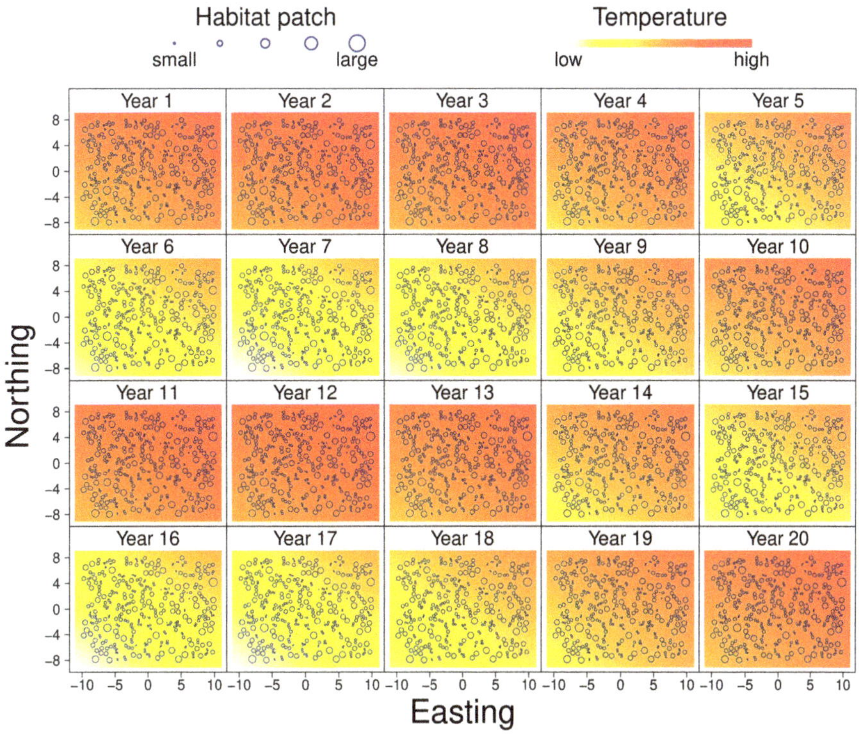

Fig. 1.4 The spatiotemporal variation of simulated habitat patch size and temperature in an imaginary landscape that forms the background of the models introduced in this book. While habitat patch size has no obvious spatial gradient and varies little over time, temperature has a moderate northeast-southwest gradient and a strong temporal oscillation

```
for (t in 1:nyear) {
 temp[,t] <- (temp_mean + sin(t / 1.5) * 2) / 0.6 - 4
} # t
```

We then use the following code to visualize the generated covariates in Fig. 1.4:

```
ngrid <- 10000
x <- rep(seq(-11,11,length.out=sqrt(ngrid)), times=sqrt(ngrid))
y <- rep(seq(- 9, 9,length.out=sqrt(ngrid)), each =sqrt(ngrid))
temp_area_mean <- numeric(ngrid)
temp_area <- matrix(, ngrid, nyear)
for (i in 1:ngrid) {
 temp_area_mean[i] <- inv.logit(x[i] / 30 + y[i] / 20) * 10
 for (t in 1:nyear) {
  temp_area[i,t] <- (temp_area_mean[i] + sin(t / 1.5) * 2) / 0.6 - 4
 } # t
} # i
```

```
nscale <- 100
scale <- round((temp_area - min(temp_area)) / (max(temp_area) - min
(temp_area)) * (nscale - 1)) + 1
xsize <- (max(x) - min(x)) / (sqrt(ngrid)-1) / 2
ysize <- (max(y) - min(y)) / (sqrt(ngrid)-1) / 2

laymat <- matrix(
 c(rep(rep( 1: 2, each=5), times=2),
  rep(rep( 3: 7, each=2), times=3),
  rep(rep( 8:12, each=2), times=3),
  rep(rep(13:17, each=2), times=3),
  rep(rep(18:22, each=2), times=3)),
 nrow=14, ncol=10, byrow=TRUE)
layout(mat=laymat)
par(mar=c(1,0,0,0))
par(oma=c(6,7,0,1))

plot(1, type='n', axes=F, xlab='', ylab='')
points(x=seq(0.8,1.2,length.out=5), y=rep(0.92,5), pch=21,
bg='white', col='navy', lwd=2, cex=seq(1,9,2)/2)
text(x=0.8, y=0.86, labels='small',    pos=1, cex=2.4)
text(x=1.2, y=0.86, labels='large',    pos=1, cex=2.4)
text(x=1.0, y=1.05, labels='Habitat patch', pos=3, cex=3.0)

plot(1, type='n', axes=F, xlab='', ylab='')
nscale <- 100
for (i in 1:nscale) {
 rect(seq(0.8,1.2,length.out=nscale+1)[i],  0.88,
    seq(0.8,1.2,length.out=nscale+1)[i+1], 0.96,
    col=rev(heat.colors(nscale))[i], border=NA)
} # i
text(x=0.8, y=0.86, labels='low',     pos=1, cex=2.4)
text(x=1.2, y=0.86, labels='high',    pos=1, cex=2.4)
text(x=1.0, y=1.05, labels='Temperature', pos=3, cex=3.0)

par(mar=c(0,0,0,0))
for (t in 1:nyear) {
 plot(1, xlim=c(-11,11), ylim=c(-9,12), type='n', axes=F, xlab='',
ylab='')
 for (i in 1:ngrid) {
  rect(x[i]-xsize,y[i]-ysize,x[i]+xsize,y[i]+ysize,col=rev(heat.
colors(nscale))[scale[i,t]],border=NA)
 } # i
 points(lon, lat, pch=21, cex=size[,t]*3, col='navy', lwd=0.5)
```

```
 if (t %% 5 == 1) {
  axis(2, at=seq(-08,08,length.out=5), cex.axis=1.8, las=2)
 }
 if (t > 15) {
  axis(1, at=seq(-10,10,length.out=5), cex.axis=1.8)
 }
 box()
 text(x=0, y=11, labels=paste('Year', t, sep=' '), cex=2.4)
 if (t == 11) {
  axis(2, at=12, labels='Northing', line=3.2, cex.axis=4, tick=F)
 }
 if (t == 18) {
  axis(1, at= 0, labels='Easting', line=3.4, cex.axis=4, tick=F)
 }
} # t
```

1.5 Missing Values, Model Selection, and Model Checking

Because missing values are commonly found in ecological data, dealing with missing values is an important issue for any ecological models. We will illustrate how to deal with missing values in Chap. 2, Sect. 2.4.2. In reality, it is very common that we have several candidate models that are all reasonable from a scientific perspective, and we cannot be immediately sure which one is the best. Model selection aims to pick the best candidate model(s) for the purpose of ecological hypothesis testing and prediction (Hooten and Hobbs 2015; Tredennick et al. 2021). Information-theoretic approaches (e.g., Akaike's Information Criterion or AIC; Akaike 1973, Burnham and Anderson 2002) have been broadly used in model selection as they reward good model fit while penalizing model complexity which may cause overfitting. Among information-theoretic approaches, the Watanabe-Akaike information criterion or WAIC (Watanabe 2013) is commonly used for Bayesian models and thus will be illustrated in Chap. 3, Sect. 3.4.1. When a model is implemented, we need to check if there is any discrepancy between the model and our data (Conn et al. 2018). For example, we can visualize or statistically test whether the residuals of a linear regression actually follow a Normal distribution, as the model assumes. We can also construct a Receiver Operating Characteristic (ROC) curve and calculate the area under the curve (AOC) to evaluate models with binary responses. Among many model checking approaches, the posterior predictive check (Rubin 1984; Gelman et al. 1996) is broadly applicable for data following pretty much any distribution and thus is commonly used for Bayesian models and in ecology. We will illustrate how to conduct the posterior predictive check in Chap. 4, Sect. 4.4.2. Once the approaches of dealing with missing values, calculating WAIC, and conducting the posterior predictive check are introduced, readers are highly encouraged to adapt them to other models in this book. Readers can also find valuable information about model selection and model checking in Wikle et al. (2019, Chap. 6).

1.6 Connection to Species Distribution Models

Species Distribution Models (SDMs) are statistical or machine learning approaches that characterize the spatial distributions of species occurrence or abundance (Elith and Leathwick 2009). In that sense, occupancy models and N-mixture models are both SDMs as they can be used to explain and predict the distributions of species occurrence and abundance, respectively. While non-parametric, machine learning SDMs have advantages in detecting complex non-linear relationships between species distributions and environmental drivers (e.g., Hao et al. 2020), occupancy and N-mixture models are advantageous in their capability of separating observation from true ecological processes. Furthermore, dynamic occupancy and dynamic N-mixture models can provide information for the underlying mechanisms (e.g., extinction-colonization and demography) of distributional dynamics, which are normally difficult to obtain with non-parametric approaches. Even spatially explicit integrated population models (Zhao 2020) can be considered SDMs, as they provide information for the spatial distributions and dynamics of species abundance as well as underlying demographic processes. Spatial capture-recapture models provide spatial distributions of animal habitat use, which can be further aggregated into distributions of occurrence and abundance at coarse spatial scales. Therefore, all the models introduced in this book are arguably SDMs—as a matter of fact, some of them are very advanced mechanism-driven SDMs.

1.7 Summary

In this chapter, we started by talking about the importance and key challenges in studying spatially structured population dynamics. We then reviewed the history of modeling metapopulations, and more generally, spatially structured populations. After that, we learned some basic knowledge about Bayesian modeling and skills of writing our own MCMC algorithms in R, which will be very useful for implementing the complex models introduced in this book. We then simulated environmental covariates to set the stage for later chapters. We also briefly talked about the importance of dealing with missing values, model selection and model checking, and pointed the readers to the corresponding chapters and sections for these topics. The last but not the least, let us keep in mind that the models introduced in this book can be viewed as advanced, mechanism-driven species distribution models. Now we are ready to dive into the real fun of the book. Let us go!

References

Akaike H (1973) Information theory and an extension of the maximum likelihood principle. In: Petrov BN, Csaki F (eds) Proceedings of the 2nd international symposium on information theory. Akademiai Kiado, pp 267–281

Bartlett MS (1960) Stochastic population models in ecology and epidemiology. Methuen and Wiley

Bayes T (1763) LII. An essay towards solving a problem in the doctrine of chances. By the late rev. Mr. Bayes, FRS communicated by Mr. Price, in a letter to John Canton, AMFR S. Philos Trans R Soc Lond 53:370–418

Bellier E, Kéry M, Schaub M (2016) Simulation-based assessment of dynamic N-mixture models in the presence of density dependence and environmental stochasticity. Methods Ecol Evol 7: 1029–1040

Besbeas P, Freeman SN, Morgan BJT, Catchpole EA (2002) Integrating mark-recapture-recovery and census data to estimate animal abundance and demographic parameters. Biometrics 58:540–547

Broms KM, Hooten MB, Johnson DS, Altwegg R, Conquest LL (2016) Dynamic occupancy models for explicit colonization processes. Ecology 97:194–204

Brooks SP, Gelman A (1998) General methods for monitoring convergence of iterative simulations. J Comput Graph Stat 7:434–455

Brooks SP, King R, Morgan BJT (2004) A Bayesian approach to combining animal abundance and demographic data. Anim Biodivers Conserv 27:515–529

Burnham KP, Anderson DR (2002) Model selection and inference: a practical information-theoretic approach. Springer

Cantrell RS, Cosner C (2003) Spatial ecology via reaction-diffusion equations. Wiley

Caswell H, Etter RJ (1993) Ecological interactions in patchy environments: from patch-occupancy models to cellular automata. In: Levin SA et al (eds) Patch dynamics. Springer-Verlag, Berlin, pp 93–109

Conn PB, Johnson DS, Williams PJ, Melin SR, Hooten MB (2018) A guide to Bayesian model checking for ecologists. Ecol Monogr 88:526–542

Cormack RM (1964) Estimates of survival from the sighting of marked animals. Biometrika 51: 429–438

Dail D, Madsen L (2011) Models for estimating abundance from repeated counts of an open metapopulation. Biometrics 67:577–587

de Valpine P, Turek D, Paciorek CJ, Anderson-Bergman C, Lang DT, Bodik R (2017) Programming with models: writing statistical algorithms for general model structures with NIMBLE. J Comput Graph Stat 26:103

Dorazio RM, Royle JA (2005) Estimating size and composition of biological communities by modeling the occurrence of species. J Am Stat Assoc 100:389–398

Efford MG (2004) Density estimation in live-trapping studies. Oikos 106:598–610

Elith J, Leathwick JR (2009) Species distribution models: ecological explanation and prediction across space and time. Annu Rev Ecol Evol Syst 40:677–697

Fuller AK, Sutherland CS, Royle JA, Hare MP (2016) Estimating population density and connectivity of American mink using spatial capture-recapture. Ecol Appl 26:1125–1135

Gelman A, Meng XL, Stern H (1996) Posterior predictive assessment of model fitness via realized discrepancies. Stat Sin 6:733–760

Gelman A, Carlin JB, Stern HS, Dunson DB, Vehtari A, Rubin DB (2014) Bayesian data analysis. Chapman & Hall/CRC

Geman S, Geman D (1984) Stochastic relaxation, gibbs distributions, and the Bayesian restoration of images. IEEE Trans Pattern Anal Mach Intell PAMI-6:721–741

Hanski I (1994) A practical model of metapopulation dynamics. J Anim Ecol 63:151–162

Hanski I (1998) Metapopulation dynamics. Nature 396:41–49

Hanski I, Gilpin ME (1997) Metapopulation biology: ecology, genetics, and evolution. Academic Press

Hao T, Elith J, Lahoz-Monfort JJ, Guillera-Arroita G (2020) Testing whether ensemble modelling is advantageous for maximising predictive performance of species distribution models. Ecography 43:549–558

Hassell MP, Comins HN, Mayt RM (1991) Spatial structure and chaos in insect population dynamics. Nature 353:255–258

Hooten MB, Hefley TJ (2019) Bringing Bayesian models to life. CRC Press

Hooten MB, Hobbs NT (2015) A guide to Bayesian model selection for ecologists. Ecol Monogr 85:3–28

Hostetter NJ, Regehr EV, Wilson RR, Royle JA, Converse SJ (2022) Modeling spatiotemporal abundance and movement dynamics using an integrated spatial capture–recapture movement model. Ecology 103:e3772

Howell PE, Muths E, Hossack BR, Sigafus BH, Chandler RB (2018) Increasing connectivity between metapopulation ecology and landscape ecology. Ecology 99:1119–1128

Jolly GM (1965) Explicit estimates from capture-recapture data with both death and immigration-stochastic model. Biometrika 52:225–248

Kéry M, Royle JA (2016) Applied hierarchical modeling in ecology: analysis of distribution, abundance and species richness in R and BUGS, Volume 1: Prelude and static models. Elsevier

Kéry M, Royle JA (2021) Applied hierarchical modeling in ecology: analysis of distribution, abundance and species richness in R and BUGS, Volume 2: Dynamic and advanced models. Elsevier

Levins R (1969) Some demographic and genetic consequences of environmental heterogeneity for biological control. Bull Entomol Soc Am 15:237–240

Levins R (1970) Extinction. In: Gerstenhaber M (ed) Some mathematical problems in biology. American Mathematical Society, Providence, pp 75–107

Link WA, Eaton MJ (2012) On thinning of chains in MCMC. Methods Ecol Evol 3:112–115

MacKenzie DI, Nichols JD, Lachman GB, Droege S, Royle AA, Langtimm CA (2002) Estimating site occupancy rates when detection probabilities are less than one. Ecology 83:2248–2255

MacKenzie DI, Nichols JD, Hines JE, Knutson MG, Franklin AB (2003) Estimating site occupancy, colonization, and local extinction when a species is detected imperfectly. Ecology 84: 2200–2207

Moilanen A (2002) Implications of empirical data quality to metapopulation model parameter estimation and application. Oikos 96:516–530

Ovaskainen, O. and Abrego, N. 2020. Joint species distribution modelling

Ovaskainen O, Saastamoinen M (2018) Frontiers in metapopulation biology: the legacy of Ilkka Hanski. Annu Rev Ecol Evol Syst 49:231–252

Ovaskainen O, Tikhonov G, Norberg A, Guillaume Blanchet F, Duan L, Dunson D, Roslin T, Abrego N (2017) How to make more out of community data? A conceptual framework and its implementation as models and software. Ecol Lett 20:561–576

Pulliam HR (1988) Sources, sinks and population regulation. Am Nat 132:652–661

R Development Core Team (2019) R: a language and environment for statistical computing. R Foundation for Statistical Computing

Rakhimberdiev E, Karagicheva J, Saveliev A, Loonstra AHJ, Verhoeven MA, Hooijmeijer JCEW, Schaub M, Piersma T (2022) Misidentification errors in reencounters result in biased estimates of survival probability from CJS models: evidence and a solution using the robust design. Methods Ecol Evol 13:1106–1118

Rossman S, Yackulic CB, Saunders SP, Reid J, Davis R, Zipkin EF (2016) Dynamic N-occupancy models: estimating demographic rates and local abundance from detection-nondetection data. Ecology 97:3300

Royle JA (2004) N-mixture models for estimating population size from spatially replicated counts. Biometrics 60:108–115

Royle J, Dorazio R (2008) Hierarchical modeling and inference in ecology: the analysis of data from populations, metapopulations and communities. Elsevier

Royle JA, Young KV (2008) A hierarchical model for spatial capture recapture data. Ecology 89: 2281–2289

Royle JA, Fuller AK, Sutherland C (2018) Unifying population and landscape ecology with spatial capture–recapture. Ecography 41:444–456

Rubin DB (1984) Bayesianly justifiable and relevant frequency calculations for the applied statistician. Ann Stat 12:1151–1172

Seber GAF (1965) A note on the multiple-recapture census. Biometrika 52:249–260

Thomas CD, Kunin WE (1999) The spatial structure of populations. J Anim Ecol 68:647–657

Tredennick AT, Hooker G, Ellner SP, Adler PB (2021) A practical guide to selecting models for exploration, inference, and prediction in ecology. Ecology 102:e03336

Turing AM (1952) The chemical basis of morphogenesis. Philos Trans R Soc Lond Ser B Biol Sci 237:37–72

Tyre AJ, Tenhumberg B, Field SA, Niejalke D, Parris K, Possingham HP (2003) Improving precision and reducing bias in biological surveys: estimating false-negative error rates. Ecol Appl 13:1790–1801

Watanabe S (2013) A widely applicable Bayesian information criterion. J Mach Learn Res 14:867–897

Wikle CK (2003) Hierarchical Bayesian models for predicting the spread of ecological processes. Ecology 84:1382–1394

Wikle CK, Zammit-Mangion A, Cressie N (2019) Spatio-temporal statistics with R. CRC Press

Williams BK, Nichols JD, Conroy MJ (2002) Analysis and management of animal populations: modeling, estimation, and decision making. Elsevier

Zhao Q (2020) On the sampling design of spatially explicit integrated population models. Methods Ecol Evol 11:1207–1220

Occupancy Models 2

2.1 Introduction

In this chapter, we will introduce occupancy models (MacKenzie et al. 2002; Tyre et al. 2003) and extensions that use detection/non-detection data to estimate the true occupancy status of species. Occupancy is probably the simplest way to measure population status and thus provides limited information for the spatial structures and dynamics of populations. Occupancy status, which is often represented by 1's (presence) and 0's (absence), is a reduced form of abundance in the sense that the study species will be considered present at a site as long as there is at least one individual of the species at that site (i.e., local abundance ≥ 1), and absent only when there is no individual at all at that site (i.e., local abundance $= 0$). Dynamic occupancy models (MacKenzie et al. 2003; Royle and Kéry 2007) further explain the temporal changes of occupancy status with local extinction (change from 1 to 0) and colonization (change from 0 to 1), which again is a reduced form of the BIDE processes in the sense that a local population goes extinct when all individuals at that site either die (D) or emigrate (E) without giving birth (B) to any new individual, and an unoccupied site is colonized when at least one individual immigrates (I) into this site. Despite the limited information occupancy models can provide, they are nonetheless one of the most broadly used family of models in population ecology, due to the relative ease with which detection/non-detection data can be collected.

Occupancy models work only when there are both 1's and 0's in the true occupancy status (although this may not be immediately clear from the 1's and 0's in the data due to imperfect detection, which we will detail in Sect. 2.2). Dynamic occupancy models work only when there is a substantial rate of turnover (extinction and colonization); in that sense, dynamic occupancy models more closely represent classic metapopulation models than other models that will be introduced later in this

Supplementary Information The online version contains supplementary material available at https://doi.org/10.1007/978-3-031-64518-1_2.

Q. Zhao, *Bayesian Analysis of Spatially Structured Population Dynamics*, Analysis and Synthesis 253, https://doi.org/10.1007/978-3-031-64518-1_2

book. Dynamic occupancy models are nonetheless generalized forms of classic metapopulation models in their ability of accounting for imperfect detection and dealing with patchy to continuous habitats. For continuous habitats, we need to divide the study area into irregular (e.g., counties in the USA) or regular polygons (e.g., grids) that cover the entire landscape. Natural habitat patches or artificially defined polygons will then serve as basic spatial units or "sites" in occupancy modeling.

Regardless of how these "sites" are defined, let us consider the situation when they are larger than the territory of the study species so that individuals will not move in or out of a site during a primary occasion (e.g., a breeding season in a year); we still want to allow movement among sites between years. In this way, the true occupancy status (in fact local abundance) at each site does not change within a year, which is called the closed population, or simply closure, assumption. When such an assumption is violated, it is no longer "occupancy" that is modeled; instead, 1 represents "use" of the site, and 0 otherwise. This will also change the ecological interpretation of dynamic occupancy models; for example, a change from 1 in 1 year to 0 in the next year will represent stopping using a site by the study species rather than local extinction. To allow modeling occupancy, extinction, and colonization, we will follow the within-year closed population assumption in this chapter. We will follow the same assumption in Chap. 3 for N-mixture models and Chap. 4 for integrated population models, and release the assumption in Chap. 5 for spatial capture-recapture models.

Next, we will introduce a series of occupancy models with increasing ecological realism, and thus model complexity, to provide a friendly learning curve. We will start with the original "static" occupancy model (MacKenzie et al. 2002; Tyre et al. 2003), which considers neither space nor time. We then will introduce a dynamic occupancy model without considering spatial dependency between sites (MacKenzie et al. 2003; Royle and Kéry 2007) to help the readers understand the temporal aspects of these models. After that, we will introduce dynamic occupancy models that account for spatial dependency (Bled et al. 2011; Sutherland et al. 2014; Chandler et al. 2015; Broms et al. 2016; Howell et al. 2018) to further help the readers to understand the spatial aspects of these models. At last, we will introduce a spatial dynamic occupancy model that accounts for interspecific interactions (Mackenzie et al. 2004; Waddle et al. 2010; Rota et al. 2016; Fidino et al. 2019). Within increasing ecological realism, we hope to prepare the readers for a broad range of situations they may encounter in their studies, a philosophy we will also adopt in later chapters. We will end this chapter with a brief summary of other extensions of occupancy models.

2.2 Static Occupancy Models

The static occupancy model (MacKenzie et al. 2002; Tyre et al. 2003) is a hierarchical Bayesian model that contains three sub-models, a process sub-model, an observation sub-model, and a parameter sub-model. The process sub-model describes the true occupancy status of a species given environmental conditions, the observation sub-model describes how detection/non-detection data are obtained given the true occupancy status, and the parameter sub-model describes the prior distributions of the parameters. Because we use vague priors throughout the book, we will focus on describing the process and observation sub-models.

The process sub-model assumes that the true occupancy status or presence/absence of a species follows a Bernoulli distribution with a success probability that is a function of covariates. For example, we can imagine some fictitious butterfly species living in the landscape described in Fig. 1.4, in which patchy meadows contain their food plant and are located along a temperature gradient. If meadow size has a positive effect on the occupancy probability of this butterfly species, that means the butterfly is more likely to live in larger meadows at some given temperature. Likewise, if temperature has a negative effect on occupancy probability, the butterfly is less likely to live in warmer meadows, for a given meadow size. Using statistical language to describe the process sub-model, we have

$$\left.\begin{array}{c} z_i \sim Bernoulli(\psi_i) \\ \psi_i = logit^{-1}\left(\beta_0 + \beta_1 \times x_{i,1} + \ldots + \beta_{n_{covs}} \times x_{i,n_{covs}}\right) \end{array}\right\}, \tag{2.1}$$

in which z_i represents the presence ($z_i = 1$) or absence ($z_i = 0$) of the study species at the ith site among a total of n_{site} sites, ψ_i is occupancy probability, x's are n_{covs} covariates ($n_{covs} = 2$ if we consider two covariates, meadow size, and temperature), and β's are intercept and slope parameters representing the effects of x's on ψ at a logit scale. At this point, you may notice the similarity between Eqs. (2.1) and (1.3). Indeed, the process sub-model of the occupancy model can be considered a logistic regression, except that data y in Eq. (1.3) is replaced by the true but latent occupancy status z here.

The observation sub-model of the occupancy model links the observed detection/non-detection data to the true occupancy status. Let us continue with the example of the fictitious butterfly. When our field observers visit a meadow to determine if the butterfly lives there or not, they can make mistakes. If the butterfly lives in a meadow, they may find it and correctly write down "detection" or "1" for that meadow, or miss it and mistakenly write down "non-detection" or "0." If the butterfly does not live in a given meadow, we assume that our observers are trained well enough not to mistakenly find it, and thus will always correctly write down a "0." Again, using statistical language, we have

$$\left.\begin{array}{c} y_{i,j} \sim Bernoulli\left(z_i \times p_{i,j}\right) \\ p_{i,j} = logit^{-1}\left(\alpha_0 + \alpha_1 \times w_{i,j,1} + \ldots + \alpha_{n_{pcvs}} \times w_{i,j,n_{pcvs}}\right) \end{array}\right\}, \tag{2.2}$$

in which $y_{i,j}$ represents the detection ($y_{i,j} = 1$) or non-detection ($y_{i,j} = 0$) of the study species during the jth visit or replicated observation occasion (among a total of n_{reps} visits) of site i, $p_{i,j}$ is detection probability, w's are n_{pcvs} observation covariates for each site and visit, and α's are intercept and slope parameters governing the effects of w's on $p_{i,j}$ at a logit scale. Note that the number of visits, n_{reps}, can differ among sites and thus be written as $n_{reps,\,i}$. We can still let any $n_{reps,i} = \max_i (n_{reps,i})$, in which case we will have missing values in y (and likely also in w). We will talk about how to deal with missing values in Sect. 2.4.2.

The observation sub-model has two main assumptions. First, it assumes that the n_{reps} visits for each site i are made over a time period that is short enough so that z_i does not change among the visits. This is consistent with the closed population assumption we described earlier in Sect. 2.1 in the sense that, the longer the time period is, the larger the territory of the species is likely to be, and eventually a site can no longer contain the territory of the study species. Having multiple visits to a site during a period of closed population such as a breeding season is called the robust sampling design (Pollock 1982; Kendall et al. 1995), which improves model inference and thus is adopted in this and later models. It is worth noting though, that some researchers have found it possible to estimate occupancy based on a single visit to each site (Lele et al. 2012; but see Doser and Stoudt 2024). The second main assumption is that observers do not mistakenly detect a species that is not actually present at a site. This assumption is represented by having $y_{i,\,j} \sim Bernoulli(z_i \times p_{i,\,j})$, which means that we have a probability of $p_{i,\,j}$ to detect the study species ($y_{i,\,j} = 1$) if the species exists at the site ($z_i = 1$ leads to $z_i \times p_{i,\,j} = p_{i,\,j}$), and we will never detect the species ($y_{i,\,j} = 0$) if it does not exist at that site ($z_i = 0$ leads to $z_i \times p_{i,\,j} = 0$). In other words, the model in its current form only allows for false-negative observation errors (not detecting a species when it exists) but not false positive observation errors (detecting a species when it does not exist). We note that this second assumption has been relaxed in occupancy models that account for both false-positive and false-negative observation errors (Royle and Link 2006; Miller et al. 2011; Chambert et al. 2015; Kéry and Royle 2021; Chap. 7).

Next, let us simulate data for the model described above using the following code. In particular, we assume that the two covariates simulated in Chap. 1, meadow size and temperature, have positive ($\beta_{meadow} = 1$) and negative ($\beta_{temperature} = -0.4$) effects on the occupancy probability of the target species at a logit scale, respectively.

```
library(boot) # for using logit and inv.logit functions

set.seed(2)

# Basic values
nsite <- 360 # number of sites
nreps <- 5  # number of within-season replicates
ncovs <- 2  # number of environmental covariates
npcvs <- 2  # number of observational covariates

beta <- c(0.4, 1, -0.4)  #intercept and slopes for true occupancy status
alpha <- c(0.3, -0.6, 0.3) # intercept and slopes for detection
probability
```

```
# Read in environmental covariates
load('data/covariates.RData')
x1 <- (log(cov$size) - mean(log(cov$size))) / sd(log(cov$size)) #
standardized log habitat patch size
x2 <- (cov$temp - mean(cov$temp)) / sd(cov$temp) # standardized
temperature
x <- cbind(x1[,1], x2[,2]) # environmental covariates

# Simulate data
psi <- inv.logit(cbind(1,x) %*% beta) # occupancy probability
z <- rbinom(nsite, 1, psi) # true occupancy status

w <- array(rnorm(nsite * nreps * npcvs, 0, 1), dim=c(nsite, nreps,
npcvs)) # observational covariates

p <- matrix(, nsite, nreps) # detection probability
for (i in 1:nsite) {
 p[i,] <- inv.logit(cbind(1,w[i,,]) %*% alpha)
} # i

y <- matrix(, nsite, nreps) # detection/non-detection data
for (i in 1:nsite) {
 y[i,] <- rbinom(nreps, 1, z[i] * p[i,])
} # i
```

The simulated true occupancy status of the first five sites looks like the following. Here, we can see that the species is absent at site 1, and present at sites 2–5.

	Occupancy
Site 1	0
Site 2	1
Site 3	1
Site 4	1
Site 5	1

The simulated detection/non-detection data for the first five sites looks like the following. The species is not detected at site 1 during any of the visits, which is not surprising because the species is absent at this site and we assume no false positive observation errors. However, for sites 2–5, we have missed the species for quite a few times, despite the presence of the species at these sites. In particular, we missed the species at site 5 during all five visits, which is indeed a rare case but nonetheless has happened.

	Detection/non-detection				
	Visit 1	Visit 2	Visit 3	Visit 4	Visit 5
Site 1	0	0	0	0	0
Site 2	0	1	0	0	1
Site 3	1	0	1	1	0
Site 4	1	0	1	1	1
Site 5	0	0	0	0	0

We can then develop a custom MCMC algorithm to implement the model, as shown below. The interest is to estimate parameters β, α, and z given data y, x, and w.

```
occu_mcmc <- function(y, x, w, nmcmc) {

 # Setup variables
 nsite <- dim(y) [1]
 nreps <- dim(y) [2]
 ncovs <- dim(x) [2]
 npcvs <- dim(w) [3]
 ysum <- rowSums(y)

 beta_save <- matrix(0, ncovs + 1, nmcmc)
 alpha_save <- matrix(0, npcvs + 1, nmcmc)
 z_save <- matrix(0, nsite, 2000)

 # Priors
 beta_mean <- rep(0, ncovs + 1)
 beta_sd <- 2
 alpha_mean <- rep(0, npcvs + 1)
 alpha_sd <- 2

 # Starting values
 beta <- rep(0, ncovs + 1)
 alpha <- rep(0, npcvs + 1)
 p <- matrix(0.5, nsite, nreps)
 z <- ifelse(ysum==0, 0, 1)

 # Tuning factors
 beta_tune <- 0.15
 alpha_tune <- 0.1

 # Begin MCMC loop
 for (k in 1:nmcmc) {
  ### Sample beta
  beta_star <- rnorm(ncovs + 1, beta, beta_tune)
  mh1 <- sum(dbinom(z, 1, inv.logit(cbind(1,x) %*% beta_star),
log=TRUE)) +
      sum(dnorm(beta_star, beta_mean, beta_sd, log=TRUE))
```

```
  mh2 <- sum(dbinom(z, 1, inv.logit(cbind(1,x) %*% beta   ), log=TRUE)) +
      sum(dnorm(beta    , beta_mean, beta_sd, log=TRUE))
  mh <- exp(mh1 - mh2)
  if (mh > runif(1)) {
   beta <- beta_star
  }

  ### Sample alpha
  alpha_star <- rnorm(npcvs + 1, alpha, alpha_tune)
  p_star <- matrix(, nsite, nreps)
  for (i in 1:nsite) {
   p_star[i,] <- inv.logit(cbind(1,w[i,,]) %*% alpha_star)
  } # i
 mh1 <- sum(dbinom(y[which(z==1),], 1, p_star[which(z==1),], log=TRUE)) +
       sum(dnorm(alpha_star, alpha_mean, alpha_sd, log=TRUE))
 mh2 <- sum(dbinom(y[which(z==1),], 1, p   [which(z==1),], log=TRUE)) +
       sum(dnorm(alpha    , alpha_mean, alpha_sd, log=TRUE))
  mh <- exp(mh1 - mh2)
  if (mh > runif(1)) {
   alpha <- alpha_star
   p <- p_star
  }

  ### Sample z
  s_temp <- inv.logit(cbind(1,x) %*% beta)
  num_temp <- s_temp * apply((1 - p), 1, prod)
  psi_temp <- num_temp / (num_temp + (1 - s_temp))
  z[which(ysum==0)] <- rbinom(length(which(ysum==0)), 1, psi_temp
[which(ysum==0)])

  ### Save samples
  beta_save[,k] <- beta
  alpha_save[,k] <- alpha
  if (nmcmc > 2000) {
   if (k > nmcmc-2000) {
    z_save[,k-nmcmc+2000] <- z
   }
  }
 } # k

 # Write output
 list(beta_save=beta_save, alpha_save=alpha_save, z_save=z_save)
} # occu_mcmc
```

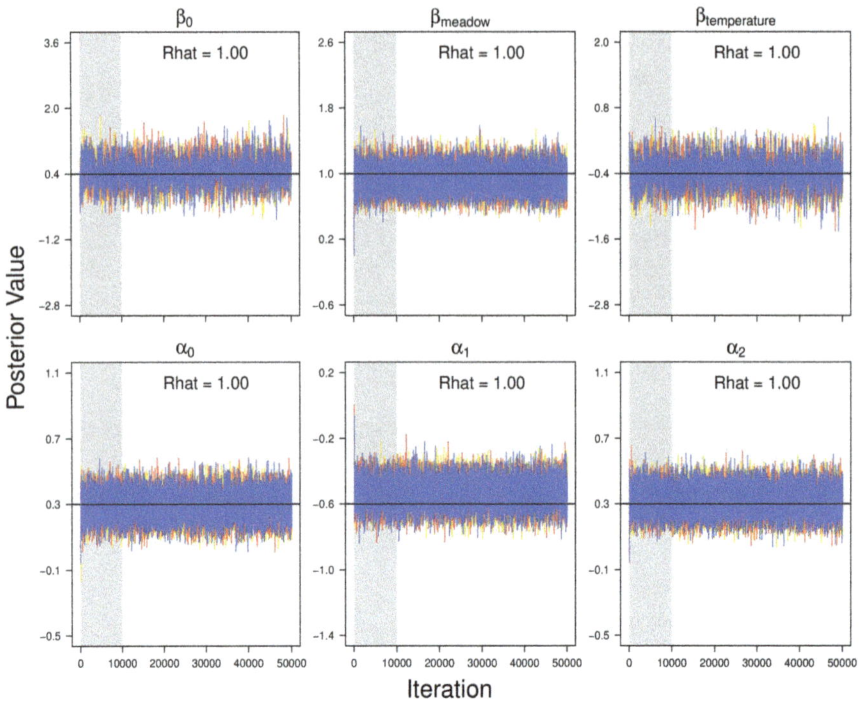

Fig. 2.1 Traceplots showing the posterior chains for β and α in the static occupancy model. Rhat values are calculated after discarding the first 10,000 iterations. Horizontal lines represent the true values used in data simulation

Next, let us run the following code to implement the algorithm. The computing takes a little bit less than 3 min. Using code in Appendix 2.1, we create some traceplots to check model convergence; the model converges well, as indicated by the well-mixed chains and Rhat values that are output as 1.00 (Fig. 2.1). The traceplots also show consistency between the estimates of β and α and their true values that are used in data simulation (Fig. 2.1).

```
library(foreach)  # for parallel computing
library(doParallel) # for parallel computing

numCores <- round(detectCores() / 2) # only use half of the cores for
parallel computing
registerDoParallel(numCores) # setup parallel computing

nmcmc <- 50000 # number of iterations of MCMC computing
chain <- 3   # number of chains
```

```
start_time <- Sys.time() # start time of computing
out <- foreach (i=1:chain, .packages='boot') %dopar% {
 occu_mcmc(y=y, x=x, w=w, nmcmc=nmcmc)
} # i
stopImplicitCluster() # clean up the cluster
end_time <- Sys.time() # end time of computing
time_taken <- end_time - start_time # computing time
```

We can also use the following code to plot the estimated occupancy status, z, which again shows consistency with the simulated true occupancy status (Fig. 2.2). The estimated number of occupied sites ($\sum_{i=1}^{n_{site}} z_i$) is also consistent with the simulated true value.

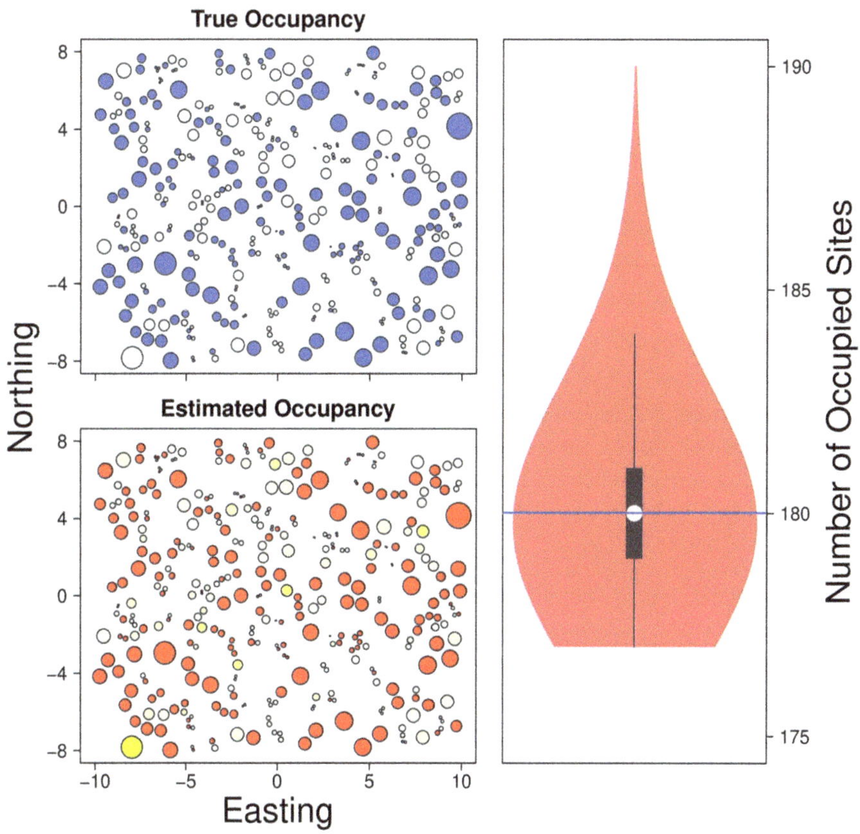

Fig. 2.2 True occupancy status z in the simulated butterfly data (upper-left panel; blue represents presence and white represents absence) and its estimates (lower-left panel; a deeper color represents a higher occupancy probability, and vice versa) as well as the true (horizontal line) and estimated (violin) number of occupied sites among a total of 360 sites ($\sum_{i=1}^{n_{site}} z_i$; right panel)

```
library(vioplot) # for making violin plots

for (i in 1:chain) {
 if (i == 1) {
  z_post <- out[[i]]$z_save
 } else {
  z_post <- cbind(z_post, out[[i]]$z_save)
 }
} # i

nscale <- 100
z_scale <- round(apply(z_post, 1, mean) * (nscale - 1)) + 1

layout(mat=matrix(c(rep(c(1,1,2,2),times=3),rep(3,8)), nrow=4,
ncol=5, byrow=FALSE))
par(mar=c(1,2,3,1))
par(oma=c(4,4,0,6))

plot(cov$lon, cov$lat, xlim=c(-10,10), ylim=c(-8,8), axes=F,
xlab='', ylab='',
   pch=21, cex=cov$size[,1]*6, col='grey18', bg=ifelse(z==1,
'royalblue', 'white'))
axis(1, at=seq(-10,10,5), labels=rep('',5))
axis(2, at=seq(-08,08,4), cex.axis=1.6, las=2)
box()
title(main='True Occupancy', cex.main=2)

plot(cov$lon, cov$lat, xlim=c(-10,10), ylim=c(-8,8), axes=F,
xlab='', ylab='',
   pch=21, cex=cov$size[,1]*6, col='grey18', bg=rev(heat.colors
(nscale))[z_scale])
axis(1, at=seq(-10,10,5), cex.axis=1.6)
axis(2, at=seq(-08,08,4), cex.axis=1.6, las=2)
box()
title(main='Estimated Occupancy', cex.main=2)

axis(1, at=0, labels='Easting', tick=F, line=2.5, cex.axis=3)
title(ylab='Northing', cex.lab=3, line=1.5, outer=T)

par(mar=c(1,1,3,1))
plot(1, ylim=c(175, 190), type='n', axes=F, xlab='', ylab='')
vioplot(apply(z_post,2,sum), col='lightcoral', rectCol='grey36',
lineCol='grey36',
    border=NA, cex=3, add=T)
abline(h=sum(z), col='royalblue', lwd=1.8)
```

```
axis(4, at=seq(175, 190, 5), cex.axis=1.6, las=2)
axis(4, at=182.5, label='Number of Occupied Sites', line=4.5, cex.
axis=3, tick=F)
```

2.3 Dynamic Occupancy Models

The original occupancy model of MacKenzie et al. (2002) and Tyre et al. (2003) is a static model in the sense that it only models a snapshot of the system. By specifying that occupancy status is only a function of environmental conditions at the same time (although time is not specified in the model and thus is only implied), the model basically assumes that species occupancy always tracks environmental changes instantaneously (or there is no environmental change at all). However, this is hardly the case in reality, as the state of a population at a given time is arguably always related to its status in the past, and it normally takes some time for a population to respond to environmental changes. The dynamic occupancy model (MacKenzie et al. 2003; Royle and Kéry 2007) deals with such an issue by modeling changes in occupancy as a function of covariates, rather than linking occupancy to covariates directly (except for the first year). The dynamic occupancy model is a model for discrete time steps (also called primary occasions), and these time steps in practice are often years, or breeding seasons. But note that the time steps necessary to describe occupancy dynamics in an adequate manner will have to be adapted to the life-cycle of a study species: a time step may be 1 day for a metapopulation of some damselfly species or successive 10 years for a tree species.

The dynamic occupancy model contains a process sub-model that describes the true occupancy status and an observation sub-model that links detection/non-detection data with true occupancy. The process sub-model first describes the initial occupancy status (i.e., occupancy at the first year) using basically Eq. (2.1) for the static occupancy model; however, now z, ψ and x's are all indexed by 1 to indicate that they are for $t = 1$. Thus, we have

$$\left.\begin{aligned} z_{i,1} &\sim Bernoulli\left(\psi_{i,1}\right) \\ \psi_{i,1} &= logit^{-1}\left(\beta_0^{[0]} + \beta_1^{[0]} \times x_{i,1,1} + \ldots + \beta_{n_{covs}}^{[0]} \times x_{i,1,n_{covs}}\right) \end{aligned}\right\}, \tag{2.3}$$

in which $z_{i,\ 1}$ represents the presence ($z_{i,\ 1} = 1$) or absence ($z_{i,\ 1} = 0$) of the study species at site i in the first year, $\psi_{i,\ 1}$ is initial occupancy probability, x's are covariates, and $\beta^{[0]}$'s are intercept and slope parameters governing the environmental effects on initial occupancy. The superscript [0] in the regression coefficients β's is meant to distinguish them from other regression coefficients used in other sub-models (see below). At first glance, the sub-model for initial occupancy status seems to contradict our wish of modeling changes in occupancy, but it is a must-have. This formulation is a typical example of a hidden Markov model, where the dynamics of a system in discrete time is first described at $t = 1$ as a static model, and then allowed to evolve from t to $t + 1$ with a probabilistic, recursive rule. The initial status in a dynamic occupancy model is often estimated less reliably, but this usually has little influence on the inferences for later time steps.

From the second year (i.e., $t \geq 2$), the model assumes that a site that is occupied in year t has a probability of remaining occupied in year $t + 1$ (persistence, the opposite of extinction), and a site that is unoccupied in year t has a probability to become occupied in year $t + 1$ (colonization). Furthermore, in the model we can specify persistence and colonization probabilities as functions of covariates. Using our little butterfly as an example, instead of assuming that the butterfly is more likely to live in a larger meadow, we now assume that when the butterfly lives in a meadow, it is more likely to continue living there the next year if the meadow is larger, and when the butterfly does not live in a meadow, they are more likely to move into that meadow next year if the meadow is larger, and vice versa. Using statistical language, we have

$$\left.\begin{array}{r}z_{i,t} \sim Bernoulli\left(\psi_{i,t}\right)\\ \psi_{i,t} = z_{i,t-1} \times \phi_{i,t-1} + \left(1 - z_{i,t-1}\right) \times \gamma_{i,t-1}\\ \phi_{i,t-1} = logit^{-1}\left(\beta_0^{[\phi]} + \beta_1^{[\phi]} \times x_{i,t,1} + \ldots + \beta_{n_{covs}}^{[\phi]} \times x_{i,t,n_{covs}}\right)\\ \gamma_{i,t-1} = logit^{-1}\left(\beta_0^{[\gamma]} + \beta_1^{[\gamma]} \times x_{i,t,1} + \ldots + \beta_{n_{covs}}^{[\gamma]} \times x_{i,t,n_{covs}}\right)\end{array}\right\}, \tag{2.4}$$

in which $\phi_{i,\,t-1}$ is persistence probability for site i from year $t - 1$ to year t, $\gamma_{i,\,t-1}$ is colonization probability, $\beta^{[\phi]}$'s are intercept and slope parameters governing environmental effects on persistence, and $\beta^{[\gamma]}$ are intercept and slope parameters governing environmental effects of colonization. The occupancy probability, $\psi_{i,\,t}$, equals $\phi_{i,\,t-1}$ when $z_{i,\,t-1} = 1$ (i.e., site i is occupied in year $t - 1$), and equals $\gamma_{i,\,t-1}$ when $z_{i,\,t-1} = 0$ (i.e., site i is unoccupied in year $t - 1$), a trick to comprise persistent and colonization in a single Bernoulli distribution.

Then, the observation sub-model links the observed detection/non-detection data to the true occupancy status in the same way as in Eq. (2.2), except that all variables are now indexed by t. Thus, we have

$$\left.\begin{array}{r}y_{i,t,j} \sim Bernoulli\left(z_{i,t} \times p_{i,t,j}\right)\\ p_{i,t,j} = logit^{-1}\left(\alpha_0 + \alpha_1 \times w_{i,t,j,1} + \ldots + \alpha_{n_{pcvs}} \times w_{i,t,j,n_{pcvs}}\right)\end{array}\right\}. \tag{2.5}$$

Using the code below, let us simulate data for the model specified above. We assume that meadow size has positive effects on initial occupancy ($\beta_{meadow}^{[0]} = 1$), persistence ($\beta_{meadow}^{[\phi]} = 0.6$) and colonization ($\beta_{meadow}^{[\gamma]} = 0.4$), and temperature has negative effects on initial occupancy ($\beta_{temperature}^{[0]} = -0.4$), persistence ($\beta_{temperature}^{[\phi]} = -0.3$), and colonization ($\beta_{temperature}^{[\gamma]} = -0.2$) at a logit scale. The simulated data look quite similar to the simulated data for the static occupancy model, except that now the data have an additional dimension for a year.

```
library(boot) # for using logit and inv.logit functions

set.seed(3)

# Basic values
nsite <- 360 # number of sites
nyear <- 20  # number of years
nreps <- 5   # number of within-season replicates
ncovs <- 2   # number of environmental covariates
npcvs <- 2   # number of observational covariates

beta0 <- c(0.4, 1, -0.4)       # intercept and slopes for initial occupancy
beta_phi   <- c(-0.2, 0.6, -0.3) # intercept and slopes for persistence
beta_gamma <- c(0.2, 0.4, -0.2)  # intercept and slopes for colonization
alpha <- c(0.3, -0.6, 0.3)       # intercept and slopes for detection
probability

# Read in environmental covariates
load('data/covariates.RData')
x1 <- (log(cov$size) - mean(log(cov$size))) / sd(log(cov$size)) #
standardized log habitat patch size
x2 <- (cov$temp - mean(cov$temp)) / sd(cov$temp) # standardized
temperature
x <- array(, dim=c(nsite, nyear, ncovs)) # environmental covariates
x[,,1] <- x1
x[,,2] <- x2

# Simulate data
psi <- matrix(, nsite, nyear) # occupancy probability
z <- matrix(, nsite, nyear) # occupancy status
psi[,1] <- inv.logit(cbind(1,x[,1,]) %*% beta0) # initial occupancy
probability
z[,1] <- rbinom(nsite, 1, psi[,1]) # initial occupancy status

phi   <- matrix(, nsite, nyear-1) # probability of persistence
gamma <- matrix(, nsite, nyear-1) # probability of colonization
for (t in 2:nyear) {
  phi  [,t-1] <- inv.logit(cbind(1,x[,t,]) %*% beta_phi  )
  gamma[,t-1] <- inv.logit(cbind(1,x[,t,]) %*% beta_gamma)
  psi[,t] <- z[,t-1] * phi[,t-1] + (1 - z[,t-1]) * gamma[,t-1]
  z[,t] <- rbinom(nsite, 1, psi[,t])
} # t
```

```
w <- array(rnorm(nsite * nyear * nreps * npcvs, 0, 1), dim=c(nsite,
nyear, nreps, npcvs)) # observational covariates

p <- array(, dim=c(nsite, nyear, nreps)) # detection probability
for (i in 1:nsite) {
 for (t in 1:nyear) {
  p[i,t,] <- inv.logit(cbind(1,w[i,t,,]) %*% alpha)
 } # t
} # i

y <- array(, dim=c(nsite, nyear, nreps)) # detection/non-detection
data
for (i in 1:nsite) {
 for (t in 1:nyear) {
  y[i,t,] <- rbinom(nreps, 1, z[i,t] * p[i,t,])
 } # t
} # i
```

Next, let us develop a custom MCMC algorithm to estimate parameters $\beta^{[0]}$, $\beta^{[\phi]}$, $\beta^{[\gamma]}$, α, and z using data y, x, and w, as shown below:

```
dynam_occu_mcmc <- function(y, x, w, nmcmc) {

 # Setup variables
 nsite <- dim(y)[1]
 nyear <- dim(y)[2]
 nreps <- dim(y)[3]
 ncovs <- dim(x)[3]
 npcvs <- dim(w)[4]
 ysum <- apply(y, 1:2, sum)

 beta0_save <- matrix(0, ncovs + 1, nmcmc)
 beta_phi_save <- matrix(0, ncovs + 1, nmcmc)
 beta_gamma_save <- matrix(0, ncovs + 1, nmcmc)
 alpha_save <- matrix(0, npcvs + 1, nmcmc)
 zsum_save <- matrix(0, nyear, nmcmc)

 # Priors
 beta0_mean <- rep(0, ncovs + 1)
 beta0_sd <- 2
 beta_phi_mean <- rep(0, ncovs + 1)
 beta_phi_sd <- 2
 beta_gamma_mean <- rep(0, ncovs + 1)
 beta_gamma_sd <- 2
```

```
alpha_mean <- rep(0, npcvs + 1)
alpha_sd <- 2

# Starting values
beta0 <- rep(0, ncovs + 1)
beta_phi  <- rep(0, ncovs + 1)
beta_gamma <- rep(0, ncovs + 1)
alpha <- rep(0, npcvs + 1)
phi  <- matrix(.5, nsite, nyear-1)
gamma <- matrix(.5, nsite, nyear-1)
p <- array(.5, dim=c(nsite, nyear, nreps))
z <- ifelse(ysum==0, 0, 1)

# Tuning factors
beta0_tune <- 0.2
beta_phi_tune <- 0.05
beta_gamma_tune <- 0.05
alpha_tune <- 0.035

# Begin MCMC loop
for (k in 1:nmcmc) {
  ### Sample beta0
  beta0_star <- rnorm(ncovs + 1, beta0, beta0_tune)
  mh1 <- sum(dbinom(z[,1], 1, inv.logit(cbind(1,x[,1,]) %*%
beta0_star), log=TRUE)) +
        sum(dnorm(beta0_star, beta0_mean, beta0_sd, log=TRUE))
  mh2 <- sum(dbinom(z[,1], 1, inv.logit(cbind(1,x[,1,]) %*% beta0   ),
log=TRUE)) +
        sum(dnorm(beta0   , beta0_mean, beta0_sd, log=TRUE))
  mh <- exp(mh1 - mh2)
  if (mh > runif(1)) {
   beta0 <- beta0_star
  }

  ### Sample beta_phi
  beta_phi_star <- rnorm(ncovs + 1, beta_phi, beta_phi_tune)
  phi_star <- matrix(, nsite, nyear-1)
  for (t in 2:nyear) {
   phi_star[,t-1] <- inv.logit(cbind(1,x[,t,]) %*% beta_phi_star)
  } # t
  mh1 <- sum(dbinom(z[,-1], 1, z[,-nyear] * phi_star + (1 - z[,-nyear]) *
gamma, log=TRUE)) +
        sum(dnorm(beta_phi_star, beta_phi_mean, beta_phi_sd, log=TRUE))
  mh2 <- sum(dbinom(z[,-1], 1, z[,-nyear] * phi   + (1 - z[,-nyear]) *
gamma, log=TRUE)) +
        sum(dnorm(beta_phi   , beta_phi_mean, beta_phi_sd, log=TRUE))
```

```
  mh <- exp(mh1 - mh2)
  if (mh > runif(1)) {
   beta_phi <- beta_phi_star
   phi <- phi_star
  }

  ### Sample beta_gamma
  beta_gamma_star <- rnorm(ncovs + 1, beta_gamma, beta_gamma_tune)
  gamma_star <- matrix(, nsite, nyear-1)
  for (t in 2:nyear) {
   gamma_star[,t-1] <- inv.logit(cbind(1,x[,t,]) %*% beta_gamma_star)
  } # t
  mh1 <- sum(dbinom(z[,-1], 1, z[,-nyear] * phi + (1 - z[,-nyear]) *
gamma_star, log=TRUE)) +
         sum(dnorm(beta_gamma_star, beta_gamma_mean, beta_gamma_sd,
log=TRUE))
  mh2 <- sum(dbinom(z[,-1], 1, z[,-nyear] * phi + (1 - z[,-nyear]) * gamma
, log=TRUE)) +
         sum(dnorm(beta_gamma    , beta_gamma_mean, beta_gamma_sd,
log=TRUE))
  mh <- exp(mh1 - mh2)
  if (mh > runif(1)) {
   beta_gamma <- beta_gamma_star
   gamma <- gamma_star
  }

  ### Sample alpha
  alpha_star <- rnorm(npcvs + 1, alpha, alpha_tune)
  p_star <- array(, dim=c(nsite, nyear, nreps))
  for (i in 1:nsite) {
   for (t in 1:nyear) {
    p_star[i,t,] <- inv.logit(cbind(1,w[i,t,,]) %*% alpha_star)
   } # t
  } # i
  mh1t <- mh2t <- numeric(nyear)
  for (t in 1:nyear) {
   mh1t[t] <- sum(dbinom(y[which(z[,t]==1),t,], 1, p_star[which(z[,
t]==1),t,], log=TRUE))
   mh2t[t] <- sum(dbinom(y[which(z[,t]==1),t,], 1, p    [which(z[,t]
==1),t,], log=TRUE))
  } # t
  mh1 <- sum(mh1t) +
         sum(dnorm(alpha_star, alpha_mean, alpha_sd, log=TRUE))
  mh2 <- sum(mh2t) +
         sum(dnorm(alpha    , alpha_mean, alpha_sd, log=TRUE))
```

```
  mh <- exp(mh1 - mh2)
  if (mh > runif(1)) {
   alpha <- alpha_star
   p <- p_star
  }

  ### Sample z
  s_temp <- matrix(, nsite, nyear)
  s_temp[,1] <- inv.logit(cbind(1,x[,1,]) %*% beta0)
  for (t in 2:nyear) {
   s_temp[,t] <- z[,t-1] * phi[,t-1] + (1 - z[,t-1]) * gamma[,t-1]
  } # t
  num_temp <- s_temp * apply((1 - p), 1:2, prod)
  psi_temp <- num_temp / (num_temp + (1 - s_temp))
  z[which(ysum==0)] <- rbinom(length(which(ysum==0)), 1, psi_temp
[which(ysum==0)])

  ### Save samples
  beta0_save[,k] <- beta0
  beta_phi_save[,k] <- beta_phi
  beta_gamma_save[,k] <- beta_gamma
  alpha_save[,k] <- alpha
  zsum_save[,k] <- colSums(z)
 } # k

 # Write output
 list(beta0_save=beta0_save,
   beta_phi_save=beta_phi_save, beta_gamma_save=beta_gamma_save,
   alpha_save=alpha_save, zsum_save=zsum_save)
} # dynam_occu_mcmc
```

When the algorithm is developed, we can use the following code to implement it, which takes about 1 h to run. Then, using code in Appendix 2.2, we can create traceplots that show good model convergence and consistency between parameter estimates of $\beta^{[0]}$, $\beta^{[\phi]}$, $\beta^{[\gamma]}$, and α and their true values used in data simulation (Fig. 2.3).

```
library(foreach)  # for parallel computing
library(doParallel) # for parallel computing

numCores <- round(detectCores() / 2) # only use half of the cores for
parallel computing
registerDoParallel(numCores) # setup parallel computing
```

```
nmcmc <- 50000 # number of iterations
chain <- 3    # number of chains

start_time <- Sys.time() # start time of computing
out <- foreach (i=1:chain, .packages='boot') %dopar% {
 dynam_occu_mcmc(y=y, x=x, w=w, nmcmc=nmcmc)
} # i
stopImplicitCluster() # clean up the cluster
end_time <- Sys.time() # end time of computing
time_taken <- end_time - start_time # computing time
```

We can also use the following code to create Fig. 2.4, which shows that the model can represent the temporal variation in the number of occupied sites (i.e., $\sum_{i=1}^{n_{site}} z_{i,t}$), which is mainly driven by the oscillation of temperature. Recall that temperature negatively influences both persistence and colonization. Thus, we have more sites occupied in colder years, and vice versa.

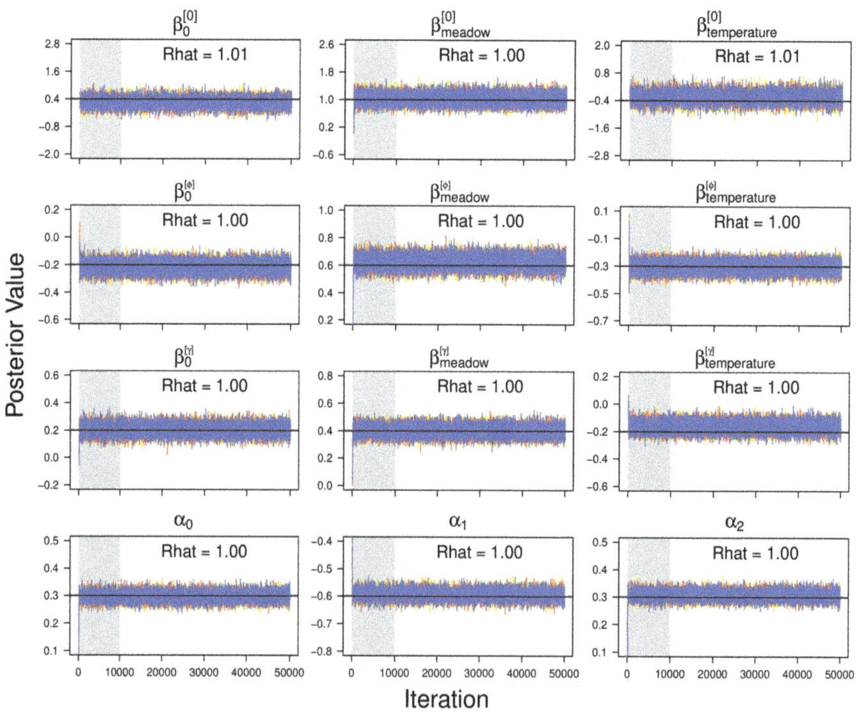

Fig. 2.3 Traceplots showing posterior chains for $\beta^{[0]}$, $\beta^{[\phi]}$, $\beta^{[\gamma]}$, and α in the non-spatial dynamic occupancy model. Rhat values are calculated after discarding the first 10,000 iterations. Horizontal lines represent the true values used in data simulation

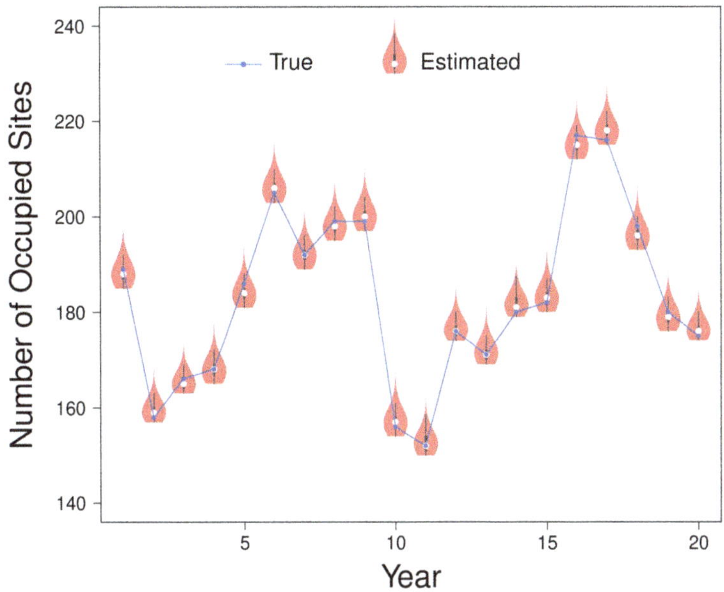

Fig. 2.4 Violin plots showing an excellent agreement between the estimated number of occupied sites ($\sum_{i=1}^{n_{site}} z_{i,t}$; violin) and the simulated true values (blue points)

```
library(vioplot) # for making violin plots

for (i in 1:chain) {
 if (i == 1) {
  zsum_post <- out[[i]]$zsum_save[, (nmcmc*0.2+1):nmcmc]
 } else {
  zsum_post <- cbind(zsum_post, out[[i]]$zsum_save[, (nmcmc*0.2+1):
nmcmc])
 }
} # i

par(mfrow=c(1,1))
par(mar=c(5,7,1,1))

plot(1, xlim=c(1,nyear), ylim=c(140,240), type='n', xlab='',
ylab='', axes=F)
vioplot(t(zsum_post), add=T, col='lightcoral', rectCol='grey36',
lineCol='grey36', border=NA)
lines(colSums(z) ~ c(1:nyear), type='o', pch=16, cex=.8, lwd=1.2,
col='royalblue')
axis(1, at=seq(0,nyear,5), cex.axis=1.5)
```

```
axis(2, at=seq(140,240,20), cex.axis=1.5, las=2)
title(xlab='Year', cex.lab=2.5, line=3.5)
title(ylab='Number of Occupied Sites', cex.lab=2.5, line=4.5)

points(x=5, y=232, pch=16, cex=1, col='royalblue')
lines(x=c(4.4,5.6), y=c(232,232), lwd=1.2, col='royalblue')
vioplot(zsum_post[11,]-median(zsum_post[11,])+232, at=10, add=T,
col='lightcoral', rectCol='grey36', lineCol='grey36', border=NA)
text(x= 5.6, y=232, labels='True', pos=4, cex=1.8)
text(x=10.6, y=232, labels='Estimated', pos=4, cex=1.8)
```

2.4 Spatial Dynamic Occupancy Models

While the dynamic occupancy model of MacKenzie et al. (2003) considers temporal changes in occupancy status, it only allows the parameters governing such changes, persistence, and colonization, to be functions of local environmental covariates but not occupancy status at other sites. In that sense, the dynamic occupancy model is non-spatial because the occupancy dynamics at a given site are unrelated to that at any other site, making the spatial configuration of the sites irrelevant. This is obviously unrealistic and misses something very important, as we have learned from all those metapopulation studies that population dynamics at a site are driven not only by the environmental conditions of that given site but also the connectivity (i.e., probability of moving) between that site and other sites (Hanski 1998). For instance, one might imagine that a site nearer an occupied site may have a higher probability of being occupied than one that is further away, simply because the potential of exchanging animals from the nearer site is higher. Therefore, researchers have developed spatial dynamic occupancy models (e.g., Bled et al. 2011; Broms et al. 2016) to account for connectivity among sites.

The first part of the process sub-model of spatial dynamic occupancy models describes the initial occupancy status of a species as a function of covariates such that

$$\left.\begin{array}{c} z_{i,1} \sim Bernoulli(\psi_{i,1}) \\ \psi_{i,1} = logit^{-1}\left(\beta_0^{[0]} + \beta_1^{[0]} \times x_{i,1,1} + \ldots + \beta_{n_{covs}}^{[0]} \times x_{i,1,n_{covs}}\right) \end{array}\right\}, \quad (2.6)$$

which is no different than what we have in the non-spatial dynamic occupancy model (Eq. 2.3). The second part of the process sub-model describes the inter-annual transitions of occupancy status, and again assumes that occupancy after the first year is driven by persistence and colonization, and persistence is a function of covariates such that

$$\left.\begin{array}{c} z_{i,t} \sim Bernoulli(\psi_{i,t}) \\ \psi_{i,t} = z_{i,t-1} \times \phi_{i,t-1} + (1 - z_{i,t-1}) \times \gamma_{i,t-1} \\ \phi_{i,t-1} = logit^{-1}\left(\beta_0^{[\phi]} + \beta_1^{[\phi]} \times x_{i,t,1} + \ldots + \beta_{n_{covs}}^{[\phi]} \times x_{i,t,n_{covs}}\right) \\ \gamma_{i,t-1} = \ldots \end{array}\right\}. \tag{2.7}$$

These assumptions are the same as what we can see in the non-spatial version of the model (Eq. 2.4). However, the spatial dynamic occupancy model assumes that colonization at an unoccupied site, $\gamma_{i,\,t\,-\,1}$, is influenced by the occupancy status of other sites given the connectivity between these sites; the connectivity can further be a function of adjacency or distance between these sites as well as environmental conditions in these sites, which we will describe in detail a little bit later. Therefore, the process sub-model of spatial dynamic occupancy models is very similar to the incidence function model (Hanski 1994) in metapopulation studies (Risk et al. 2011). Note that persistence at a site can also be influenced by other sites (e.g., Chandler et al. 2015), but here we will focus on colonization to illustrate how to incorporate spatial dependency in dynamic occupancy models. Later, we will detail how colonization can be specified in a few different ways.

Spatial dynamic occupancy models also have an observation sub-model that links detection/non-detection data to the true occupancy status given some observation covariates such that

$$\left.\begin{array}{c} y_{i,t,j} \sim Bernoulli(z_{i,t} \times p_{i,t,j}) \\ p_{i,t,j} = logit^{-1}\left(\alpha_0 + \alpha_1 \times w_{i,t,j,1} + \ldots + \alpha_{n_{pcvs}} \times w_{i,t,j,n_{pcvs}}\right) \end{array}\right\}. \tag{2.8}$$

which is the same as in the non-spatial version of the model (Eq. 2.5). However, note that, to account for spatial dependency among sites, the model requires us to model all the sites, even though we normally wouldn't be able to survey every site. One solution is to sign missing values to $y_{i,\,t,\,j}$ for unsurvey sites, but of course locations and covariates of these sites are still needed. An approach to dealing with missing values will be introduced in Sect. 2.4.2; model performance when missing values exist requires further investigation.

2.4.1 With Adjacency-Based Movement

As mentioned above, colonization at a site is related to the occupancy status at other sites based on their connectivity with this site. The simplest way of specifying connectivity is to assume that only adjacent sites (also called neighbors) are connected. For a continuous habitat, adjacency can be defined as the case where two sites (polygons) share some of their boundaries. For a patchy habitat such as the one we consider in this book, adjacency can be defined as the case where two sites are within a given distance. When adjacency is defined, the model assumes that an unoccupied site can be colonized only if any of its adjacent sites are occupied, and the probability of colonization is higher when more of its adjacent sites are occupied.

Using our butterfly as an example, this basically says if the butterfly does not live in a given meadow in a year, it has a chance to move into this meadow next year only if the species lives in at least one of the adjacent meadows. Furthermore, the species is more likely to move into this empty meadow if it lives in more of the adjacent meadows, and vice versa. To give the model greater flexibility and make it more realistic, some researchers further allow a small probability of colonization even when no adjacent sites are occupied to represent a potential effect of long-distance movement (Broms et al. 2016). Here, we introduce a model structure that is motivated by, but slightly different from what Broms et al. (2016) specify, which gives us

$$\gamma_{i_1,t-1} = 1 - (1-\zeta)^{\sum_{i_2=1}^{n_{site}} a_{i_1,i_2} \times z_{i_2,t-1}} \times (1-\delta), \tag{2.9}$$

in which ζ is the baseline probability of colonization from adjacent sites, a_{i_1,i_2} is a binary indicator of adjacency between any two sites i_1 and i_2 ($a_{i_1,i_2} = 1$ if sites i_1 and i_2 are adjacent to each other and $a_{i_1,i_2} = 0$ otherwise), and δ is a probability of colonization from non-adjacent sites that are considered to be due to long-distance movement (Broms et al. 2016). Because $\sum_{i_2=1}^{n_{site}} a_{i_1,i_2} \times z_{i_2,t-1}$ gives the number of sites that are adjacent to the site i_1 and are occupied in year $t-1$, ζ can be considered the probability of colonization if only one adjacent site is occupied and no colonization comes from non-adjacent sites.

Using the code below, let us simulate data while assuming positive effects of meadow size on initial occupancy ($\beta_{meadow}^{[0]} = 1$) and persistence ($\beta_{meadow}^{[\phi]} = 0.6$) and negative effects of temperature on initial occupancy ($\beta_{temperature}^{[0]} = -0.4$) and persistence ($\beta_{temperature}^{[\phi]} = -0.3$). We also assume a moderate colonization probability from adjacent sites ($\zeta = 0.4$) and a low colonization probability due to long-distance movement ($\delta = 0.1$). We consider that two sites are adjacent if the distance between their centroids is less than 1; the choice of the cutoff point here is arbitrary but should be adequate for an illustrative purpose.

```
library(boot)   # for using logit and inv.logit functions

set.seed(3)

# Basic values
nsite <- 360 # number of sites
nyear <- 20 # number of years
nreps <- 5  # number of within-season replicates
ncovs <- 2  # number of environmental covariates
npcvs <- 3  # number of observational covariates

beta0 <- c(0.4, 1, -0.4)      # intercept and slopes for initial occupancy
beta_phi <- c(-0.2, 0.6, -0.3)  # intercept and slopes for persistence
zeta <- 0.4                 # probability of neighborhood colonization
```

```
delta <- 0.1            # probability of long-distance colonization
alpha <- c(0.3, -0.6, 0.4, -0.2) # intercept and slopes for detection
probability

# Read in environmental covariates
load('data/covariates.RData')
x1 <- (log(cov$size) - mean(log(cov$size))) / sd(log(cov$size)) #
standardized log habitat patch size
x2 <- (cov$temp - mean(cov$temp)) / sd(cov$temp) # standardized
temperature
x <- array(, dim=c(nsite, nyear, ncovs)) # environmental covariates
x[,,1] <- x1
x[,,2] <- x2

# Simulate data
psi <- matrix(, nsite, nyear) # occupancy probability
z <- matrix(, nsite, nyear) # occupancy status
psi[,1] <- inv.logit(cbind(1,x[,1,]) %*% beta0) # initial occupancy
probability
z[,1] <- rbinom(nsite, 1, psi[,1]) # initial occupancy status

lon <- cov$lon # longitude or easting
lat <- cov$lat # latitude or northing
d <- matrix(, nsite, nsite) # distance matrix
for (i1 in 1:nsite) {
 for (i2 in 1:nsite) {
  d[i1,i2] <- sqrt((lon[i1] - lon[i2]) ^ 2 + (lat[i1] - lat[i2]) ^ 2)
 } # i2
} # i1
a <- ifelse(d < 1, 1, 0) # adjacency matrix
diag(a) <- 0 # a site is not a "neighbor" to ifself

phi   <- matrix(, nsite, nyear-1) # probability of persistence
gamma <- matrix(, nsite, nyear-1) # probability of colonization
for (t in 2:nyear) {
 for (i in 1:nsite) {
  phi  [i,t-1] <- inv.logit(c(1,x[i,t,]) %*% beta_phi)
  gamma[i,t-1] <- 1 - (1 - zeta) ^ sum(a[i,] * z[,t-1]) * (1 - delta)
  psi[i,t] <- z[i,t-1] * phi[i,t-1] + (1 - z[i,t-1]) * gamma[i,t-1]
  z[i,t] <- rbinom(1, 1, psi[i,t])
 } # i
} # t

w <- array(rnorm(nsite * nyear * nreps * npcvs, 0, 1), dim=c(nsite,
nyear, nreps, npcvs)) # observational covariates
```

```
p <- array(, dim=c(nsite, nyear, nreps)) # detection probability
for (i in 1:nsite) {
 for (t in 1:nyear) {
  p[i,t,] <- inv.logit(cbind(1,w[i,t,,]) %*% alpha)
 } # t
} # i

y <- array(, dim=c(nsite, nyear, nreps)) # detection/non-detection
data
for (i in 1:nsite) {
 for (t in 1:nyear) {
  y[i,t,] <- rbinom(nreps, 1, z[i,t] * p[i,t,])
 } # t
} # i
```

Next, we can develop a custom MCMC algorithm to estimate parameters $\beta^{[0]}$, $\beta^{[\phi]}$, ζ, δ, α, and z using data y, x, w, and a, as shown below:

```
adj_dynam_occu_mcmc <- function(y, x, w, a, nmcmc) {

 # Setup variables
 nsite <- dim(y) [1]
 nyear <- dim(y) [2]
 nreps <- dim(y) [3]
 ncovs <- dim(x) [3]
 npcvs <- dim(w) [4]
 ysum <- apply(y, 1:2, sum)

 beta0_save <- matrix(0, ncovs + 1, nmcmc)
 beta_phi_save <- matrix(0, ncovs + 1, nmcmc)
 zeta_save <- rep(0, nmcmc)
 delta_save <- rep(0, nmcmc)
 alpha_save <- matrix(0, npcvs + 1, nmcmc)
 z_save <- array(0, dim=c(nsite, nyear, 2000))

 # Priors
 beta0_mean <- rep(0, ncovs + 1)
 beta0_sd <- 2
 beta_phi_mean <- rep(0, ncovs + 1)
 beta_phi_sd <- 2
 logit_zeta_mean <- 0
 logit_zeta_sd <- 2
 logit_delta_mean <- 0
 logit_delta_sd <- 2
 alpha_mean <- rep(0, npcvs + 1)
 alpha_sd <- 2
```

```
# Starting values
beta0 <- rep(0, ncovs + 1)
beta_phi <- rep(0, ncovs + 1)
zeta <- 0.5
delta <- 0.5
alpha <- rep(0, npcvs + 1)
z <- ifelse(ysum==0, 0, 1)
phi  <- matrix(0.5, nsite, nyear-1)
gamma <- matrix(, nsite, nyear-1)
for (t in 2:nyear) {
 for (i in 1:nsite) {
  gamma[i,t-1] <- 1 - (1 - zeta) ^ sum(a[i,] * z[,t-1]) * (1 - delta)
 } # i
} # t
p <- array(0.5, dim=c(nsite, nyear, nreps))

# Tuning factors
beta0_tune <- 0.3
beta_phi_tune <- 0.05
zeta_tune <- 0.05
delta_tune <- 0.05
alpha_tune <- 0.025

# Begin MCMC loop
for (k in 1:nmcmc) {
 ### Sample beta0
 beta0_star <- rnorm(ncovs + 1, beta0, beta0_tune)
 mh1 <- sum(dbinom(z[,1], 1, inv.logit(cbind(1,x[,1,]) %*%
beta0_star), log=TRUE)) +
        sum(dnorm(beta0_star, beta0_mean, beta0_sd, log=TRUE))
 mh2 <- sum(dbinom(z[,1], 1, inv.logit(cbind(1,x[,1,]) %*% beta0   ),
log=TRUE)) +
        sum(dnorm(beta0   , beta0_mean, beta0_sd, log=TRUE))
 mh <- exp(mh1 - mh2)
 if (mh > runif(1)) {
  beta0 <- beta0_star
 }

 ### Sample beta_phi
 beta_phi_star <- rnorm(ncovs + 1, beta_phi, beta_phi_tune)
 phi_star <- matrix(, nsite, nyear-1)
 for (t in 2:nyear) {
  phi_star[,t-1] <- inv.logit(cbind(1,x[,t,]) %*% beta_phi_star)
 } # t
 mh1 <- sum(dbinom(z[,-1], 1, z[,-nyear] * phi_star + (1 - z[,-nyear]) *
```

```
gamma, log=TRUE)) +
        sum(dnorm(beta_phi_star, beta_phi_mean, beta_phi_sd,
log=TRUE))
  mh2 <- sum(dbinom(z[,-1], 1, z[,-nyear] * phi     + (1 - z[,-nyear]) *
gamma, log=TRUE)) +
        sum(dnorm(beta_phi     , beta_phi_mean, beta_phi_sd, log=TRUE))
  mh <- exp(mh1 - mh2)
  if (mh > runif(1)) {
   beta_phi <- beta_phi_star
   phi <- phi_star
  }

  ### Sample zeta
  zeta_star <- inv.logit(rnorm(1, logit(zeta), zeta_tune))
  gamma <- gamma_star <- matrix(, nsite, nyear-1)
  for (t in 2:nyear) {
   for (i in 1:nsite) {
    gamma     [i,t-1] <- 1 - (1 - zeta     ) ^ sum(a[i,] * z[,t-1]) * (1 - delta)
    gamma_star[i,t-1] <- 1 - (1 - zeta_star) ^ sum(a[i,] * z[,t-1]) * (1 -
delta)
   } # i
  } # t
  mh1 <- sum(dbinom(z[,-1], 1, z[,-nyear] * phi + (1 - z[,-nyear]) *
gamma_star, log=TRUE)) +
      dnorm(logit(zeta_star), logit_zeta_mean, logit_zeta_sd,
log=TRUE)
  mh2 <- sum(dbinom(z[,-1], 1, z[,-nyear] * phi + (1 - z[,-nyear]) * gamma
, log=TRUE)) +
      dnorm(logit(zeta     ), logit_zeta_mean, logit_zeta_sd, log=TRUE)
  mh <- exp(mh1 - mh2)
  if (mh > runif(1)) {
   zeta <- zeta_star
  }

  ### Sample delta
  delta_star <- inv.logit(rnorm(1, logit(delta), delta_tune))
  gamma <- gamma_star <- matrix(, nsite, nyear-1)
  for (t in 2:nyear) {
   for (i in 1:nsite) {
    gamma     [i,t-1] <- 1 - (1 - zeta) ^ sum(a[i,] * z[,t-1]) * (1 - delta     )
    gamma_star[i,t-1] <- 1 - (1 - zeta) ^ sum(a[i,] * z[,t-1]) * (1 -
delta_star)
   } # i
  } # t
  mh1 <- sum(dbinom(z[,-1], 1, z[,-nyear] * phi + (1 - z[,-nyear]) *
gamma_star, log=TRUE)) +
```

```
      dnorm(logit(delta_star), logit_delta_mean, logit_delta_sd,
log=TRUE)
  mh2 <- sum(dbinom(z[,-1], 1, z[,-nyear] * phi + (1 - z[,-nyear]) * gamma
, log=TRUE)) +
      dnorm(logit(delta   ), logit_delta_mean, logit_delta_sd,
log=TRUE)
  mh <- exp(mh1 - mh2)
  if (mh > runif(1)) {
   delta <- delta_star
  }

  ### Sample alpha
  alpha_star <- rnorm(npcvs + 1, alpha, alpha_tune)
  p_star <- array(, dim=c(nsite, nyear, nreps))
  for (i in 1:nsite) {
   for (t in 1:nyear) {
    p_star[i,t,] <- inv.logit(cbind(1,w[i,t,,]) %*% alpha_star)
   } # t
  } # i
  mh1t <- mh2t <- numeric(nyear)
  for (t in 1:nyear) {
   mh1t[t] <- sum(dbinom(y[which(z[,t]==1),t,], 1, p_star[which(z[,
t]==1),t,], log=TRUE))
   mh2t[t] <- sum(dbinom(y[which(z[,t]==1),t,], 1, p   [which(z[,t]
==1),t,], log=TRUE))
  } # t
  mh1 <- sum(mh1t) +
      sum(dnorm(alpha_star, alpha_mean, alpha_sd, log=TRUE))
  mh2 <- sum(mh2t) +
      sum(dnorm(alpha,    alpha_mean, alpha_sd, log=TRUE))
  mh <- exp(mh1 - mh2)
  if (mh > runif(1)) {
   alpha <- alpha_star
   p <- p_star
  }

  ### Sample z
  s_temp <- matrix(, nsite, nyear)
  s_temp[,1] <- inv.logit(cbind(1,x[,1,]) %*% beta0)
  for (t in 2:nyear) {
   s_temp[,t] <- z[,t-1] * phi[,t-1] + (1 - z[,t-1]) * gamma[,t-1] + 1e-10
  } # t
  num_temp <- s_temp * apply((1 - p), 1:2, prod)
  psi_temp <- num_temp / (num_temp + (1 - s_temp))
  z[which(ysum==0)] <- rbinom(length(which(ysum==0)), 1, psi_temp
[which(ysum==0)])
```

```
  ### Save samples
  beta0_save[,k] <- beta0
  beta_phi_save[,k] <- beta_phi
  zeta_save[k] <- zeta
  delta_save[k] <- delta
  alpha_save[,k] <- alpha
  if (nmcmc > 2000) {
   if (k > nmcmc-2000) {
    z_save[,,k-nmcmc+2000] <- z
   }
  }
 } # k

 # Write output
 list(beta0_save=beta0_save,
    beta_phi_save=beta_phi_save, zeta_save=zeta_save,
delta_save=delta_save,
    alpha_save=alpha_save,
    z_save=z_save)
} # adj_dynam_occu_mcmc
```

We then can use the code below to implement the above-mentioned algorithm, which takes 4.7 h to run. Using the code in Appendix 2.3 to create some traceplots, we can see good model convergence indicated by the well-mixed chains and Rhat statistics as well as consistency between parameter estimates of $\beta^{[0]}$, $\beta^{[\phi]}$, ζ, δ, α, and their true values used in data simulation (Fig. 2.5).

```
library(foreach)   # for parallel computing
library(doParallel) # for parallel computing

numCores <- round(detectCores() / 2) # only use half of the cores for
parallel computing
registerDoParallel(numCores) # setup parallel computing

nmcmc <- 50000 # number of iterations
chain <- 3    # number of chains

start_time <- Sys.time() # start time of computing
out <- foreach (i=1:chain, .packages='boot') %dopar% {
 adj_dynam_occu_mcmc(y=y, x=x, w=w, a=a, nmcmc=nmcmc)
} # i
stopImplicitCluster() # clean up the cluster
end_time <- Sys.time() # end time of computing
time_taken <- end_time - start_time # computing time
```

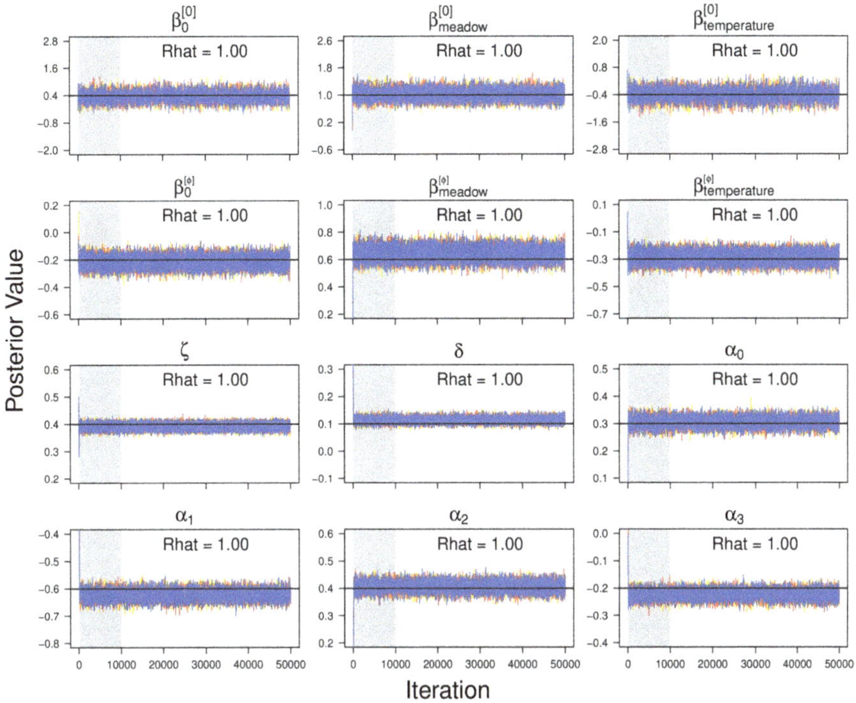

Fig. 2.5 Traceplots showing the posterior chains for $\beta^{[0]}$, $\beta^{[\phi]}$, ζ, δ, and α in the spatial dynamic occupancy model with adjacency-based movement. Rhat values are calculated after discarding the first 10,000 iterations. Horizontal lines represent the true values used in data simulation

We can also use the following code to calculate the proportion of colonized sites when the total area of their neighbors is large (> 0.656, the median of total neighboring area) vs. small. Note that we consider to have a colonization event for site i when we have $z_{i,\,t-1} = 0$ and $z_{i,\,t} = 1$. We can see that sites with large total neighboring area are more likely to be colonized than sites with small total neighboring area (Fig. 2.6). This is because, larger sites are more likely to be occupied,

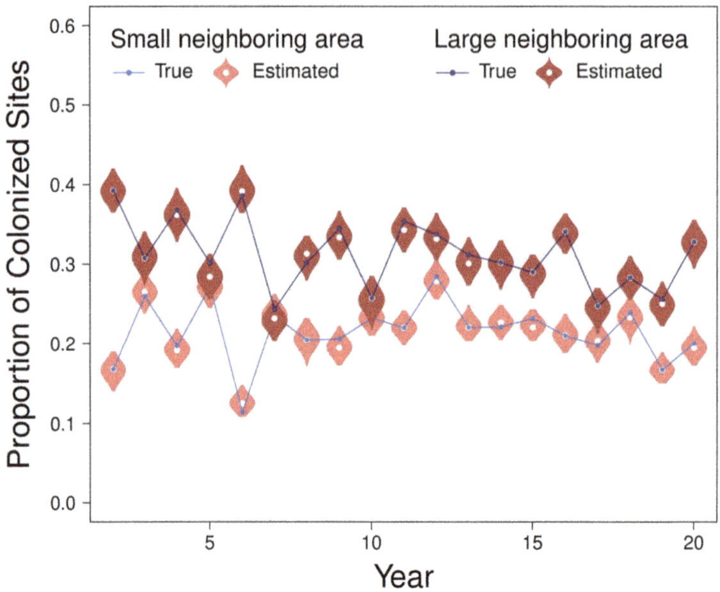

Fig. 2.6 Violin plots showing an excellent agreement between the estimated proportion of colonized sites (violin) and the simulated true values (blue points) when the total area of their neighbors is large (deeper colors) vs. small (lighter colors)

and thus, sites with large total neighboring area tend to have more occupied sites in their adjacency to boost their probability of being colonized.

```
library(vioplot)   # for making violin plots

nb_size <- matrix(, nsite, nyear)
for (i in 1:nsite) {
 for (t in 1:nyear) {
  nb_size[i,t] <- sum(cov$size[,t] * a[i,])
 } # t
} # i
nb_large <- ifelse(nb_size > median(nb_size), 1, 0)

colon_small <- colon_large <- numeric(nyear-1)
for (t in 1:(nyear-1)) {
 colon_small[t] <- length(which(z[which(nb_large[,t]==0),t]==0 & z
[which(nb_large[,t]==0),t+1]==1)) / length(which(nb_large[,t]
==0))
 colon_large[t] <- length(which(z[which(nb_large[,t]==1),t]==0 & z
[which(nb_large[,t]==1),t+1]==1)) / length(which(nb_large[,t]
==1))
} # i
```

```
z_post <- array(, dim=c(nsite, nyear, 6000))
for (i in 1:chain) {
 z_post[,,2000*(i-1)+c(1:2000)] <- out[[i]]$z_save
} # i

colon_post_small <- colon_post_large <- matrix(, 6000, nyear-1)
for (t in 1:(nyear-1)) {
 colon_post_small[,t] <- colMeans(ifelse(z_post[which(nb_large[,t]
==0),t,]==0 & z_post[which(nb_large[,t]==0),t+1,]==1, 1, 0))
 colon_post_large[,t] <- colMeans(ifelse(z_post[which(nb_large[,t]
==1),t,]==0 & z_post[which(nb_large[,t]==1),t+1,]==1, 1, 0))
} # i

par(mfrow=c(1,1))
par(mar=c(5,6.5,1,1))

plot(1, xlim=c(1,nyear-1), ylim=c(0,0.6), type='n', axes=F,
xlab='', ylab='')
vioplot(colon_post_small, col='lightcoral', rectCol=NA, lineCol=NA,
border=NA, add=T)
lines(colon_small, type='o', pch=16, cex=0.8, lwd=1.2,
col='royalblue')
vioplot(colon_post_large, col='firebrick', rectCol=NA, lineCol=NA,
border=NA, add=T)
lines(colon_large, type='o', pch=16, cex=0.8, lwd=1.2, col='navy')
axis(1, at=seq(4,nyear,5), labels=seq(4,nyear,5)+1, cex.axis=1.5)
axis(2, at=seq(0, 0.6, 0.1), cex.axis=1.5, las=2)
title(xlab='Year', cex.lab=2.5, line=3.5)
title(ylab='Proportion of Colonized Sites', cex.lab=2.5, line=4)

points(x=1.5, y=0.54, pch=16, cex=1, col='royalblue')
lines(x=c(1, 2), y=c(0.54,0.54), lwd=1.2, col='royalblue')
vioplot(rnorm(1e4,0.54,0.005), at=4.5, col='lightcoral',
rectCol=NA, lineCol=NA, border=NA, add=T)
text(x=2.1, y=0.54, labels='True', pos=4, cex=1.6)
text(x=5.1, y=0.54, labels='Estimated', pos=4, cex=1.6)
text(x=0.7, y=0.58, labels='Small neighboring area', pos=4, cex=2)

points(x=11.5, y=.54, pch=16, cex=1, col='navy')
lines(x=c(11,12), y=c(0.54,0.54), lwd=1.2, col='navy')
vioplot(rnorm(1e4,0.54,0.005), at=14.5, col='firebrick', rectCol=NA,
lineCol=NA, border=NA, add=T)
text(x=12.1, y=0.54, labels='True', pos=4, cex=1.6)
text(x=15.1, y=0.54, labels='Estimated', pos=4, cex=1.6)
text(x=10.7, y=0.58, labels='Large neighboring area', pos=4, cex=2)
```

2.4.2 With Distance-Based Movement

The definition of adjacency can be arbitrary if sites are not naturally adjacent to each other in patchy habitats. Furthermore, it may be an oversimplification to assume that only adjacent sites are connected. More realistically, we can assume that connectivity between two sites is related to their distance, typically with a negative relationship. Note that distance can be calculated even for sites (polygons) defined in continuous habitats. Imagine a butterfly population, if the species does not live in a meadow in a year, it has a better chance to move into this meadow from a closer meadow than from a more distant one if they live in both. At the end of the day, whether the butterfly will move into a meadow or not depends on the number of meadows they live in and the distance from these meadows to the empty meadow. To specify the relationship between colonization probability and distance, we often use something called a decay function, i.e., a monotone declining function of distance. Here we use a simple and widely used decay function, which is

$$\left.\begin{aligned} \gamma_{i_1,t-1} &= \sum_{i_2=1}^{n_{site}} \theta_{i_1,i_2} \times z_{i_2,t-1} \\ \theta_{i_1,i_2} &= \frac{\eta_{i_1,i_2}}{\sum_{i_2=1}^{n_{site}} \eta_{i_1,i_2}} \\ \eta_{i_1,i_2} &= exp(-1 \times \kappa \times d_{i_1,i_2}) \end{aligned}\right\}, \tag{2.10}$$

in which θ_{i_1,i_2} is a parameter representing (standardized) colonization rate between sites i_1 and i_2, η_{i_1,i_2} is unstandardized colonization rate, and κ is a decay parameter governing the relationship between colonization and distance, denoted d_{i_1,i_2}. A larger κ represents a more rapid rate of decay, meaning that the colonization rate decreases faster with increasing distance, and vice versa.

Next, let us use the code below to simulate data under this model specification. We assume positive effects of meadow size on initial occupancy ($\beta^{[0]}_{meadow} = 1$) and persistence ($\beta^{[\phi]}_{meadow} = 0.6$) and negative effects of temperature on initial occupancy ($\beta^{[0]}_{temperature} = -0.4$) and persistence ($\beta^{[\phi]}_{temperature} = -0.3$). We also assume a relatively weak rate of decay, $\kappa = 0.5$, to allow a high probability of long-distance colonization.

```
library(boot) # for using logit and inv.logit functions

set.seed(3)

# Basic values
nsite <- 360 # number of sites
nyear <- 20 # number of years
nreps <- 5  # number of within-season replicates
ncovs <- 2  # number of environmental covariates
npcvs <- 1  # number of observational covariates
```

```
beta0 <- c(0.4, 1, -0.4)      # intercept and slopes for initial occupancy
beta_phi <- c(-0.2, 0.6, -0.3) # intercept and slopes for persistence
kappa <- 0.5           # decay parameter for colonization
alpha <- c(0.3, -0.6)       # intercept and slopes for detection
probability

# Read in environmental covariates
load('data/covariates.RData')
x1 <- (log(cov$size) - mean(log(cov$size))) / sd(log(cov$size)) #
standardized log habitat patch size
x2 <- (cov$temp - mean(cov$temp)) / sd(cov$temp) # standardized
temperature
x <- array(, dim=c(nsite, nyear, ncovs)) # environmental covariates
x[,,1] <- x1
x[,,2] <- x2

# Simulate data
psi <- matrix(, nsite, nyear) # occupancy probability
z <- matrix(, nsite, nyear) # occupancy status
psi[,1] <- inv.logit(cbind(1,x[,1,]) %*% beta0) # initial occupancy
probability
z[,1] <- rbinom(nsite, 1, psi[,1]) # initial occupancy status

lon <- cov$lon # longitude or easting
lat <- cov$lat # latitude or northing
d <- matrix(, nsite, nsite) # distance matrix
for (i1 in 1:nsite) {
 for (i2 in 1:nsite) {
  d[i1,i2] <- sqrt((lon[i1] - lon[i2]) ^ 2 + (lat[i1] - lat[i2]) ^ 2)
 } # i2
} # i1
eta <- exp(-1 * kappa * d) # unstandardized colonization rate
theta <- eta / rowSums(eta) #  standardized colonization rate

phi  <- matrix(, nsite, nyear-1) # probability of persistence
gamma <- matrix(, nsite, nyear-1) # probability of colonization
for (t in 2:nyear) {
 for (i in 1:nsite) {
  phi  [i,t-1] <- inv.logit(c(1,x[i,t,]) %*% beta_phi)
  gamma[i,t-1] <- sum(theta[i,] * z[,t-1])
  psi[i,t] <- z[i,t-1] * phi[i,t-1] + (1 - z[i,t-1]) * gamma[i,t-1]
  z[i,t] <- rbinom(1, 1, psi[i,t])
 } # i
} # t
```

```
w <- array(rnorm(nsite * nyear * nreps * npcvs, 0, 1), dim=c(nsite,
nyear, nreps, npcvs)) # observational covariates

p <- array(, dim=c(nsite, nyear, nreps)) # detection probability
for (i in 1:nsite) {
 for (t in 1:nyear) {
  p[i,t,] <- inv.logit(cbind(1,w[i,t,,]) %*% alpha)
 } # t
} # i

y <- array(, dim=c(nsite, nyear, nreps)) # detection/non-detection
data
for (i in 1:nsite) {
 for (t in 1:nyear) {
  y[i,t,] <- rbinom(nreps, 1, z[i,t] * p[i,t,])
 } # t
} # i
```

Remember that we also use this model to illustrate how to deal with missing values. Therefore, we include several additional lines as shown below to create missing values (NA's) in 20% of the observations.

```
pmiss <- 0.2 # proportion of observations with missing values
nmiss <- nsite * nyear * nreps * pmiss # number of observations with
missing values
miss <- array(sample(c(rep(0,nmiss),rep(1,nsite*nyear*nreps-
nmiss)), nsite*nyear*nreps, replace=F), dim=c(nsite, nyear, nreps)) #
identifier for missing values (0 represents missing value)
y[which(miss == 0)] <- NA # create missing values in data
```

The data from the first three sites in year one are shown below. This is to say, in year one, site 1 was actually visited three times ($n_{reps,\ 1,\ 1} = 3$) instead of five times, site 2 five times ($n_{reps,\ 2,\ 1} = 5$), and site 3 four times ($n_{reps,\ 3,\ 1} = 4$), but we put NA's for the missed visits (visits 1 & 2 for site 1 and visit 3 for site 3) to collate the data into a "balanced" manner. Later, we will see how to deal with these NA's in the MCMC algorithm.

	Detection/non-detection				
	Visit 1	Visit 2	Visit 3	Visit 4	Visit 5
Site 1	NA	NA	0	0	0
Site 2	0	0	0	0	0
Site 3	0	1	NA	1	1

Next, let us develop a custom MCMC algorithm to estimate parameters $\beta^{[0]}$, $\beta^{[\phi]}$, κ, and α using data y, x, w, and d. Remember that we have missing values in y. There are two major tricks to deal with these missing values in the MCMC algorithm. First, detection probability p needs to be forced to 0 whenever there is a missing value in y, which is important for sampling z. Second, we need to make sure that missing values do not contribute to the likelihood of α, the only parameter whose probability is calculated directly from y. Code of dealing with missing values is highlighted in **bold** below:

```
dist_dynam_occu_mcmc <- function(y, x, w, d, nmcmc) {

 # Setup variables
 nsite <- dim(y)[1]
 nyear <- dim(y)[2]
 nreps <- dim(y)[3]
 ncovs <- dim(x)[3]
 npcvs <- dim(w)[4]
 ysum <- apply(y, 1:2, sum, na.rm=T) # need na.rm=T to deal with missing
values in y

 beta0_save <- matrix(0, ncovs + 1, nmcmc)
 beta_phi_save <- matrix(0, ncovs + 1, nmcmc)
 kappa_save <- rep(0, nmcmc)
 alpha_save <- matrix(0, npcvs + 1, nmcmc)

 # Priors
 beta0_mean <- rep(0, ncovs + 1)
 beta0_sd <- 2
 beta_phi_mean <- rep(0, ncovs + 1)
 beta_phi_sd <- 2
 log_kappa_mean <- 0
 log_kappa_sd <- 2
 alpha_mean <- rep(0, npcvs + 1)
 alpha_sd <- 2

 # Starting values
 beta0 <- rep(0, ncovs + 1)
 beta_phi <- rep(0, ncovs + 1)
 kappa <- 1
 alpha <- rep(0, npcvs + 1)
 phi <- matrix(0.5, nsite, nyear-1)
 eta <- exp(-1 * kappa * d)
 theta <- eta / rowSums(eta)
 z <- ifelse(ysum==0, 0, 1)
 gamma <- matrix(, nsite, nyear-1)
```

```
 for (t in 2:nyear) {
  for (i in 1:nsite) {
   gamma[i,t-1] <- sum(theta[i,] * z[,t-1])
  } # i
 } # t
 p <- array(0.5, dim=c(nsite, nyear, nreps))
 p[which(is.na(y))] <- 0 # force detection to be 0 for missing values;
this is important for sampling z

 # Tuning factors
 beta0_tune <- 0.4
 beta_phi_tune <- 0.05
 kappa_tune <- 0.1
 alpha_tune <- 0.03

 # Begin MCMC loop
 for (k in 1:nmcmc) {
  ### Sample beta0
  beta0_star <- rnorm(ncovs + 1, beta0, beta0_tune)
  mh1 <- sum(dbinom(z[,1], 1, inv.logit(cbind(1,x[,1,]) %*%
beta0_star), log=TRUE)) +
      sum(dnorm(beta0_star, beta0_mean, beta0_sd, log=TRUE))
  mh2 <- sum(dbinom(z[,1], 1, inv.logit(cbind(1,x[,1,]) %*% beta0    ),
log=TRUE)) +
      sum(dnorm(beta0    , beta0_mean, beta0_sd, log=TRUE))
  mh <- exp(mh1 - mh2)
  if (mh > runif(1)) {
   beta0 <- beta0_star
  }

  ### Sample beta_phi
  beta_phi_star <- rnorm(ncovs + 1, beta_phi, beta_phi_tune)
  phi_star <- matrix(, nsite, nyear-1)
  for (t in 2:nyear) {
   phi_star[,t-1] <- inv.logit(cbind(1,x[,t,]) %*% beta_phi_star)
  } # t
  mh1 <- sum(dbinom(z[,-1], 1, z[,-nyear] * phi_star + (1 - z[,-nyear]) *
gamma, log=TRUE)) +
     sum(dnorm(beta_phi_star, beta_phi_mean, beta_phi_sd, log=TRUE))
  mh2 <- sum(dbinom(z[,-1], 1, z[,-nyear] * phi    + (1 - z[,-nyear]) *
gamma, log=TRUE)) +
      sum(dnorm(beta_phi    , beta_phi_mean, beta_phi_sd, log=TRUE))
  mh <- exp(mh1 - mh2)
  if (mh > runif(1)) {
   beta_phi <- beta_phi_star
```

```
    phi <- phi_star
  }

  ### Sample kappa
  kappa_star <- exp(rnorm(1, log(kappa), kappa_tune))
  eta_star <- exp(-1 * kappa_star * d)
  theta_star <- eta_star / rowSums(eta_star)
  gamma <- gamma_star <- matrix(, nsite, nyear-1)
  for (t in 2:nyear) {
   for (i in 1:nsite) {
    gamma   [i,t-1] <- sum(theta   [i,] * z[,t-1])
    gamma_star[i,t-1] <- sum(theta_star[i,] * z[,t-1])
   } # i
  } # t
  mh1 <- sum(dbinom(z[,-1], 1, z[,-nyear] * phi + (1 - z[,-nyear]) *
gamma_star, log=TRUE)) +
      dnorm(log(kappa_star), log_kappa_mean, log_kappa_sd, log=TRUE)
  mh2 <- sum(dbinom(z[,-1], 1, z[,-nyear] * phi + (1 - z[,-nyear]) * gamma
, log=TRUE)) +
      dnorm(log(kappa   ), log_kappa_mean, log_kappa_sd, log=TRUE)
  mh <- exp(mh1 - mh2)
  if (mh > runif(1)) {
   kappa <- kappa_star
   eta <- eta_star
   theta <- theta_star
  }

  ### Sample alpha
  alpha_star <- rnorm(npcvs + 1, alpha, alpha_tune)
  p_star <- array(, dim=c(nsite, nyear, nreps))
  for (i in 1:nsite) {
   for (t in 1:nyear) {
    p_star[i,t,] <- inv.logit(cbind(1,w[i,t,,]) %*% alpha_star)
   } # t
  } # i
  p_star[which(is.na(y))] <- 0 # again, force detection to be 0 for
missing values
  mh1t <- mh2t <- numeric(nyear)
  for (t in 1:nyear) {
   mh1t[t] <- sum(dbinom(y[which(z[,t]==1),t,], 1, p_star[which(z[,t]
==1),t,], log=TRUE), na.rm=T) # missing values don't contribute to the
likelihood
    mh2t[t] <- sum(dbinom(y[which(z[,t]==1),t,], 1, p    [which(z[,t]
==1),t,], log=TRUE), na.rm=T) # missing values don't contribute to the
likelihood
```

```
  } # t
  mh1 <- sum(mh1t) +
      sum(dnorm(alpha_star, alpha_mean, alpha_sd, log=TRUE))
  mh2 <- sum(mh2t) +
      sum(dnorm(alpha     , alpha_mean, alpha_sd, log=TRUE))
  mh <- exp(mh1 - mh2)
  if (mh > runif(1)) {
   alpha <- alpha_star
   p <- p_star
  }

  ### Sample z
  s_temp <- matrix(, nsite, nyear)
  s_temp[,1] <- inv.logit(cbind(1,x[,1,]) %*% beta0)
  for (t in 2:nyear) {
   s_temp[,t] <- z[,t-1] * phi[,t-1] + (1 - z[,t-1]) * gamma[,t-1]
  } # t
  num_temp <- s_temp * apply((1 - p), 1:2, prod)
  psi_temp <- num_temp / (num_temp + (1 - s_temp))
  z[which(ysum==0)] <- rbinom(length(which(ysum==0)), 1, psi_temp
[which(ysum==0)])

  ### Save samples
  beta0_save[,k] <- beta0
  beta_phi_save[,k] <- beta_phi
  kappa_save[k] <- kappa
  alpha_save[,k] <- alpha
 } # k

 # Write output
 list(beta0_save=beta0_save,
    beta_phi_save=beta_phi_save, kappa_save=kappa_save,
    alpha_save=alpha_save)
} # dist_dynam_occu_mcmc
```

We can then use the following code to implement the algorithm, which takes 2.8 h to run. Using the code in Appendix 2.4 to create traceplots, we can see good model convergence indicated by the well-mixed chains and Rhat statistics and consistency between parameter estimates of $\beta^{[0]}$, $\beta^{[\phi]}$, κ, and α and their true values used in data simulation (Fig. 2.7).

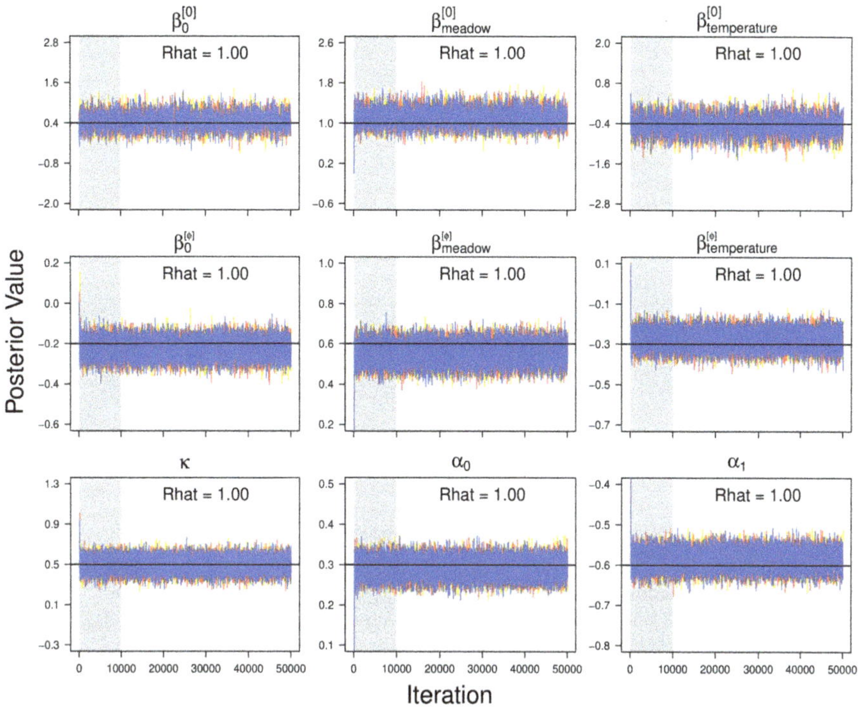

Fig. 2.7 Traceplots showing the posterior chains for $\beta^{[0]}$, $\beta^{[\phi]}$, κ, and α in the spatial dynamic occupancy model with distance-based movement. Rhat values are calculated after discarding the first 10,000 iterations. Horizontal lines represent the true values used in data simulation

```
library(foreach)  # for parallel computing
library(doParallel) # for parallel computing

numCores <- round(detectCores() / 2) # only use half of the cores for
parallel computing
registerDoParallel(numCores) # setup parallel computing

nmcmc <- 50000 # number of iterations
chain <- 3   # number of chains

start_time <- Sys.time() # start time of computing
out <- foreach (i=1:chain, .packages='boot') %dopar% {
 dist_dynam_occu_mcmc(y=y, x=x, w=w, d=d, nmcmc=nmcmc)
} # i
stopImplicitCluster() # clean up the cluster
end_time <- Sys.time() # end time of computing
time_taken <- end_time - start_time # computing time
```

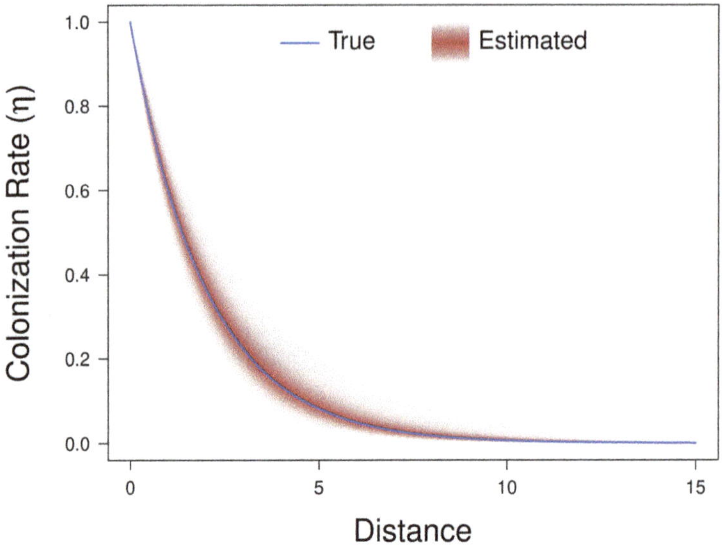

Fig. 2.8 A response density curve (red) showing the estimated relationship between unstandardized colonization rate η and the distance between two sites, governed by κ. The blue line represents the true relationship used in data simulation

Using the code below, we can further construct a response density curve to visualize the relationship between the unstandardized colonization rate η and the distance between any pair of sites, governed by κ, which again shows satisfying agreement with the true relationship (Fig. 2.8).

```
kappa_post <- matrix(, nmcmc*0.8, chain)
for (j in 1:chain) {
 kappa_post[,j] <- out[[j]]$kappa_save[(nmcmc*0.2+1):nmcmc]
} # j
kappa_post <- as.vector(kappa_post)
npred <- 100
d_pred <- seq(0, 15, length.out=npred)
eta_pred <- matrix(, npred, nmcmc*0.8*chain)
for (i in 1:npred) {
 eta_pred[i,] <- exp(-1 * kappa_post * d_pred[i])
} # i
nqt <- 200
eta_qt <- apply(eta_pred, 1, quantile, probs=seq(0.001, 0.999, length.
out=nqt+1))
cols <- colorRampPalette(c('white','firebrick'))(nqt/2)
cols <- c(cols, rev(cols))
```

```
par(mfrow=c(1,1))
par(mar=c(5,6,1,1))

plot(1, xlim=c(0,15), ylim=c(0,1), type='n', xlab='', ylab='',
axes=F)
for (i in 1:nqt) {
 polygon(x=c(d_pred,rev(d_pred)), y=c(eta_qt[i,],rev(eta_qt[i
+1,])),
     col=cols[i], border=NA)
} # i
curve(exp(-1 * kappa * x), col='royalblue', lwd=2, add=T)
axis(1, at=seq(0,15,5.0), cex.axis=1.2)
axis(2, at=seq(0, 1,0.2), cex.axis=1.2, las=2)
box()
lines(x=c(4,5), y=c(0.95,0.95), col='royalblue', lwd=2)
for (i in 1:nqt) {
 rect(8, seq(0.9,1,length.out=nqt+1)[i], 9, seq(0.9,1,length.
out=nqt+1)[i+1],
    col=cols[i], border=NA)
} # i
text(x=5, y=0.95, labels='True', pos=4, cex=1.6)
text(x=9, y=0.95, labels='Estimated', pos=4, cex=1.6)

title(xlab='Distance', line=3.5, cex.lab=2)
title(ylab=expression(paste('Colonization Rate (', eta, ')',
sep='')), line=3.8, cex.lab=2)
```

2.4.3 With Distance- and Environment-Based Movement

More realistically, colonization at an unoccupied site is not only related to the distance from other occupied sites but also affected by environmental conditions (Sutherland et al. 2012; Drake et al. 2022). Imagine an area with just three meadows, where our butterfly lives in one of them but not the other two. Also, imagine that the two empty meadows are equally far from the meadow that the species live in, but one empty meadow is much larger than the other. Based on the model introduced in Sect. 2.4.2, the chance that the butterfly will move into either of these empty meadows is the same. However, here we want to consider the case when the butterfly is more likely to move into the larger empty meadow. Using statistical language, we have

$$\left.\begin{aligned} \gamma_{i_1,t-1} &= logit^{-1}\left(\beta_0^{[\gamma]} + \beta_1^{[\gamma]} \times x_{i_1,t,1} + \ldots\right) \times \left(\sum_{i_2=1}^{n_{site}} \theta_{i_1,i_2} \times z_{i_2,t-1}\right) \\ \theta_{i_1,i_2} &= \frac{\eta_{i_1,i_2}}{\sum_{i_2=1}^{n_{site}} \eta_{i_1,i_2}} \\ \eta_{i_1,i_2} &= exp(-1 \times \kappa \times d_{i_1,i_2}) \end{aligned}\right\}, \quad (2.11)$$

in which $\beta^{[\gamma]}$'s are the intercept and slope parameters governing the effects of covariates (x's) on colonization, θ_{i_1,i_2} is the standardized colonization rate between sites i_1 and i_2, η_{i_1,i_2} is unstandardized colonization rate, and κ is a decay parameter governing the relationship between colonization and distance (d_{i_1,i_2}). Equation (2.11) can be considered a combination of Eqs. (2.4) and (2.10). Here, we only consider the situation when colonization is influenced by the environmental conditions at the target sites; colonization can also be influenced by the environmental conditions at the source sites, or a function (e.g., difference) of environmental conditions between the source and target sites, a situation that we will consider in the next chapter (Sect. 3.4.2). Also, note that some researchers have developed alternative functional forms for colonization processes in spatial dynamic occupancy models (Broms et al. 2016), which are worth checking out.

Next, let us use the following code to simulate data. We assume positive effects of meadow size on initial occupancy ($\beta_{meadow}^{[0]} = 1$), persistence ($\beta_{meadow}^{[\phi]} = 0.6$) and colonization ($\beta_{meadow}^{[\gamma]} = 0.4$), negative effects of temperature on initial occupancy ($\beta_{temperature}^{[0]} = -0.4$), persistence ($\beta_{temperature}^{[\phi]} = -0.3$), and colonization ($\beta_{temperature}^{[\gamma]} = -0.2$), as well as a relatively weak rate of decay ($\kappa = 0.5$).

```
library(boot) # for using logit and inv.logit functions

set.seed(3)

# Basic values
nsite <- 360 # number of sites
nyear <- 20  # number of years
nreps <- 5  # number of within-season replicates
ncovs <- 2  # number of environmental covariates
npcvs <- 1  # number of observational covariates

beta0 <- c(0.4, 1, -0.4)           # intercept and slopes for initial
occupancy
beta_phi  <- c(-0.2, 0.6, -0.3) # intercept and slopes for persistence
beta_gamma <- c(0.2, 0.4, -0.2) # intercept and slopes for colonization
kappa <- 0.5                       # decay parameter for colonization
alpha <- c(0.3, -0.6)              # intercept and slopes for detection
probability

# Read in environmental covariates
load('data/covariates.RData')
```

```
x1 <- (log(cov$size) - mean(log(cov$size))) / sd(log(cov$size)) #
standardized log habitat patch size
x2 <- (cov$temp - mean(cov$temp)) / sd(cov$temp) # standardized
temperature
x <- array(, dim=c(nsite, nyear, ncovs)) # environmental covariates
x[,,1] <- x1
x[,,2] <- x2

# Simulate data
psi <- matrix(, nsite, nyear) # occupancy probability
z <- matrix(, nsite, nyear) # occupancy status
psi[,1] <- inv.logit(cbind(1,x[,1,]) %*% beta0) # initial occupancy
probability
z[,1] <- rbinom(nsite, 1, psi[,1]) # initial occupancy status

lon <- cov$lon # longitude or easting
lat <- cov$lat # latitude or northing
d <- matrix(, nsite, nsite) # distance matrix
for (i1 in 1:nsite) {
 for (i2 in 1:nsite) {
  d[i1,i2] <- sqrt((lon[i1] - lon[i2]) ^ 2 + (lat[i1] - lat[i2]) ^ 2)
 } # i2
} # i1
eta <- exp(-1 * kappa * d)    # unstandardized colonization rate
theta <- eta / rowSums(eta) #  standardized colonization rate

phi  <- matrix(, nsite, nyear-1)  # probability of persistence
gamma <- matrix(, nsite, nyear-1) # probability of colonization
for (t in 2:nyear) {
 for (i in 1:nsite) {
  phi [i,t-1] <- inv.logit(c(1, x[i,t,]) %*% beta_phi )
  gamma[i,t-1] <- inv.logit(c(1, x[i,t,]) %*% beta_gamma) * sum(theta
[i,] * z[,t-1])
  psi[i,t] <- z[i,t-1] * phi[i,t-1] + (1 - z[i,t-1]) * gamma[i,t-1]
  z[i,t] <- rbinom(1, 1, psi[i,t])
 } # i
} # t

w <- array(rnorm(nsite * nyear * nreps * npcvs, 0, 1), dim=c(nsite,
nyear, nreps, npcvs)) # observational covariates

p <- array(, dim=c(nsite, nyear, nreps)) # detection probability
for (i in 1:nsite) {
 for (t in 1:nyear) {
  p[i,t,] <- inv.logit(cbind(1,w[i,t,,]) %*% alpha)
```

```
  } # t
} # i

y <- array(, dim=c(nsite, nyear, nreps)) # detection/non-detection
data
for (i in 1:nsite) {
  for (t in 1:nyear) {
    y[i,t,] <- rbinom(nreps, 1, z[i,t] * p[i,t,])
  } # t
} # i
```

We then develop a custom MCMC algorithm to estimate parameters $\beta^{[0]}$, $\beta^{[\phi]}$, $\beta^{[\gamma]}$, κ, and α using data y, x, w, and d.

```
dist_cov_dynam_occu_mcmc <- function(y, x, w, d, nmcmc) {

  # Setup variables
  nsite <- dim(y)[1]
  nyear <- dim(y)[2]
  nreps <- dim(y)[3]
  ncovs <- dim(x)[3]
  npcvs <- dim(w)[4]
  ysum <- apply(y, 1:2, sum)

  beta0_save <- matrix(0, ncovs + 1, nmcmc)
  beta_phi_save <- matrix(0, ncovs + 1, nmcmc)
  beta_gamma_save <- matrix(0, ncovs + 1, nmcmc)
  kappa_save <- rep(0, nmcmc)
  alpha_save <- matrix(0, npcvs + 1, nmcmc)

  # Priors
  beta0_mean <- rep(0, ncovs + 1)
  beta0_sd <- 2
  beta_phi_mean <- rep(0, ncovs + 1)
  beta_phi_sd <- 2
  beta_gamma_mean <- rep(0, ncovs + 1)
  beta_gamma_sd <- 2
  log_kappa_mean <- 0
  log_kappa_sd <- 2
  alpha_mean <- rep(0, npcvs + 1)
  alpha_sd <- 2

  # Starting values
  beta0 <- rep(0, ncovs + 1)
  beta_phi <- rep(0, ncovs + 1)
```

```
beta_gamma <- rep(0, ncovs + 1)
kappa <- 1
alpha <- rep(0, npcvs + 1)
phi <- matrix(0.5, nsite, nyear-1)
eta <- exp(-1 * kappa * d)
theta <- eta / rowSums(eta)
z <- ifelse(ysum==0, 0, 1)
gamma <- matrix(, nsite, nyear-1)
for (t in 2:nyear) {
  for (i in 1:nsite) {
    gamma[i,t-1] <- inv.logit(c(1, x[i,t,]) %*% beta_gamma) * sum(theta
[i,] * z[,t-1])
  } # i
} # t
p <- array(.5, dim=c(nsite, nyear, nreps))

# Tuning factors
beta0_tune <- 0.2
beta_phi_tune <- 0.05
beta_gamma_tune <- 0.05
kappa_tune <- 0.1
alpha_tune <- 0.035

# Begin MCMC loop
for (k in 1:nmcmc) {
  ### Sample beta0
  beta0_star <- rnorm(ncovs + 1, beta0, beta0_tune)
  mh1 <- sum(dbinom(z[,1], 1, inv.logit(cbind(1,x[,1,]) %*%
beta0_star), log=TRUE)) +
      sum(dnorm(beta0_star, beta0_mean, beta0_sd, log=TRUE))
  mh2 <- sum(dbinom(z[,1], 1, inv.logit(cbind(1,x[,1,]) %*% beta0    ),
log=TRUE)) +
      sum(dnorm(beta0    , beta0_mean, beta0_sd, log=TRUE))
  mh <- exp(mh1 - mh2)
  if (mh > runif(1)) {
    beta0 <- beta0_star
  }

  ### Sample beta_phi
  beta_phi_star <- rnorm(ncovs + 1, beta_phi, beta_phi_tune)
  phi_star <- matrix(, nsite, nyear-1)
  for (t in 2:nyear) {
    phi_star[,t-1] <- inv.logit(cbind(1,x[,t,]) %*% beta_phi_star)
  } # t
  mh1 <- sum(dbinom(z[,-1], 1, z[,-nyear] * phi_star + (1 - z[,-nyear]) *
```

```
gamma, log=TRUE)) +
     sum(dnorm(beta_phi_star, beta_phi_mean, beta_phi_sd, log=TRUE))
  mh2 <- sum(dbinom(z[,-1], 1, z[,-nyear] * phi    + (1 - z[,-nyear]) *
gamma, log=TRUE)) +
      sum(dnorm(beta_phi    , beta_phi_mean, beta_phi_sd, log=TRUE))
  mh <- exp(mh1 - mh2)
  if (mh > runif(1)) {
   beta_phi <- beta_phi_star
   phi <- phi_star
  }

  ### Sample beta_gamma
  beta_gamma_star <- rnorm(ncovs + 1, beta_gamma, beta_gamma_tune)
  gamma <- gamma_star <- matrix(, nsite, nyear-1)
  for (t in 2:nyear) {
   for (i in 1:nsite) {
    gamma   [i,t-1] <- inv.logit(c(1, x[i,t,]) %*% beta_gamma    ) * sum
(theta[i,] * z[,t-1])
   gamma_star[i,t-1] <- inv.logit(c(1, x[i,t,]) %*% beta_gamma_star) *
sum(theta[i,] * z[,t-1])
   } # i
  } # t
  mh1 <- sum(dbinom(z[,-1], 1, z[,-nyear] * phi + (1 - z[,-nyear]) *
gamma_star, log=TRUE)) +
      sum(dnorm(beta_gamma_star, beta_gamma_mean, beta_gamma_sd,
log=TRUE))
  mh2 <- sum(dbinom(z[,-1], 1, z[,-nyear] * phi + (1 - z[,-nyear]) * gamma
, log=TRUE)) +
      sum(dnorm(beta_gamma    , beta_gamma_mean, beta_gamma_sd,
log=TRUE))
  mh <- exp(mh1 - mh2)
  if (mh > runif(1)) {
   beta_gamma <- beta_gamma_star
  }

  ### Sample kappa
  kappa_star <- exp(rnorm(1, log(kappa), kappa_tune))
  eta_star <- exp(-1 * kappa_star * d)
  theta_star <- eta_star / rowSums(eta_star)
  gamma <- gamma_star <- matrix(, nsite, nyear-1)
  for (t in 2:nyear) {
   for (i in 1:nsite) {
    gamma   [i,t-1] <- inv.logit(c(1, x[i,t,]) %*% beta_gamma) * sum
(theta   [i,] * z[,t-1])
    gamma_star[i,t-1] <- inv.logit(c(1, x[i,t,]) %*% beta_gamma) * sum
```

```
(theta_star[i,] * z[,t-1])
    } # i
  } # t
  mh1 <- sum(dbinom(z[,-1], 1, z[,-nyear] * phi + (1 - z[,-nyear]) *
gamma_star, log=TRUE)) +
      dnorm(log(kappa_star), log_kappa_mean, log_kappa_sd, log=TRUE)
  mh2 <- sum(dbinom(z[,-1], 1, z[,-nyear] * phi + (1 - z[,-nyear]) * gamma
, log=TRUE)) +
      dnorm(log(kappa)    , log_kappa_mean, log_kappa_sd, log=TRUE)
  mh <- exp(mh1 - mh2)
  if (mh > runif(1)) {
   kappa <- kappa_star
   eta <- eta_star
   theta <- theta_star
  }

  ### Sample alpha
  alpha_star <- rnorm(npcvs + 1, alpha, alpha_tune)
  p_star <- array(, dim=c(nsite, nyear, nreps))
  for (i in 1:nsite) {
   for (t in 1:nyear) {
    p_star[i,t,] <- inv.logit(cbind(1,w[i,t,,]) %*% alpha_star)
   } # t
  } # i
  mh1t <- mh2t <- numeric(nyear)
  for (t in 1:nyear) {
   mh1t[t] <- sum(dbinom(y[which(z[,t]==1),t,], 1, p_star[which(z[,
t]==1),t,], log=TRUE))
   mh2t[t] <- sum(dbinom(y[which(z[,t]==1),t,], 1, p    [which(z[,t]
==1),t,], log=TRUE))
  } # t
  mh1 <- sum(mh1t) +
      sum(dnorm(alpha_star, alpha_mean, alpha_sd, log=TRUE))
  mh2 <- sum(mh2t) +
      sum(dnorm(alpha    , alpha_mean, alpha_sd, log=TRUE))
  mh <- exp(mh1 - mh2)
  if (mh > runif(1)) {
   alpha <- alpha_star
   p <- p_star
  }

  ### Sample z
  s_temp <- matrix(, nsite, nyear)
  s_temp[,1] <- inv.logit(cbind(1,x[,1,]) %*% beta0)
  for (t in 2:nyear) {
```

```
    s_temp[,t] <- z[,t-1] * phi[,t-1] + (1 - z[,t-1]) * gamma[,t-1]
   } # t
   num_temp <- s_temp * apply((1 - p), 1:2, prod)
   psi_temp <- num_temp / (num_temp + (1 - s_temp))
   z[which(ysum==0)] <- rbinom(length(which(ysum==0)), 1, psi_temp
[which(ysum==0)])

   ### Save samples
   beta0_save[,k] <- beta0
   beta_phi_save[,k] <- beta_phi
   beta_gamma_save[,k] <- beta_gamma
   kappa_save[k] <- kappa
   alpha_save[,k] <- alpha
  } # k

 # Write output
 list(beta0_save=beta0_save,
    beta_phi_save=beta_phi_save, beta_gamma_save=beta_gamma_save,
kappa_save=kappa_save,
    alpha_save=alpha_save)
} # dist_cov_dynam_occu_mcmc
```

Then we can use the following code to implement the algorithm, which takes a little bit less than 6 h for the model to run. Using code in Appendix 2.5, we can create traceplots that show good model convergence and consistency between parameter estimates of $\beta^{[0]}$, $\beta^{[\phi]}$, $\beta^{[\gamma]}$, κ, and α and their true values used in data simulation (Fig. 2.9).

```
   library(foreach)   # for parallel computing
library(doParallel) # for parallel computing

numCores <- round(detectCores() / 2) # only use half of the cores for
parallel computing
registerDoParallel(numCores) # setup parallel computing

nmcmc <- 50000 # number of iterations
chain <- 3    # number of chains

start_time <- Sys.time() # start time of computing
out <- foreach (i=1:chain, .packages='boot') %dopar% {
 dist_cov_dynam_occu_mcmc(y=y, x=x, w=w, d=d, nmcmc=nmcmc)
} # i
stopImplicitCluster() # clean up the cluster
end_time <- Sys.time() # end time of computing
time_taken <- end_time - start_time # computing time
```

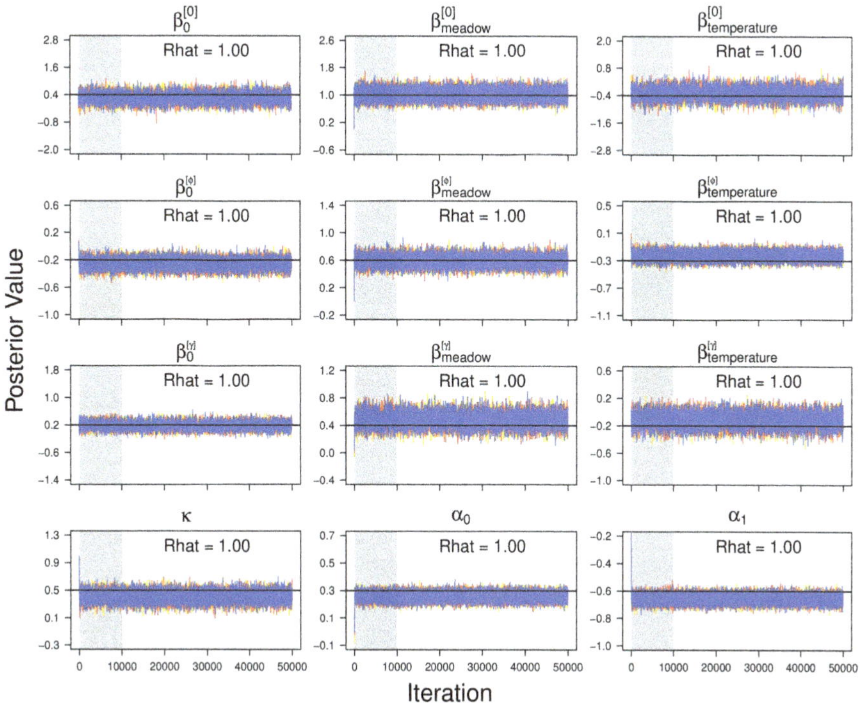

Fig. 2.9 Traceplots showing the posterior chains for $\beta^{[0]}$, $\beta^{[\phi]}$, $\beta^{[\gamma]}$, κ, and α in the spatial dynamic occupancy model with distance- and environment-based movement. Rhat values are calculated after discarding the first 10,000 iterations. Horizontal lines represent the true values used in data simulation

We can also use the following code to create some response surface, which shows the positive effect of meadow size and negative effect of temperature on colonization probability, γ, given everything else equal (Fig. 2.10).

```
 beta_gamma_post <- cbind(
 out[[1]]$beta_gamma[,(nmcmc-1999):nmcmc],
 out[[2]]$beta_gamma[,(nmcmc-1999):nmcmc],
 out[[3]]$beta_gamma[,(nmcmc-1999):nmcmc])

npred <- 500
size_pred <- rep(seq(0.5, 12, length.out=npred), each=npred)
temp_pred <- rep(seq(1.5, 8, length.out=npred), times=npred)
x1_pred <- (log(size_pred) - mean(log(cov$size))) / sd(log(cov$size))
x2_pred <- (temp_pred - mean(cov$temp)) / sd(cov$temp)
gamma_pred <- matrix(, 6000, npred^2)
for (i in 1:(npred^2)) {
```

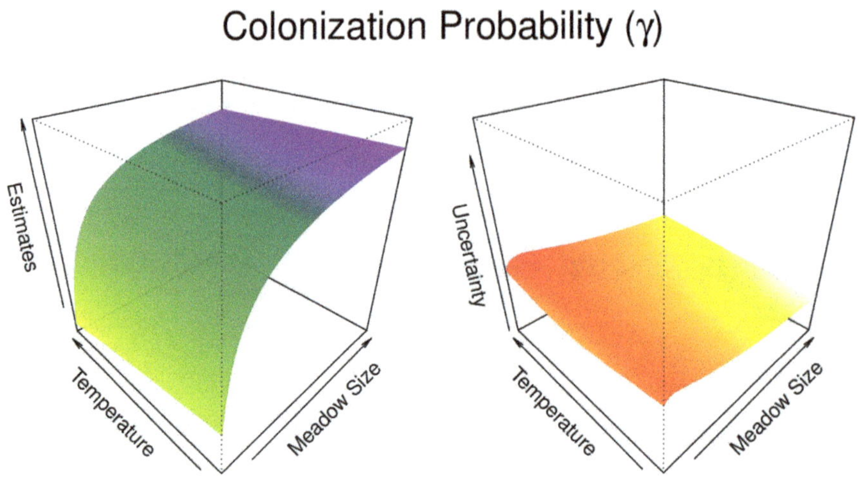

Fig. 2.10 Response surface showing the estimated positive effect of meadow size and negative effect of temperature on colonization probability, γ, given everything else equal (left). The corresponding uncertainty (width of 80% Credible Interval) is also shown (right)

```
 gamma_pred[,i] <- inv.logit(cbind(1,x1_pred[i],x2_pred[i]) %*%
beta_gamma_post)
} # i

gamma_med <- apply(gamma_pred, 2, quantile, probs=0.5)
gamma_wid <- apply(gamma_pred, 2, quantile, probs=0.9) - apply
(gamma_pred, 2, quantile, probs=0.1)
rm(gamma_pred)

nscale <- 500
gamma_med_mat <- matrix(gamma_med, npred, npred, byrow=T)
gamma_wid_mat <- matrix(gamma_wid, npred, npred, byrow=T)
gamma_med_col <- gamma_wid_col <- matrix(, npred-1, npred-1)
for (i in 1:(npred-1)) {
 for (j in 1:(npred-1)) {
  gamma_med_col[i,j] <- log(mean(gamma_med_mat[i:(i+1),j:(j+1)]))
  gamma_wid_col[i,j] <- log(mean(gamma_wid_mat[i:(i+1),j:(j+1)]))
 } # j
} # i
gamma_med_scale <- round((gamma_med_col - min(gamma_med_col)) / (max
(gamma_med_col) - min(gamma_med_col)) * (nscale - 1)) + 1
gamma_wid_scale <- round((gamma_wid_col - min(gamma_wid_col)) / (max
(gamma_wid_col) - min(gamma_wid_col)) * (nscale - 1)) + 1
cols <- c(colorRampPalette(c('yellow','seagreen'))(nscale*0.9),
colorRampPalette(c('seagreen','purple'))(nscale*0.1))
```

```
layout(mat=matrix(c(1,1,rep(c(2,3),times=5)), nrow=6, ncol=2,
byrow=T))
par(oma=c(0,1,0,0))

par(mar=c(1,0,0,0))
plot(1, type='n', xlab='', ylab='', axes=F)
text(x=1, y=0.8, labels=expression(paste('Colonization Probability
(', gamma, ')', sep='')), cex=3)

par(mar=c(0,1,1,1))
persp(x=sort(unique(size_pred)), y=sort(unique(temp_pred)),
z=gamma_med_mat, zlim=c(.65,1),
   theta=-45, phi=30, col=cols[gamma_med_scale], cex.lab=1.6,
border=NA,
   xlab='\nMeadow Size', ylab='\nTemperature', zlab='\nEstimates')

persp(x=sort(unique(size_pred)), y=sort(unique(temp_pred)),
z=gamma_wid_mat, zlim=c(0,.35),
   theta=-45, phi=30, col=rev(heat.colors(nscale))
[gamma_wid_scale], cex.lab=1.6, border=NA,
  xlab='\nMeadow Size', ylab='\nTemperature', zlab='\nUncertainty')
```

2.5 Spatial Dynamic Occupancy Models for Interacting Species

Understanding the coexistence of interacting species has long been a focus in metapopulation studies (Hanski and Gilpin 1997; Hanski 1998). In this case, the occupancy status of a species is influenced not only by its abiotic environment but also by the presence or absence of other interacting species. For example, competition and predator–prey interactions can influence persistence and colonization, and thus the stability of local populations (Bengtsson 1989; Bonsall et al. 2002). The capability of moving among habitat patches is also key for metapopulation stability (Holyoak and Lawler 1996; Briggs and Hoopes 2004; Cooper et al. 2012; Wisz et al. 2013). Therefore, studies have been motivated to incorporate interspecific interactions in static occupancy models (Mackenzie et al. 2004; Waddle et al. 2010; Rota et al. 2016) as well as in their dynamic counterparts (Fidino et al. 2019).

Here, we consider a model in which persistence and colonization of one species are functions of abiotic environmental covariates as well as occupancy of other species. For illustration, imagine that our little butterfly lives in the same area as some ant species that is a fierce predator of the butterfly. If the butterfly lives in a meadow, it will be less likely to continue living there if the ant also lives in the same meadow (because all of the butterflies may simply be eaten!). Conversely, if the ant lives in a meadow, it will be more likely to continue living there if the butterfly is there (because the ant gets to eat the butterfly). Meanwhile, we can also imagine that the butterfly is less likely to move into a meadow where the ant lives in, and the ant is more likely to move into a meadow if the butterfly is there. Using statistical language, we have the following for the first year:

$$\left.\begin{array}{c} z_{i,1,s} \sim Bernoulli\left(\psi_{i,1,s}\right) \\ \psi_{i,1,s} = logit^{-1}\left(\beta_{0,s}^{[0]} + \beta_{1,s}^{[0]} \times x_{i,1,1} + \ldots + \beta_{n_{covs},s}^{[0]} \times x_{i,1,n_{covs}}\right) \end{array}\right\}, \tag{2.12}$$

in which $z_{i,\ 1,\ s}$ is the occupancy status of species s at site i and year 1, $\psi_{i,\ 1,\ s}$ is the initial occupancy probability, and $\beta_{\cdot,s}^{[0]}$'s are species-specific intercepts and slopes governing the effects of covariates (x's) on initial occupancy. Here, if we consider a system with two species (i.e., $n_{spec} = 2$), we will have $s = ant$ or *butterfly*.

From the second year, we have

$$\left.\begin{array}{c} z_{i_1,t,s} \sim Bernoulli\left(\psi_{i_1,t,s}\right) \\ \psi_{i_1,t,s} = z_{i_1,t-1,s} \times \phi_{i_1,t-1,s} + \left(1 - z_{i_1,t-1,s}\right) \times \gamma_{i_1,t-1,s} \\ \phi_{i_1,t-1,s} = logit^{-1}\left[\beta_{0,s,z_{-s}=0}^{[\phi]} \times \left(1 - z_{i_1,t-1,-s}\right) + \beta_{0,s,z_{-s}=1}^{[\phi]} \times z_{i_1,t-1,-s} + \beta_{1,s}^{[\phi]} \times x_{i_1,t,1} + \ldots\right] \\ \gamma_{i_1,t-1,s} = \sum_{i_2=1}^{n_{site}} \theta_{i_1,i_2,s} \times z_{i_2,t-1,s} \\ \theta_{i_1,i_2,s} = \dfrac{\eta_{i_1,i_2,s}}{\sum_{i_2=1}^{n_{site}} \eta_{i_1,i_2,s}} \\ \eta_{i_1,i_2,s} = exp\left\{-1 \times \left[\kappa_{s,z_{-s}=0} \times \left(1 - z_{i_1,t-1,-s}\right) + \kappa_{s,z_{-s}=1} \times z_{i_1,t-1,-s}\right] \times d_{i_1,i_2}\right\} \end{array}\right\}, \tag{2.13}$$

This model also has an observation sub-model that links the observed detection/non-detection data to the true occupancy status such that

$$\left.\begin{array}{c} y_{i,t,j,s} \sim Bernoulli\left(z_{i,t,s} \times p_{i,t,j,s}\right) \\ p_{i,t,j,s} = logit^{-1}\left(\alpha_{0,s} + \alpha_{1,s} \times w_{i,t,j,1} + \ldots + \alpha_{n_{pcvs},s} \times w_{i,t,j,n_{pcvs}}\right) \end{array}\right\}. \tag{2.14}$$

This is similar to Eq. (2.5) except that the state variables and parameters are species specific and thus indexed by s.

Next, we can simulate data for this multispecies model using the code below. Let us choose values from a predator–prey system, in which the predator ant has a negative effect on the prey butterfly's persistence and colonization, and the prey butterfly has a positive effect on the predator ant's persistence and colonization. The model is nonetheless useful for other kinds of interspecific interactions such as competition, for which both species have negative effects on each other's persistence and colonization. More specifically, we assume that the predator ant has a lower baseline persistence when the prey butterfly is absent ($\beta_{0,ant,z_{butterfly}=0}^{[\phi]} = -1$) vs. when the butterfly is present ($\beta_{0,ant,z_{butterfly}=1}^{[\phi]} = 1$), while the butterfly has a higher baseline persistence when the ant is absent ($\beta_{0,butterfly,z_{ant}=0}^{[\phi]} = 0.3$) vs. when the ant is present ($\beta_{0,butterfly,z_{ant}=1}^{[\phi]} = -0.3$). We also assume that the ant is less likely to colonize a site from long distance when the butterfly is absent at the site ($\kappa_{ant,z_{butterfly}=0} = 3$) vs. when the butterfly is present ($\kappa_{ant,z_{butterfly}=1} = 1$), while the butterfly is more likely to colonize a site from long distance when the ant is absent at a site ($\kappa_{butterfly,z_{ant}=0} =$

1.5) vs. when the ant is present ($\kappa_{butterfly, z_{ant}=1} = 2.5$). We assume that meadow size has a positive effect on the persistence of the butterfly ($\beta^{[\phi]}_{meadow, butterfly} = 0.8$) but has no effect on the persistence of the ant ($\beta^{[\phi]}_{meadow, ant} = 0$). We also assume that temperature has a negative effect on the persistence of the butterfly ($\beta^{[\phi]}_{temperature, butterfly} = -0.4$) but has no effect on the persistence of the ant ($\beta^{[\phi]}_{temperature, ant} = 0$). Note that in the code below $\beta^{[0]}$, $\beta^{[\phi]}$, κ, and α are written in two-column matrices, in which the first column is for the predator ant and the second for the prey butterfly.

```
library(boot) # for using logit and inv.logit functions

set.seed(4)

# Basic values
nsite <- 360 # number of sites
nyear <- 20  # number of years
nreps <- 5   # number of within-season replicates
nspec <- 2   # number of species
ncovs <- 2   # number of environmental covariates
npcvs <- 2   # number of observational covariates

### For following parameters, the first column is for the predator
species, and the second column is for the prey species
beta0 <- matrix(c(-1, 0.4, -0.2, 1, 1, -0.5), ncovs+1, nspec)
# intercept and slopes for initial occupancy
beta_phi <- matrix(c(-1, 1, 0, 0, 0.3, -0.3, 0.8, -0.4), ncovs+2, nspec)
# intercept and slopes for persistence
kappa <- matrix(c(3, 1, 1.5, 2.5), nspec, nspec)
# decay parameter for colonization
alpha <- matrix(c(0.8, -0.6, 0.3, 0.6, -0.4, 0.2), npcvs+1, nspec)
# intercept and slopes for detection probability

# Read in environmental covariates
load('data/covariates.RData')
x1 <- (log(cov$size) - mean(log(cov$size))) / sd(log(cov$size))
# standardized log habitat patch size
x2 <- (cov$temp - mean(cov$temp)) / sd(cov$temp)
# standardized temperature
x <- array(, dim=c(nsite, nyear, ncovs)) # environmental covariates
x[,,1] <- x1
x[,,2] <- x2

# Simulate data
psi <- array(, dim=c(nsite, nyear, nspec)) # occupancy probability
z <- array(, dim=c(nsite, nyear, nspec)) # occupancy status
psi[,1,] <- inv.logit(cbind(1,x[,1,]) %*% beta0) # initial occupancy
```

```
probability
z[,1,] <- rbinom(nsite*nspec, 1, psi[,1,]) # initial occupancy status

lon <- cov$lon # longitude or easting
lat <- cov$lat # latitude or northing
d <- matrix(, nsite, nsite) # distance matrix
for (i1 in 1:nsite) {
 for (i2 in 1:nsite) {
  d[i1,i2] <- sqrt((lon[i1] - lon[i2]) ^ 2 + (lat[i1] - lat[i2]) ^ 2)
 } # i2
} # i1
eta <- array(, dim=c(nsite, nsite, nspec, nspec)) # unstandardized
colonization rate
theta <- array(, dim=c(nsite, nsite, nspec, nspec)) # standardized
colonization rate
for (s1 in 1:nspec) {
 for (s2 in 1:nspec) {
  eta[,,s1,s2] <- exp(-1 * kappa[s1,s2] * d)
  theta[,,s1,s2] <- eta[,,s1,s2] / rowSums(eta[,,s1,s2])
 } # s2
} # s1

phi  <- array(, dim=c(nsite, nyear-1, nspec)) # probability of
persistence
gamma <- array(, dim=c(nsite, nyear-1, nspec)) # probability of
colonization
for (t in 2:nyear) {
 for (i in 1:nsite) {
  for (s in 1:nspec) {
   phi  [i,t-1,s] <- inv.logit(c(1 - z[i,t-1,-s], z[i,t-1,-s], x[i,t,])
%*% beta_phi[,s])
   gamma[i,t-1,s] <- sum(theta[i,,,s] %*% c(1 - z[i,t-1,-s], z[i,t-1,-
s]) * z[,t-1,s])
   psi[i,t,s] <- z[i,t-1,s] * phi[i,t-1,s] + (1 - z[i,t-1,s]) * gamma[i,
t-1,s]
   z[i,t,s] <- rbinom(1, 1, psi[i,t,s])
  } # s
 } # i
} # t

w <- array(rnorm(nsite * nyear * nreps * npcvs, 0, 1), dim=c(nsite,
nyear, nreps, npcvs)) # observational covariates

p <- array(, dim=c(nsite, nyear, nreps, nspec)) # detection probability
for (t in 1:nyear) {
```

```
 for (j in 1:nreps) {
  p[,t,j,] <- inv.logit(cbind(1,w[,t,j,]) %*% alpha)
 } # j
} # t

y <- array(, dim=c(nsite, nyear, nreps, nspec)) # detection/non-
detection data
for (t in 1:nyear) {
 for (j in 1:nreps) {
  for (s in 1:nspec) {
   y[,t,j,s] <- rbinom(nsite, 1, z[,t,s] * p[,t,j,s])
  } # s
 } # j
} # t
```

We then can develop a custom MCMC algorithm to estimate parameters $\beta^{[0]}$, $\beta^{[\phi]}$, κ, and α using data y, x, w, and d, as shown below:

```
sdom_ii_mcmc <- function(y, x, w, d, nmcmc) {

 # Setup variables
 nsite <- dim(y)[1]
 nyear <- dim(y)[2]
 nreps <- dim(y)[3]
 nspec <- dim(y)[4]
 ncovs <- dim(x)[3]
 npcvs <- dim(w)[4]
 ysum <- apply(y, c(1,2,4), sum)

 beta0_save <- array(0, dim=c(ncovs + 1, nspec, nmcmc))
 beta_phi_save <- array(0, dim=c(ncovs + 2, nspec, nmcmc))
 kappa_save <- array(0, dim=c(nspec, nspec, nmcmc))
 alpha_save <- array(0, dim=c(npcvs + 1, nspec, nmcmc))

 # Priors
 beta0_mean <- matrix(0, ncovs + 1, nspec)
 beta0_sd <- 2
 beta_phi_mean <- matrix(0, ncovs + 2, nspec)
 beta_phi_sd <- 2
 log_kappa_mean <- matrix(0, nspec, nspec)
 log_kappa_sd <- 2
 alpha_mean <- matrix(0, npcvs + 1, nspec)
 alpha_sd <- 2
```

```
# Starting values
beta0 <- matrix(0, ncovs + 1, nspec)
beta_phi <- matrix(0, ncovs + 2, nspec)
kappa <- matrix(1, nspec,nspec)
alpha <- matrix(0, npcvs + 1, nspec)
phi  <- array(0.5, dim=c(nsite, nyear-1, nspec))
gamma <- array(0.5, dim=c(nsite, nyear-1, nspec))
eta <- theta <- array(, dim=c(nsite, nsite, nspec, nspec))
for (s1 in 1:nspec) {
 for (s2 in 1:nspec) {
  eta[,,s1,s2] <- exp(-1 * kappa[s1,s2] * d)
  theta[,,s1,s2] <- eta[,,s1,s2] / rowSums(eta[,,s1,s2])
 } # s2
} # s1
p <- array(0.5, dim=c(nsite, nyear, nreps, nspec))
z <- ifelse(ysum==0, 0, 1)

# Tuning factors
beta0_tune <- matrix(0.4, ncovs + 1, nspec)
beta_phi_tune <- matrix(0.05, ncovs + 2, nspec)
kappa_tune <- matrix(0.1, nspec, nspec)
alpha_tune <- matrix(0.05, npcvs + 1, nspec)

# Begin MCMC loop
for (k in 1:nmcmc) {
 ### Sample beta0
 beta0_star <- matrix(rnorm((ncovs + 1) * nspec, beta0, beta0_tune),
ncovs + 1, nspec)
 for (s in 1:nspec) {
  mh1 <- sum(dbinom(z[,1,s], 1, inv.logit(cbind(1,x[,1,]) %*%
beta0_star[,s]), log=TRUE)) +
     sum(dnorm(beta0_star[,s], beta0_mean[,s], beta0_sd, log=TRUE))
  mh2 <- sum(dbinom(z[,1,s], 1, inv.logit(cbind(1,x[,1,]) %*% beta0
[,s]), log=TRUE)) +
         sum(dnorm(beta0  [,s], beta0_mean[,s], beta0_sd, log=TRUE))
  mh <- exp(mh1 - mh2)
  if (mh > runif(1)) {
   beta0[,s] <- beta0_star[,s]
  }
 } # s

 ### Sample beta_phi
 beta_phi_star <- matrix(rnorm((ncovs + 2) * nspec, beta_phi,
beta_phi_tune), ncovs + 2, nspec)
 phi <- phi_star <- array(, dim=c(nsite, nyear-1, nspec))
```

```
  for (t in 2:nyear) {
   for (s in 1:nspec) {
    phi    [,t-1,s] <- inv.logit(cbind(1 - z[,t-1,-s], z[,t-1,-s], x[,
t,]) %*% beta_phi     [,s])
   phi_star[,t-1,s] <- inv.logit(cbind(1 - z[,t-1,-s], z[,t-1,-s], x[,
t,]) %*% beta_phi_star[,s])
   } # k
  } # t
  for (s in 1:nspec) {
   mh1 <- sum(dbinom(z[,-1,s], 1, z[,-nyear,s] * phi_star[,,s] + (1 - z
[,-nyear,s]) * gamma[,,s], log=TRUE)) +
       sum(dnorm(beta_phi_star[,s], beta_phi_mean[,s], beta_phi_sd,
log=TRUE))
   mh2 <- sum(dbinom(z[,-1,s], 1, z[,-nyear,s] * phi    [,,s] + (1 - z
[,-nyear,s]) * gamma[,,s], log=TRUE)) +
       sum(dnorm(beta_phi    [,s], beta_phi_mean[,s], beta_phi_sd,
log=TRUE))
   mh <- exp(mh1 - mh2)
   if (mh > runif(1)) {
    beta_phi[,s] <- beta_phi_star[,s]
   }
  } # s

  ### Sample kappa
  kappa_star <- matrix(exp(rnorm(nspec * nspec, log(kappa),
kappa_tune)), nspec, nspec)
  eta_star <- theta_star <- array(, dim=c(nsite, nsite, nspec, nspec))
  for (s1 in 1:nspec) {
   for (s2 in 1:nspec) {
    eta_star[,,s1,s2] <- exp(-1 * kappa_star[s1,s2] * d)
    theta_star[,,s1,s2] <- eta_star[,,s1,s2] / rowSums(eta_star[,,s1,
s2])
   } # s2
  } # s1
  gamma <- gamma_star <- array(, dim=c(nsite, nyear-1, nspec))
  for (t in 2:nyear) {
   for (i in 1:nsite) {
    for (s in 1:nspec) {
     gamma    [i,t-1,s] <- sum(theta    [i,,,s] %*% c(1 - z[i,t-1,-s], z[i,
t-1,-s]) * z[,t-1,s])
     gamma_star[i,t-1,s] <- sum(theta_star[i,,,s] %*% c(1 - z[i,t-1,-
s], z[i,t-1,-s]) * z[,t-1,s])
    } # s
   } # i
  } # t
```

```
  for (s in 1:nspec) {
   mh1 <- sum(dbinom(z[,-1,s], 1, z[,-nyear,s] * phi[,,s] + (1 - z
[,-nyear,s]) * gamma_star[,,s], log=TRUE)) +
       sum(dnorm(log(kappa_star[,s]), log_kappa_mean[,s],
log_kappa_sd, log=TRUE))
   mh2 <- sum(dbinom(z[,-1,s], 1, z[,-nyear,s] * phi[,,s] + (1 - z
[,-nyear,s]) * gamma   [,,s], log=TRUE)) +
       sum(dnorm(log(kappa   [,s]), log_kappa_mean[,s], log_kappa_sd,
log=TRUE))
   mh <- exp(mh1 - mh2)
   if (mh > runif(1)) {
    kappa[,s] <- kappa_star[,s]
    eta[,,,s] <- eta_star[,,,s]
    theta[,,,s] <- theta_star[,,,s]
   }
  } # s

  ### Sample alpha
  alpha_star <- matrix(rnorm((npcvs + 1) * nspec, alpha, alpha_tune),
npcvs + 1, nspec)
  p_star <- array(, dim=c(nsite, nyear, nreps, nspec))
  for (t in 1:nyear) {
   for (j in 1:nreps) {
    p_star[,t,j,] <- inv.logit(cbind(1,w[,t,j,]) %*% alpha_star)
   } # j
  } # t
  for (s in 1:nspec) {
   mh1t <- mh2t <- numeric(nyear)
   for (t in 1:nyear) {
    mh1t[t] <- sum(dbinom(y[which(z[,t,s]==1),t,,s], 1, p_star[which
(z[,t,s]==1),t,,s], log=TRUE))
    mh2t[t] <- sum(dbinom(y[which(z[,t,s]==1),t,,s], 1, p   [which(z[,
t,s]==1),t,,s], log=TRUE))
   } # t
   mh1 <- sum(mh1t) +
      sum(dnorm(alpha_star[,s], alpha_mean[,s], alpha_sd, log=TRUE))
   mh2 <- sum(mh2t) +
      sum(dnorm(alpha   [,s], alpha_mean[,s], alpha_sd, log=TRUE))
   mh <- exp(mh1 - mh2)
   if (mh > runif(1)) {
    alpha[,s] <- alpha_star[,s]
    p[,,,s] <- p_star[,,,s]
   }
  } # s
```

```
  ### Sample z
  s_temp <- array(, dim=c(nsite, nyear, nspec))
  for (s in 1:nspec) {
   s_temp[,1,s] <- inv.logit(cbind(1,x[,1,]) %*% beta0[,s])
   for (t in 2:nyear) {
    s_temp[,t,s] <- z[,t-1,s] * phi[,t-1,s] + (1 - z[,t-1,s]) * gamma[,t-
1,s]
   } # t
  } # s
  num_temp <- s_temp * apply((1 - p), c(1,2,4), prod)
  psi_temp <- num_temp / (num_temp + (1 - s_temp))
  z[which(ysum==0)] <- rbinom(length(which(ysum==0)), 1, psi_temp
[which(ysum==0)])

  ### Save samples
  beta0_save[,,k] <- beta0
  beta_phi_save[,,k] <- beta_phi
  kappa_save[,,k] <- kappa
  alpha_save[,,k] <- alpha
 } # k

 # Write output
 list(beta0_save=beta0_save,
   beta_phi_save=beta_phi_save, kappa_save=kappa_save,
   alpha_save=alpha_save)
} # sdom_ii_mcmc
```

Next, let us use the code below to implement the algorithm, which takes a little bit less than 9 h for the model to run. From traceplots created with the code in Appendix 2.6, we can see good model convergence and consistency between parameter estimates of $\beta^{[0]}$, $\beta^{[\phi]}$, κ, and α and their true values used in data simulation (Fig. 2.11).

```
library(foreach)   # for parallel computing
library(doParallel) # for parallel computing

numCores <- round(detectCores() / 2) # only use half of the cores for
parallel computing
registerDoParallel(numCores) # setup parallel computing

nmcmc <- 50000 # number of iterations
chain <- 3   # number of chains

start_time <- Sys.time() # start time of computing
out <- foreach (i=1:chain, .packages='boot') %dopar% {
 sdom_ii_mcmc(y=y, x=x, w=w, d=d, nmcmc=nmcmc)
```

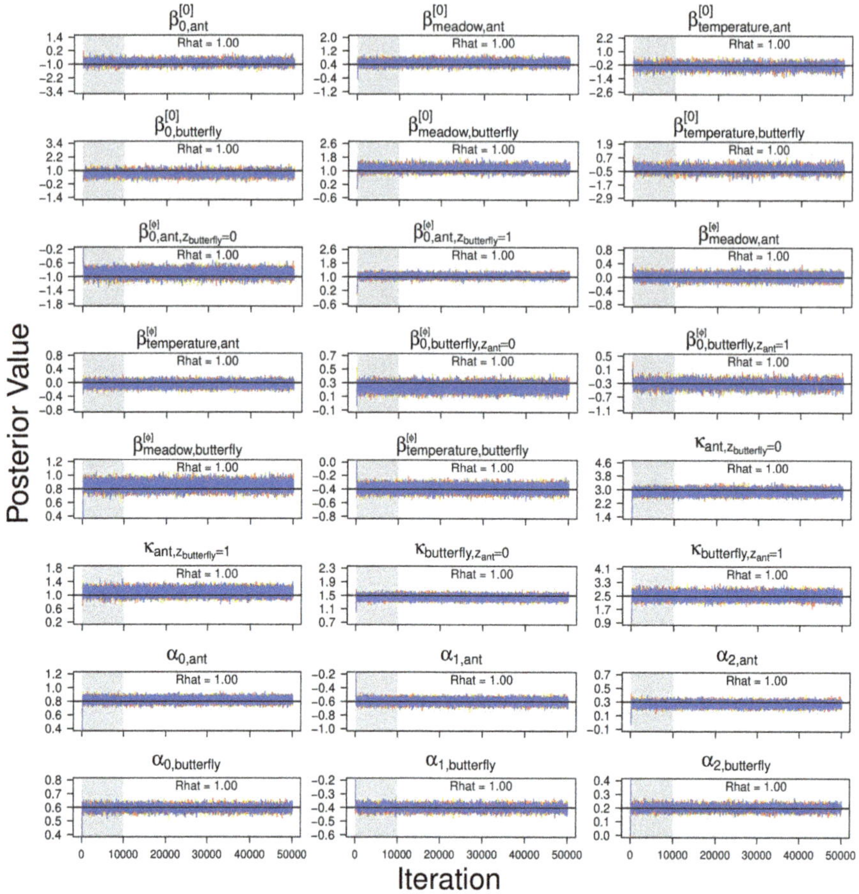

Fig. 2.11 Traceplots showing the posterior chains for $\beta^{[0]}$, $\beta^{[\phi]}$, κ, and α in the spatial dynamic occupancy model for two interacting species, a predator ant and a prey butterfly. Rhat values are calculated after discarding the first 10,000 iterations. Horizontal lines represent the true values used in data simulation

```
} # i
stopImplicitCluster() # clean up the cluster
end_time <- Sys.time() # end time of computing
time_taken <- end_time - start_time # computing time
```

We can also use the code below to plot the estimated unstandardized colonization rate, η, for the prey butterfly when the predator ant is absent or present, which shows great consistency with the true values (Fig. 2.12):

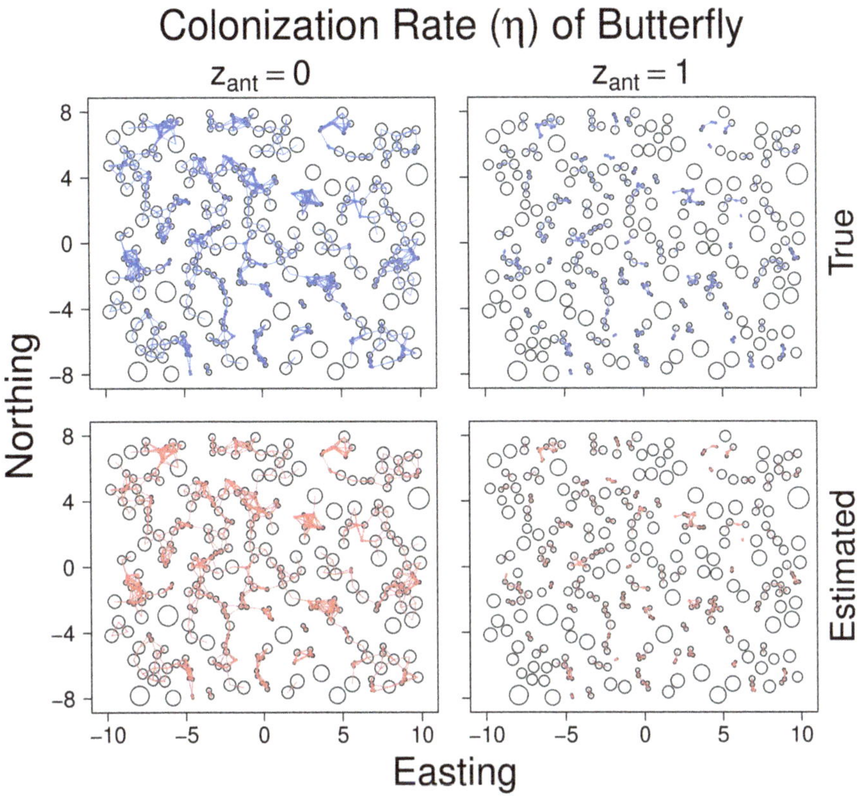

Fig. 2.12 The true (upper row) and estimated (lower row) unstandardized colonization rate, η, for the prey butterfly when the predator ant is absent (left column) or present (right column). A thicker line represents a higher colonization rate, and vice versa; only colonization rates that are greater than 0.2 are plotted

```
kappa_post <- array(, dim=c(nspec, nspec, nmcmc/2*chain))
for (s1 in 1:nspec) {
 for (s2 in 1:nspec) {
  for (i in 1:chain) {
   if (i == 1) {
    tt <- out[[i]]$kappa_save[s1,s2,(nmcmc/2+1):nmcmc]
   } else {
    tt <- c(tt, out[[i]]$kappa_save[s1,s2,(nmcmc/2+1):nmcmc])
   }
  } # i
  kappa_post[s1,s2,] <- tt
 } # s2
} # s1
kappa_med <- apply(kappa_post, 1:2, median)
```

```
eta_true <- eta_pred <- array(, dim=c(nsite, nsite, nspec, nspec))
for (s1 in 1:nspec) {
 for (s2 in 1:nspec) {
  eta_true[,,s1,s2] <- exp(-1 * kappa   [s1,s2] * d)
  eta_pred[,,s1,s2] <- exp(-1 * kappa_med[s1,s2] * d)
 } # s2
} # s1

min_eta <- 0.2
max_lwd <- 3
titles <- c(expression(z["ant"]==0), expression(z["ant"]==1))

par(mfrow=c(2,2))
par(mar=c(1,1,1,1))
par(oma=c(4,5,5,2))

for (s in 1:nspec) {

 plot(cov$lon, cov$lat, xlim=c(-10.2,10.2), ylim=c(-8.2,8.2),
axes=F, xlab='', ylab='',
    pch=21, cex=rowMeans(cov$size)*4.6, col='grey18', bg='white')
 axis(1, at=seq(-10,10,5), labels=rep('',5))
 if (s == 1) {
  axis(2, at=seq(-8,8,4), cex.axis=1.6, las=2)
 } else {
  axis(2, at=seq(-8,8,4), labels=rep('',5))
  axis(4, at=0, labels='True', cex.axis=2.5, tick=F, line=.5)
 }
 box()
 for (i in 1:(nsite-1)) {
  for (j in (i+1):nsite) {
   if (eta_true[i,j,s,2] >= min_eta) {
    lines(x=cov$lon[c(i,j)], y=cov$lat[c(i,j)], col='royalblue',
lwd=eta_true[i,j,s,2]*max_lwd)
   }
  } # j
 } # i
 axis(3, at=0, labels=titles[s], cex.axis=2.5, tick=F, line=-.8)

} # s

for (s in 1:nspec) {

 plot(cov$lon, cov$lat, xlim=c(-10.2,10.2), ylim=c(-8.2,8.2),
axes=F, xlab='', ylab='',
    pch=21, cex=rowMeans(cov$size)*4.6, col='grey18', bg='white')
```

```
 axis(1, at=seq(-10,10,5), cex.axis=1.6, las=1)
 if (s == 1) {
  axis(2, at=seq(-8,8,4), cex.axis=1.6, las=2)
 } else {
  axis(2, at=seq(-8,8,4), labels=rep('',5))
  axis(4, at=0, labels='Estimated', cex.axis=2.5, tick=F, line=.5)
 }
 box()
 for (i in 1:(nsite-1)) {
  for (j in (i+1):nsite) {
   if (eta_pred[i,j,s,2] >= min_eta) {
    lines(x=cov$lon[c(i,j)], y=cov$lat[c(i,j)], col='lightcoral',
lwd=eta_pred[i,j,s,2]*max_lwd)
   }
  } # j
 } # i

} # s

title(xlab='Easting' , line=2.6, cex.lab=2.8, outer=T)
title(ylab='Northing', line=2.4, cex.lab=2.8, outer=T)
title(main=expression(paste('Colonization Rate (', eta, ') of
Butterfly', sep='')), line=3.2, cex.main=3, outer=T)
```

2.6 Summary and Additional Reading

In this chapter, we talked about occupancy models which have evolved from static to dynamic (but non-spatial) and eventually spatial and dynamic. These models can be used for single species as well as a few interacting species. Undoubtedly, we have missed quite some important extensions of occupancy models, which the readers are encouraged to check out. One of such important extensions is described in Howell et al. (2018) which explicitly estimates a landscape resistance surface in spatial dynamic occupancy models. Sutherland et al. (2014) incorporate demographic information in a spatial dynamic occupancy model, which is also very interesting and important. We only talked about single-state occupancy models in this chapter, which can be extended to multiple-state versions with appropriate data (Royle 2004; Nichols et al. 2007; Kéry and Royle 2021). While we only considered false-negative observation errors in this chapter, occupancy models can deal with both false-positive and false-negative observation errors (Royle and Link 2006), in which false-positive errors can come from species misidentification (Miller et al. 2011). Presence-only data can also be used to estimate occupancy status of species, although strong assumptions have to be made (Royle et al. 2012). Alternatively, researchers can consider to jointly analyze detection/non-detection data and presence-only data to estimate occupancy status of a species under a data integration framework (Pacifici et al. 2017; Miller et al. 2019). There are numerous other extensions of occupancy models, which have been described in detail in synthesis papers and book chapters (Bailey et al. 2014; Kéry and Royle 2016, 2021). Thus, we conclude the chapter on occupancy models here. Next, we will move to N-mixture models that estimate abundance from count data for further insights into spatial structures and dynamics of populations.

References

Bailey LL, Mackenzie DI, Nichols JD (2014) Advances and applications of occupancy models. Methods Ecol Evol 5:1269–1279

Bengtsson J (1989) Interspecific competition increases local extinction rate in a metapopulation system. Nature 340:713–715

Bled F, Royle JA, Cam E (2011) Hierarchical modeling of an invasive spread: the eurasian collared-dove streptopelia decaocto in the United States. Ecol Appl 21:290–302

Bonsall MB, French DR, Hassell MP (2002) Metapopulation structures affect persistence of predator-prey interactions. J Anim Ecol 71:1075–1084

Briggs CJ, Hoopes MF (2004) Stabilizing effects in spatial parasitoid-host and predator-prey models: a review. Theor Popul Biol 65:299–315

Broms KM, Hooten MB, Johnson DS, Altwegg R, Conquest LL (2016) Dynamic occupancy models for explicit colonization processes. Ecology 97:194–204

Chambert T, Miller DAW, Nichols JD (2015) Modeling false positive detections in species occurrence data under different study designs. Ecology 96:332–339

Chandler RB, Muths E, Sigafus BH, Schwalbe CR, Jarchow CJ, Hossack BR (2015) Spatial occupancy models for predicting metapopulation dynamics and viability following reintroduction. J Appl Ecol 52:1325–1333

Cooper JK, Li J, Montagnes DJS (2012) Intermediate fragmentation per se provides stable predator-prey metapopulation dynamics. Ecol Lett 15:856–863

Doser JW, Stoudt S (2024) "Fractional replication" in single-visit multi-season occupancy models: impacts of spatiotemporal autocorrelation on identifiability. Methods Ecol Evol 15:358

Drake J, Lambin X, Sutherland C (2022) Spatiotemporal connectivity dynamics in spatially structured populations. J Anim Ecol 91:2050–2060

Fidino M, Simonis JL, Magle SB (2019) A multistate dynamic occupancy model to estimate local colonization–extinction rates and patterns of co-occurrence between two or more interacting species. Methods Ecol Evol 10:233–244

Hanski I (1994) A practical model of metapopulation dynamics. J Anim Ecol 63:151–162

Hanski I (1998) Metapopulation dynamics. Nature 396:41–49

Hanski I, Gilpin ME (1997) Metapopulation biology: ecology, genetics, and evolution. Academic Press

Holyoak M, Lawler SP (1996) The role of dispersal in predator—prey metapopulation dynamics. J Anim Ecol 65:640–652

Howell PE, Muths E, Hossack BR, Sigafus BH, Chandler RB (2018) Increasing connectivity between metapopulation ecology and landscape ecology. Ecology 99:1119–1128

Kendall WL, Pollock KH, Brownie C (1995) A likelihood-based approach to capture-recapture estimation of demographic parameters under the robust design. Biometrics 51:293–308

Kéry M, Royle JA (2016) Applied hierarchical modeling in ecology: analysis of distribution, abundance and species richness in R and BUGS, Volume 1: Prelude and static models. Elsevier

Kéry M, Royle JA (2021) Applied hierarchical modeling in ecology: analysis of distribution, abundance and species richness in R and BUGS, Volume 2: Dynamic and advanced models. Elsevier

Lele SR, Moreno M, Bayne E (2012) Dealing with detection error in site occupancy surveys: what can we do with a single survey? J Plant Ecol 5:22–31

MacKenzie DI, Nichols JD, Lachman GB, Droege S, Royle AA, Langtimm CA (2002) Estimating site occupancy rates when detection probabilities are less than one. Ecology 83:2248–2255

MacKenzie DI, Nichols JD, Hines JE, Knutson MG, Franklin AB (2003) Estimating site occupancy, colonization, and local extinction when a species is detected imperfectly. Ecology 84: 2200–2207

Mackenzie DI, Bailey LL, Nichols JD (2004) Investigating species co-occurrence patterns when species are detected imperfectly. J Anim Ecol 73:546–555

Miller DA, Nichols JD, McClintock BT, Grant EHC, Bailey LL, Weir LA (2011) Improving occupancy estimation when two types of observational error occur: non-detection and species misidentification. Ecology 92:1422–1428

Miller DAW, Pacifici K, Sanderlin JS, Reich BJ (2019) The recent past and promising future for data integration methods to estimate species' distributions. Methods Ecol Evol 10:22–37

Nichols JD, Hines JE, Mackenzie DI, Seamans ME, Gutiérrez RJ (2007) Occupancy estimation and modeling with multiple states and state uncertainty. Ecology 88:1395–1400

Pacifici K, Reich BJ, Miller DAW, Gardner B, Stauffer G, Singh S, McKerrow A, Collazo JA (2017) Integrating multiple data sources in species distribution modeling: a framework for data fusion. Ecology 98:840–850

Pollock KH (1982) A capture-recapture design robust to unequal probability of capture. J Wildl Manag 46:752–757

Risk BB, de Valpine P, Beissinger SR (2011) A robust-design formulation of the incidence function model of metapopulation dynamics applied to two species of rails. Ecology 92:462–474

Rota CT, Ferreira MAR, Kays RW, Forrester TD, Kalies EL, McShea WJ, Parsons AW, Millspaugh JJ (2016) A multispecies occupancy model for two or more interacting species. Methods Ecol Evol 7:1164–1173

Royle JA (2004) Modeling abundance index data from anuran calling surveys. Conserv Biol 18: 1378–1385

Royle JA, Kéry M (2007) A Bayesian state-space formulation of dynamic occupancy models. Ecology 88:1813–1823

Royle JA, Link WA (2006) Generalized site occupancy models allowing for false positive and false negative errors. Ecology 87:835–841

Royle JA, Chandler RB, Yackulic C, Nichols JD (2012) Likelihood analysis of species occurrence probability from presence-only data for modelling species distributions. Methods Ecol Evol 3: 545–554

Sutherland C, Elston DA, Lambin X (2012) Multi-scale processes in metapopulations: contributions of stage structure, rescue effect, and correlated extinctions. Ecology 93:2465–2473

Sutherland CS, Elston DA, Lambin X (2014) A demographic, spatially explicit patch occupancy model of metapopulation dynamics and persistence. Ecology 95:3149–3160

Tyre AJ, Tenhumberg B, Field SA, Niejalke D, Parris K, Possingham HP (2003) Improving precision and reducing bias in biological surveys: estimating false-negative error rates. Ecol Appl 13:1790–1801

Waddle JH, Dorazio RM, Walls SC, Rice KG, Beauchamp J, Schuman MJ, Mazzotti FJ (2010) A new parameterization for estimating co-occurrence of interacting species. Ecol Appl 20:1467–1475

Wisz MS, Pottier J, Kissling WD, Pellissier L, Lenoir J, Damgaard CF, Dormann CF, Forchhammer MC, Grytnes JA, Guisan A, Heikkinen RK, Høye TT, Kühn I, Luoto M, Maiorano L, Nilsson MC, Normand S, Öckinger E, Schmidt NM, Termansen M, Timmermann A, Wardle DA, Aastrup P, Svenning JC (2013) The role of biotic interactions in shaping distributions and realised assemblages of species: implications for species distribution modelling. Biol Rev 88: 15–30

3 N-Mixture Models

3.1 Introduction

In this chapter, we will introduce N-mixture models (Wyatt 2002; Royle 2004) and extensions that use count data to estimate true abundance of species. These models, in comparison to occupancy models, further move away from classic metapopulation models in the sense that they can work without 0's (or changes between 0's and non-0's in dynamic versions; Dail and Madsen 2011, Hostetler and Chandler 2015) in true abundance. The dynamic versions of N-mixture models also provide some demographic information, as the change in true abundance is exactly due to the BIDE processes. However, we will see that dynamic N-mixture models' ability to provide demographic information is still limited, due to stringent model assumptions.

More specifically, when N-mixture models evolved from static to dynamic versions, two major routes were taken. The first route was to consider population growth without further separating the contributions of birth (B) and death (D) to population growth, while population growth can be a function of density and environmental covariates (Hostetler and Chandler 2015). This version was further extended to allow distance-based movement (immigration I and emigration E; Howell et al. 2020) or even distance-, density- and environment-based movement between sites (this book, Sect. 3.3.2). Another route was to separate gain (B + I) and loss (D + E) in describing population dynamics (Dail and Madsen 2011) while allowing gain and loss to be functions of density and environmental covariates (Bellier et al. 2016). A spatial extension of the Dail-Madsen model further allows the separation of E and I from B and D, and thus accommodates the entire BIDE processes (Zhao et al. 2017). However, this spatial version is mostly conceptual and has very limited application value, as it only allows constant demographic rates and adjacency-based movement; further extensions of this model seem to be impossible

Supplementary Information The online version contains supplementary material available at https://doi.org/10.1007/978-3-031-64518-1_3.

Q. Zhao, *Bayesian Analysis of Spatially Structured Population Dynamics*, Analysis and Synthesis 253, https://doi.org/10.1007/978-3-031-64518-1_3

due to difficulties in parameter identifiability. Overall, at this point, it seems that it is better to either separate what is local (site-level population growth due to B and D) and what is not local (I and E between sites), or separate what makes the population increase (gain) and what makes the population decrease (loss) when using dynamic N-mixture models to analyze count data.

Next, we will gradually build up ecological realism and model complexity in a series of N-mixture models. We will first introduce the original "static" N-mixture model (Royle 2004). We then will introduce the non-spatial (Hostetler and Chandler 2015) and spatial versions (Howell et al. 2020) of dynamic N-mixture models with population growth, including an advanced version that allows distance-, density- and environment-based movements, which is described for the first time in this book (Sect. 3.3.2). We will also introduce a spatial dynamic N-mixture model with population growth for interacting species (Zhao et al. 2022). We then will introduce the original Dail and Madsen (2011) model as well as its age-structured, or multi-state, extension (Zipkin et al. 2014). We will end this chapter with a brief summary of other extensions of N-mixture models.

3.2 Static N-Mixture Models

The static N-mixture model for replicated counts of unmarked populations (Royle 2004) contains a process sub-model that describes the true abundance of a species given environmental conditions and an observation sub-model that describes how count data are obtained conditional on the true abundance. Now, let us imagine some fictitious pond turtle species (even though our little butterfly will do the trick just fine), which lives in the landscape described in Fig. 1.4 with ponds as habitat patches. If pond size has a positive effect on the abundance of this turtle species, there will be more turtles in larger ponds at some given water temperature. Likewise, if temperature has a negative effect on abundance, there will be fewer turtles living in warmer ponds, for a given pond size. Statistically speaking, we have

$$\left.\begin{array}{c} N_i \sim Poisson(\lambda_i) \\ \lambda_i = exp\left(\beta_0 + \beta_1 \times x_{i,1} + \ldots + \beta_{n_{covs}} \times x_{i,n_{covs}}\right) \end{array}\right\}, \tag{3.1}$$

in which N_i represents the true abundance of the study species at the ith site among a total of n_{site} sites, λ_i is the expectation of abundance, x's are n_{covs} covariates, and β's are intercept and slope parameters representing the effects of x's on the logarithm of λ.

The observation sub-model of the N-mixture model links the observed count data to the true abundance. Like in Chap. 2 for occupancy models, we only allow false-negative observation errors here and rule out potential false positive errors. That means, when our field observers visit a pond to count the number of turtles in that pond, they can (rarely) count every single turtle, or they can miss some of them, but they never count more (e.g., by counting the same turtle multiple times, which is called double counting) or misidentify something else (e.g., a rock) as a turtle.

Therefore, the number they write down is always smaller than or at most equal to the true number of turtles in that pond. If there is no turtle in a pond at all, they will always write down a 0, with no exception. Under these assumptions, we assume that the count follows a binomial distribution such that

$$\left.\begin{array}{c} y_{i,j} \sim Binomial(N_i, p_{i,j}) \\ p_{i,j} = logit^{-1}\left(\alpha_0 + \alpha_1 \times w_{i,j,1} + \ldots + \alpha_{n_{pcvs}} \times w_{i,j,n_{pcvs}}\right) \end{array}\right\}, \tag{3.2}$$

in which $y_{i,\,j}$ represents the count of the study species during the jth visit (among a total of n_{reps} visits) of site i, $p_{i,\,j}$ is detection probability, w's are n_{pcvs} observation covariates for each site and visit, and α's are intercept and slope parameters governing the effects of w's on $p_{i,\,j}$ at a logit scale. In addition to the assumption of false negative observation errors only, the model also assumes a closed population, and thus N_i does not vary among the visits j. Both assumptions can be relaxed in static N-mixture models (Stanley and Royle 2005; Chandler et al. 2011).

Next, let us use the following code to simulate data for the static N-mixture model. We assume that pond size has a positive effect ($\beta_{pond} = 1$), and temperature has a negative effect ($\beta_{temperature} = -0.8$) on the true abundance of the turtle on the log scale.

```
library(boot) # for using logit and inv.logit functions

set.seed(5)

# Basic values
nsite <- 360 # number of sites
nreps <- 5  # number of within-season replicates
ncovs <- 2  # number of environmental covariates
npcvs <- 2  # number of observational covariates

beta <- c(1.8, 1, -0.8)   # intercept and slopes for true abundance
alpha <- c(0.3, -0.6, 0.3) # intercept and slopes for detection probability

# Read in environmental covariates
load('data/covariates.RData')
x1 <- (log(cov$size) - mean(log(cov$size))) / sd(log(cov$size)) #
standardized log habitat patch size
x2 <- (cov$temp - mean(cov$temp)) / sd(cov$temp) # standardized
temperature
x <- cbind(x1[,1], x2[,2]) # environmental covariates

# Simulate data
lambda <- exp(cbind(1,x) %*% beta) # expectation of abundance
N <- rpois(nsite, lambda) # true abundance
```

```
w <- array(rnorm(nsite * nreps * npcvs, 0, 1), dim=c(nsite, nreps,
npcvs)) # observational covariates

p <- matrix(, nsite, nreps) # detection probability
for (i in 1:nsite) {
 p[i,] <- inv.logit(cbind(1,w[i,,]) %*% alpha)
} # i

y <- matrix(, nsite, nreps) # count data
for (i in 1:nsite) {
 y[i,] <- rbinom(nreps, N[i], p[i,])
} # i
```

The simulated true abundance of the first five sites looks like the following. Here, we can see that the turtle inhabits these five sites with varying abundance from 1 to 10.

	Abundance
Site 1	1
Site 2	3
Site 3	10
Site 4	6
Site 5	1

The simulated count data for the first five sites during five visits looks like the following. For sites 1 and 5, the only individual turtle there is detected during some, but not all visits. For sites 2, 3, and 4, most of the time some, but not all turtles there, are detected. Indeed, for site 3, the observer detects at most 5 out of 10 turtles that inhabit the site.

	Count				
	Visit 1	Visit 2	Visit 3	Visit 4	Visit 5
Site 1	0	1	1	0	1
Site 2	2	1	3	0	2
Site 3	3	1	5	1	3
Site 4	6	3	2	3	3
Site 5	1	1	1	0	1

We then develop a custom MCMC algorithm to fit the model, as shown below. The interest is to estimate parameters β, α and N given data y, x, and w.

```
Nmix_mcmc <- function(y, x, w, nmcmc) {

 # Setup variables
 nsite <- dim(y)[1]
```

```
nreps <- dim(y)[2]
ncovs <- dim(x)[2]
npcvs <- dim(w)[3]
ymax <- apply(y, 1, max)

beta_save <- matrix(0, ncovs + 1, nmcmc)
alpha_save <- matrix(0, npcvs + 1, nmcmc)
N_save <- matrix(0, nsite, 2000)

# Priors
beta_mean <- rep(0, ncovs + 1)
beta_sd <- 2
alpha_mean <- rep(0, npcvs + 1)
alpha_sd <- 2

# Starting values
beta <- rep(0, ncovs + 1)
alpha <- rep(0, npcvs + 1)
lambda <- exp(cbind(1,x) %*% beta)
N <- round((ymax + 1) / 0.5)
p <- matrix(0.5, nsite, nreps)

# Tuning factors
beta_tune <- 0.15
alpha_tune <- c(0.3, 0.1, 0.1)
N_tune <- 1

# Begin MCMC loop
for (k in 1:nmcmc) {
 ### Sample N
 N_star <- rpois(nsite, N + N_tune)
 mh1 <- apply(dbinom(y, N_star, p, log=T), 1, sum) +
     dpois(N_star, lambda, log=T) +
     dpois(N, N_star+N_tune, log=T)
 mh2 <- apply(dbinom(y, N    , p, log=T), 1, sum) +
     dpois(N    , lambda, log=T) +
     dpois(N_star, N+N_tune, log=T)
 mh <- exp(mh1 - mh2)
 Nkeep <- ((mh > runif(nsite)) & (N_star >= ymax))
 N[Nkeep] <- N_star[Nkeep]

 ### Sample beta
 beta_star <- rnorm(ncovs + 1, beta, beta_tune)
 lambda_star <- exp(cbind(1,x) %*% beta_star)
 mh1 <- sum(dpois(N, lambda_star, log=T)) +
```

```
    sum(dnorm(beta_star, beta_mean, beta_sd, log=T))
  mh2 <- sum(dpois(N, lambda     , log=T)) +
    sum(dnorm(beta     , beta_mean, beta_sd, log=T))
  mh <- exp(mh1 - mh2)
  if (mh > runif(1)) {
   beta <- beta_star
   lambda <- lambda_star
  }

  ### Sample alpha
  alpha_star <- rnorm(npcvs + 1, alpha, alpha_tune)
  p_star <- matrix(, nsite, nreps)
  for (i in 1:nsite) {
   p_star[i,] <- inv.logit(cbind(1,w[i,,]) %*% alpha_star)
  } # i
  mh1 <- sum(dbinom(y, N, p_star, log=TRUE)) +
    sum(dnorm(alpha_star, alpha_mean, alpha_sd, log=TRUE))
  mh2 <- sum(dbinom(y, N, p     , log=TRUE)) +
    sum(dnorm(alpha     , alpha_mean, alpha_sd, log=TRUE))
  mh <- exp(mh1 - mh2)
  if (mh > runif(1)) {
   alpha <- alpha_star
   p <- p_star
  }

  ### Save samples
  beta_save[,k] <- beta
  alpha_save[,k] <- alpha
  if (nmcmc > 2000) {
   if (k > nmcmc-2000) {
    N_save[,k-nmcmc+2000] <- N
   }
  }
 } # k

 # Write output
 list(beta_save=beta_save, alpha_save=alpha_save, N_save=N_save)
} # Nmix_mcmc
```

When using the following code to implement the algorithm, it takes about 4 min to run. Then using code in Appendix 3.1, we can create traceplots for the monitored parameters, which show good model convergence and consistency between parameter estimates of β and α and their true values used in data simulation (Fig. 3.1).

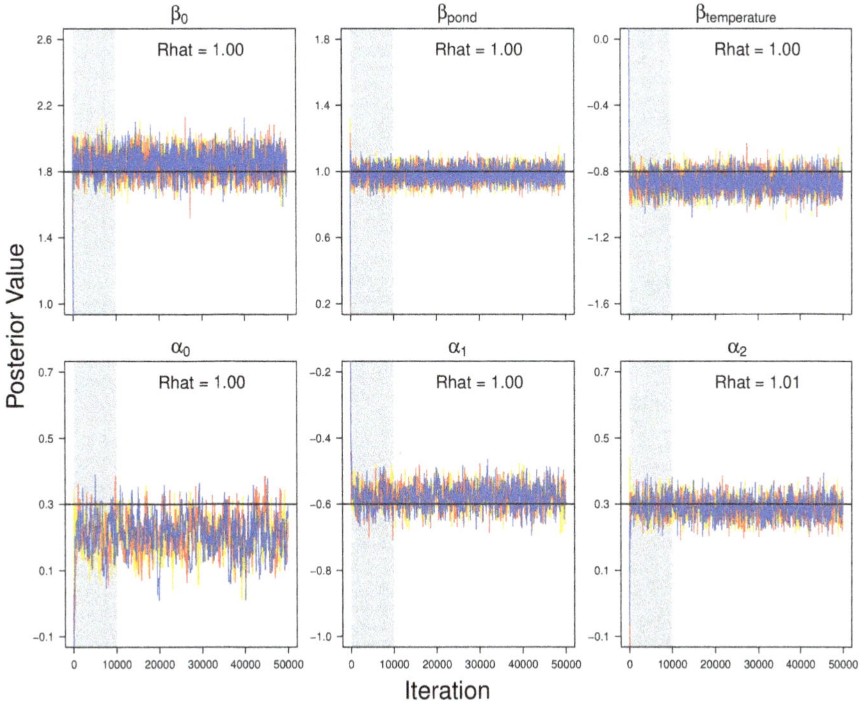

Fig. 3.1 Traceplots showing the posterior chains for β and α in the static N-mixture model. Rhat values are calculated after discarding the first 10,000 iterations. Horizontal lines represent the true values used in data simulation

```
library(foreach)   # for parallel computing
library(doParallel) # for parallel computing

numCores <- round(detectCores() / 2) # only use half of the cores for
parallel computing
registerDoParallel(numCores) # setup parallel computing

nmcmc <- 50000 # number of iterations of MCMC computing
chain <- 3    # number of chains

start_time <- Sys.time() # start time of computing
out <- foreach (i=1:chain, .packages='boot') %dopar% {
 Nmix_mcmc(y=y, x=x, w=w, nmcmc=nmcmc)
} # i
stopImplicitCluster() # clean up the cluster
end_time <- Sys.time() # end time of computing
time_taken <- end_time - start_time # computing time
```

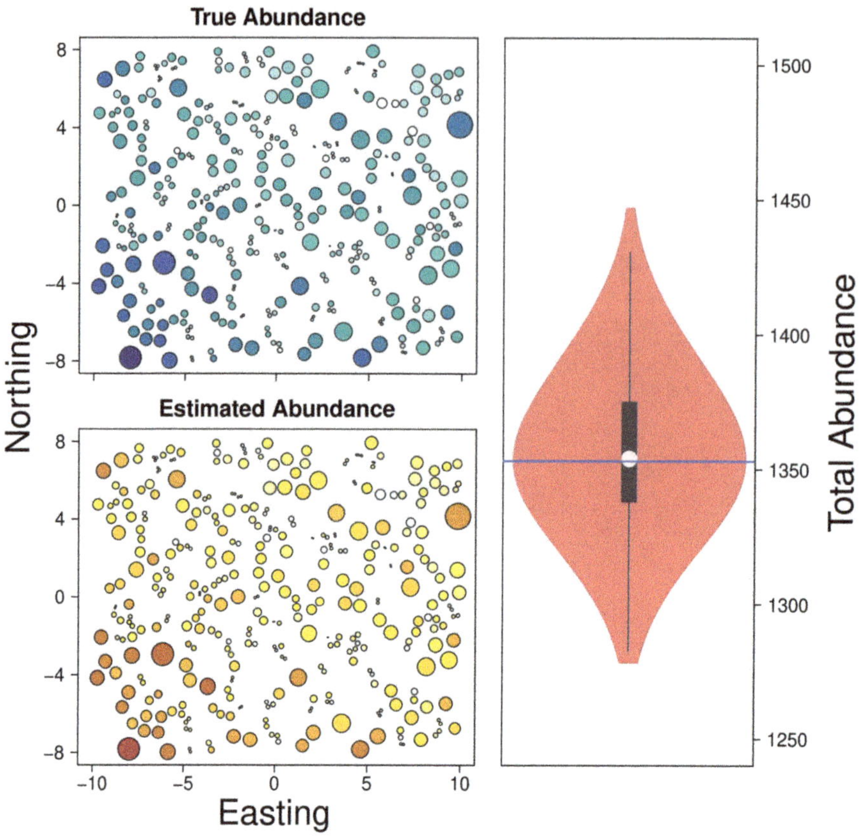

Fig. 3.2 True abundance N in the simulated pond turtle data (upper-left panel; a deeper color represents a higher abundance, and vice versa) and its estimates (lower-left panel) as well as the true (horizontal line) and estimated (violin) total abundance of the 360 sites ($\sum_{i=1}^{n_{site}} N_i$; right panel)

We can also use the following code to plot the simulated and estimated true abundance, N, and see good agreement between them (Fig. 3.2). The estimated total abundance ($\sum_{i=1}^{n_{site}} N_i$) is also consistent with the simulated true value.

```
  library(vioplot) # for making violin plots

for (i in 1:chain) {
 if (i == 1) {
  N_post <- out[[i]]$N_save
 } else {
  N_post <- cbind(N_post, out[[i]]$N_save)
 }
} # i
N_med <- apply(N_post, 1, median)
```

```
nscale <- 100
Nmax <- max(c(log(N+1), log(N_med+1)))
N_scale <- round(log(N+1) / Nmax * (nscale-1)) + 1
N_med_scale <- round(log(N_med+1) / Nmax * (nscale-1)) + 1

colt <- colorRampPalette(c('white','lightseagreen','navy'))(nscale)
cole <- colorRampPalette(c('white','gold','firebrick'))(nscale)

layout(mat=matrix(c(rep(c(1,1,2,2),times=3),rep(3,8)), nrow=4,
ncol=5, byrow=FALSE))
par(mar=c(1,2,3,1))
par(oma=c(4,4,0,6))

plot(cov$lon, cov$lat, xlim=c(-10,10), ylim=c(-8,8), axes=F,
xlab='', ylab='',
   pch=21, cex=cov$size[,1]*6, col='grey18', bg=colt[N_scale])
axis(1, at=seq(-10,10,5), labels=rep('',5))
axis(2, at=seq(-08,08,4), cex.axis=1.6, las=2)
box()
title(main='True Abundance', cex.main=2)

plot(cov$lon, cov$lat, xlim=c(-10,10), ylim=c(-8,8), axes=F,
xlab='', ylab='',
   pch=21, cex=cov$size[,1]*6, col='grey18', bg=cole[N_med_scale])
axis(1, at=seq(-10,10,5), cex.axis=1.6)
axis(2, at=seq(-08,08,4), cex.axis=1.6, las=2)
box()
title(main='Estimated Abundance', cex.main=2)

axis(1, at=0, labels='Easting', tick=F, line=2.5, cex.axis=3)
title(ylab='Northing', cex.lab=3, line=1.5, outer=T)

par(mar=c(1,1,3,2))
plot(1, ylim=c(1250, 1500), type='n', axes=F, xlab='', ylab='')
vioplot(apply(N_post,2,sum), col='lightcoral', rectCol='grey36',
lineCol='grey36',
    border=NA, cex=3, add=T)
abline(h=sum(N), col='royalblue', lwd=1.8)
axis(4, at=seq(1250, 1500, 50), cex.axis=1.6, las=2)
axis(4, at=1375, label='Total Abundance', line=5.5, cex.axis=3,
tick=F)
```

3.3 Dynamic N-Mixture Models with Population Growth

While the static N-mixture model directly links abundance to environmental conditions, dynamic N-mixture models (Dail and Madsen 2011; Hostetler and Chandler 2015; Bellier et al. 2016) describe how local abundance changes over time under certain environmental conditions. In this section, we will introduce a dynamic N-mixture model that describes such temporal changes in local abundance with population growth rates (Hostetler and Chandler 2015). Using our fictitious pond turtle as an example, the model basically says that the number of turtles in a pond is more likely to increase if (1) there are fewer turtles in that pond, (2) the pond is larger, or (3) the water is colder, and vice versa, given other conditions being the same. There is also a chance that a few turtles will move into a pond, although at this point, the model does not specify where they come from, and thus is considered non-spatial. Using statistical language, the model contains a process sub-model that first describes the initial abundance (i.e., abundance at the first year) in a way that is basically the same as the static N-mixture model (Eq. 3.1), except that variables are indexed by 1 for $t = 1$ now:

$$\left.\begin{array}{c} N_{i,1} \sim Poisson(\lambda_{i,1}) \\ \lambda_{i,1} = exp\left(\beta_0^{[0]} + \beta_1^{[0]} \times x_{i,1,1} + \ldots + \beta_{n_{covs}}^{[0]} \times x_{i,1,n_{covs}}\right) \end{array}\right\}, \tag{3.3}$$

in which $N_{i,\,1}$ is the abundance of the study species at site i in the first year, $\lambda_{i,\,1}$ is the expectation of the abundance which is a function of covariates x's governed by intercept and slope parameters $\beta^{[0]}$'s. From the second year (i.e., $t \geq 2$), the process sub-model assumes that population size is due to population growth from the previous year as well as immigration such that

$$\left.\begin{array}{c} N_{i,t} \sim Poisson(\lambda_{i,t}) \\ \lambda_{i,t} = N_{i,t-1} \times \rho_{i,t-1} + \delta \\ \rho_{i,t-1} = exp\left(\beta_0^{[\rho]} + \beta_{density}^{[\rho]} \times N_{i,t-1} + \beta_1^{[\rho]} \times x_{i,t,1} + \ldots + \beta_{n_{covs}}^{[\rho]} \times x_{i,t,n_{covs}}\right) \end{array}\right\}, \tag{3.4}$$

in which $\rho_{i,\,t-1}$ population growth rate and δ is the expected number of immigrants. Population growth rate $\rho_{i,\,t-1}$ is further a function of density and environmental covariates at a log scale, with $\beta_0^{[\rho]}$ being the intercept, $\beta_{density}^{[\rho]}$ being a slope representing the effect of density dependence process, and $\beta_1^{[\rho]}, \ldots, \beta_{n_{covs}}^{[\rho]}$ being slopes representing environmental effects on population growth. Without immigration δ, $N_{i,\,t}$ will always be 0 if $N_{i,\,t-1}$ is 0, meaning that immigration is the only way to rescue a local population after it goes extinct. Also note that $\rho_{i,\,t-1}$ is called population "growth" rate due to convention, despite the fact that it can be lower than 1 when a population is in decline.

Then, the observation sub-model links the observed count data to the true abundance in the same way as in Eq. (3.2), except that all variables are now indexed by t:

$$\left.\begin{array}{c} y_{i,t,j} \sim Binomial(N_{i,t}, p_{i,t,j}) \\ p_{i,t,j} = logit^{-1}\left(\alpha_0 + \alpha_1 \times w_{i,t,j,1} + \ldots + \alpha_{n_{pcvs}} \times w_{i,t,j,n_{pcvs}}\right) \end{array}\right\}, \tag{3.5}$$

in which detection probability $p_{i,\,t,\,j}$ is a function of observation covariates $w_{i,t,j,n_{pcvs}}$ with intercept and slope parameters α's.

Using the code below, we can simulate data while assuming that pond size has positive effects on initial abundance ($\beta^{[0]}_{pond} = 1$) and population growth rate ($\beta^{[\rho]}_{pond} = 0.6$), and water temperature has negative effects on initial abundance ($\beta^{[0]}_{temperature} = -0.8$) and population growth rate ($\beta^{[\rho]}_{temperature} = -0.4$). We also assume a negative effect of density dependence on population growth ($\beta^{[\rho]}_{density} = -0.8$) as well as an average of 0.5 immigrants per site and year (i.e., $\delta = 0.5$).

```
library(boot) # for using logit and inv.logit functions

set.seed(5)

# Basic values
nsite <- 360 # number of sites
nyear <- 20  # number of years
nreps <- 5   # number of within-season replicates
ncovs <- 2   # number of environmental covariates
npcvs <- 3   # number of observational covariates

beta0 <- c(1.8, 1, -0.8)      # intercept and slopes for initial abundance
beta_rho <- c(0, -0.8, 0.6, -0.4) # intercept and slopes for population
growth
delta <- 0.5               # expectation of the number of immigrants
alpha <- c(0.3, -0.6, 0.3, -0.3) # intercept and slopes for detection
probability

# Read in environmental covariates
load('data/covariates.RData')
x1 <- (log(cov$size) - mean(log(cov$size))) / sd(log(cov$size)) #
standardized log habitat patch size
x2 <- (cov$temp - mean(cov$temp)) / sd(cov$temp) # standardized
temperature
x <- array(, dim=c(nsite, nyear, ncovs)) # environmental covariates
x[,,1] <- x1
x[,,2] <- x2
### Standardize covariates for each site
for (i in 1:nsite) {
 for (h in 1:ncovs) {
  x[i,2:nyear,h] <- (x[i,2:nyear,h] - mean(x[i,,h])) / sd(x[i,,h])
```

```
  } # h
} # i

# Simulate data
lambda <- matrix(, nsite, nyear) # expectation of abundance
N <- matrix(, nsite, nyear) # abundance
lambda[,1] <- exp(cbind(1,x[,1,]) %*% beta0) # expectation of initial
abundance
N[,1] <- rpois(nsite, lambda[,1]) # initial abundance

rho <- matrix(0, nsite, nyear-1) # population growth rate
for (t in 2:nyear) {
  rho[,t-1] <- exp(cbind(1, (N[,t-1]-lambda[,1])/lambda[,1], x[,t,]) %
*% beta_rho)
  lambda[,t] <- N[,t-1] * rho[,t-1] + delta
  N[,t] <- rpois(nsite, lambda[,t])
} # t

w <- array(rnorm(nsite * nyear * nreps * npcvs, 0, 1), dim=c(nsite,
nyear, nreps, npcvs)) # observational covariates

p <- array(, dim=c(nsite, nyear, nreps)) # detection probability
for (i in 1:nsite) {
  for (t in 1:nyear) {
    p[i,t,] <- inv.logit(cbind(1,w[i,t,,]) %*% alpha)
  } # t
} # i

y <- array(, dim=c(nsite, nyear, nreps)) # count data
for (i in 1:nsite) {
  for (t in 1:nyear) {
    y[i,t,] <- rbinom(nreps, N[i,t], p[i,t,])
  } # t
} # i
```

Next, we develop a custom MCMC algorithm to estimate parameters $\beta^{[0]}$, $\beta^{[\rho]}$, δ, α, and N using data y, x, and w, as shown below:

```
growth_Nmix_mcmc <- function(y, x, w, nmcmc) {

  # Setup variables
  nsite <- dim(y)[1]
  nyear <- dim(y)[2]
  nreps <- dim(y)[3]
  ncovs <- dim(x)[3]
```

```
npcvs <- dim(w)[4]
ymax <- apply(y, 1:2, max)

beta0_save <- matrix(0, ncovs+1, nmcmc)
beta_rho_save <- matrix(0, ncovs+2, nmcmc)
delta_save <- rep(0, nmcmc)
alpha_save <- matrix(0, npcvs+1, nmcmc)
Nsum_save <- matrix(0, nyear, nmcmc)

# Priors
beta0_mean <- rep(0, ncovs+1)
beta0_sd <- 2
beta_rho_mean <- rep(0, ncovs+2)
beta_rho_sd <- 2
log_delta_mean <- 0
log_delta_sd <- 2
alpha_mean <- rep(0, npcvs+1)
alpha_sd <- 2

# Starting values
beta0 <- rep(0, ncovs+1)
beta_rho <- rep(0, ncovs+2)
delta <- 1
alpha <- rep(0, npcvs+1)
lambda0 <- rep(1, nsite) # expectation of initial abundance
rho <- matrix(1, nsite, nyear-1)
N <- round((ymax + 1) / 0.5)
p <- array(0.5, dim=c(nsite, nyear, nreps))

# Tuning factors
beta0_tune <- 0.08
beta_rho_tune <- 0.15
delta_tune <- 0.15
alpha_tune <- 0.01
N_tune <- 1

# Begin MCMC loop
for (k in 1:nmcmc) {
  ### Sample N
  N_star <- matrix(rpois(nsite*nyear, N + N_tune), nsite, nyear)
  mh1 <- apply(dbinom(y, N_star, p, log=T), 1:2, sum) +
      dpois(N_star, cbind(lambda0, N[,-nyear]*rho + delta), log=T) +
      dpois(N, N_star+N_tune, log=T)
  mh2 <- apply(dbinom(y, N    , p, log=T), 1:2, sum) +
      dpois(N    , cbind(lambda0, N[,-nyear]*rho + delta), log=T) +
```

```
    dpois(N_star, N+N_tune, log=T)
  mh <- exp(mh1 - mh2)
  Nkeep <- ((mh > runif(nsite*nyear)) & (N_star >= ymax))
  N[Nkeep] <- N_star[Nkeep]

  ### Sample beta0
  beta0_star <- rnorm(ncovs+1, beta0, beta0_tune)
  lambda0_star <- exp(cbind(1,x[,1,]) %*% beta0_star)
  mh1 <- sum(dpois(N[,1], lambda0_star, log=T)) +
    sum(dnorm(beta0_star, beta0_mean, beta0_sd, log=T))
  mh2 <- sum(dpois(N[,1], lambda0     , log=T)) +
    sum(dnorm(beta0     , beta0_mean, beta0_sd, log=T))
  mh <- exp(mh1 - mh2)
  if (mh > runif(1)) {
   beta0 <- beta0_star
   lambda0 <- lambda0_star
  }

  ### Sample beta_rho
  beta_rho_star <- rnorm(ncovs+2, beta_rho, beta_rho_tune)
  rho_star <- matrix(, nsite, nyear-1)
  for (t in 2:nyear) {
  rho_star[,t-1] <- exp(cbind(1, (N[,t-1]-lambda0)/lambda0, x[,t,]) %
*% beta_rho_star)
  } # t
  mh1 <- sum(dpois(N[,-1], N[,-nyear] * rho_star + delta, log=T)) +
       sum(dnorm(beta_rho_star, beta_rho_mean, beta_rho_sd, log=T))
  mh2 <- sum(dpois(N[,-1], N[,-nyear] * rho     + delta, log=T)) +
       sum(dnorm(beta_rho     , beta_rho_mean, beta_rho_sd, log=T))
  mh <- exp(mh1 - mh2)
  if (mh > runif(1)) {
   beta_rho <- beta_rho_star
   rho <- rho_star
  }

  ### Sample delta
  delta_star <- exp(rnorm(1, log(delta), delta_tune))
  mh1 <- sum(dpois(N[,-1], N[,-nyear] * rho + delta_star, log=T)) +
       dnorm(log(delta_star), log_delta_mean, log_delta_sd, log=T)
  mh2 <- sum(dpois(N[,-1], N[,-nyear] * rho + delta     , log=T)) +
       dnorm(log(delta     ), log_delta_mean, log_delta_sd, log=T)
  mh <- exp(mh1 - mh2)
  if (mh > runif(1)) {
   delta <- delta_star
  }
```

```
  ### Sample alpha
  alpha_star <- rnorm(npcvs+1, alpha, alpha_tune)
  p_star <- array(, dim=c(nsite, nyear, nreps))
  for (i in 1:nsite) {
   for (t in 1:nyear) {
    p_star[i,t,] <- inv.logit(cbind(1,w[i,t,,]) %*% alpha_star)
   } # t
  } # i
  mh1 <- sum(dbinom(y, N, p_star, log=TRUE), na.rm=T) +
         sum(dnorm(alpha_star, alpha_mean, alpha_sd, log=TRUE))
  mh2 <- sum(dbinom(y, N, p     , log=TRUE), na.rm=T) +
         sum(dnorm(alpha     , alpha_mean, alpha_sd, log=TRUE))
  mh <- exp(mh1 - mh2)
  if (mh > runif(1)) {
   alpha <- alpha_star
   p <- p_star
  }

  ### Save samples
  beta0_save[,k] <- beta0
  beta_rho_save[,k] <- beta_rho
  delta_save[k] <- delta
  alpha_save[,k] <- alpha
  Nsum_save[,k] <- colSums(N)
 } # k

 # Write output
 list(beta0_save=beta0_save,
   beta_rho_save=beta_rho_save, delta_save=delta_save,
   alpha_save=alpha_save,
   Nsum_save=Nsum_save)

} # growth_Nmix_mcmc
```

We then use the following code to implement the algorithm, which takes about 1.5 h to run. Using code in Appendix 3.2, we can create traceplots that show good model convergence and agreement between parameter estimates of $\beta^{[0]}$, $\beta^{[\rho]}$, δ, and α and their true values used in data simulation (Fig. 3.3).

```
library(foreach)   # for parallel computing
library(doParallel) # for parallel computing

numCores <- round(detectCores() / 2) # only use half of the cores for
parallel computing
registerDoParallel(numCores) # setup parallel computing
```

```
nmcmc <- 50000 # number of iterations of MCMC computing
chain <- 3    # number of chains

start_time <- Sys.time() # start time of computing
out <- foreach (i=1:chain, .packages='boot') %dopar% {
 growth_Nmix_mcmc(y=y, x=x, w=w, nmcmc=nmcmc)
} # i
stopImplicitCluster() # clean up the cluster
end_time <- Sys.time() # end time of computing
time_taken <- end_time - start_time # computing time
```

Next, we can also use the following code to plot the total abundance for each year (i.e., $\sum_{i=1}^{n_{site}} N_{i,t}$) and see that its temporal variation, which is mainly driven by the negative effect of water temperature on the population growth rate, is consistent with the simulated true values (Fig. 3.4).

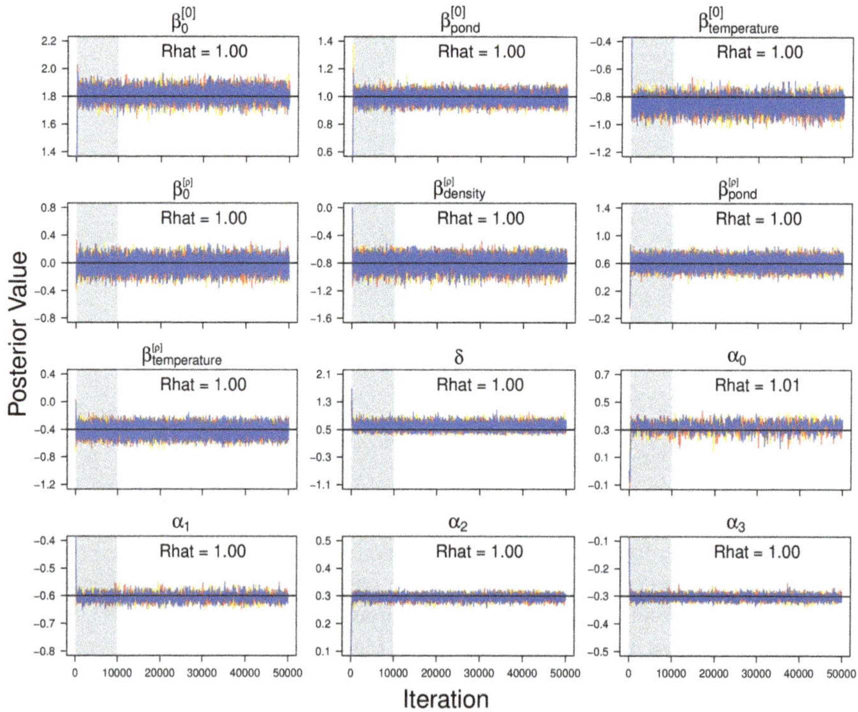

Fig. 3.3 Traceplots showing posterior chains for $\beta^{[0]}$, $\beta^{[\rho]}$, δ, and α in the non-spatial dynamic N-mixture model with population growth. Rhat values are calculated after discarding the first 10,000 iterations. Horizontal lines represent the true values used in data simulation

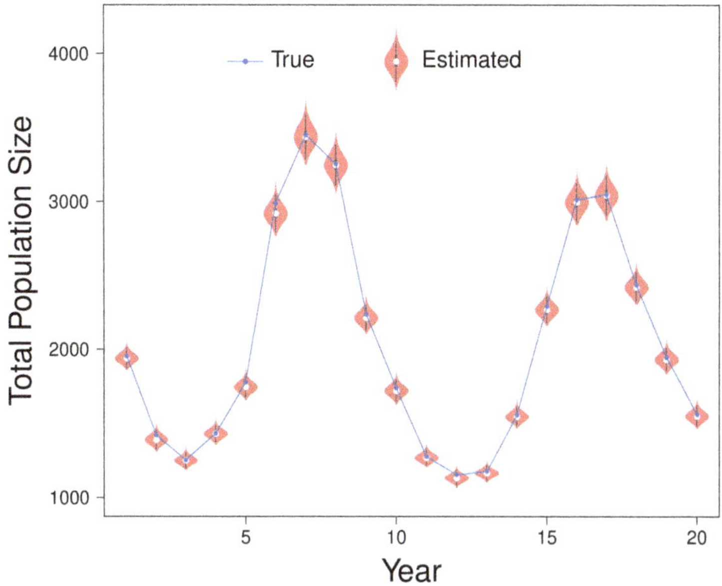

Fig. 3.4 Violin plots showing consistency between the estimated total abundance for each year ($\sum_{i=1}^{n_{site}} N_{i,t}$; violin) and its simulated true value (blue point)

```
library(vioplot) # for making violin plots

for (i in 1:chain) {
 if (i == 1) {
  Nsum_post <- out[[i]]$Nsum_save[, (nmcmc/2+1):nmcmc]
 } else {
  Nsum_post <- cbind(Nsum_post, out[[i]]$Nsum_save[, (nmcmc/2+1):
nmcmc])
 }
} # i

par(mfrow=c(1,1))
par(mar=c(5,7,1,1))

plot(1, xlim=c(1,nyear), ylim=c(1000,4200), type='n', xlab='',
ylab='', axes=F)
vioplot(t(Nsum_post), add=T, col='lightcoral', rectCol='grey36',
lineCol='grey36', border=NA)
lines(colSums(N) ~ c(1:nyear), type='o', pch=16, cex=0.8, lwd=1.2,
col='royalblue')
axis(1, at=seq(0,nyear,5), cex.axis=1.5)
```

```
axis(2, at=seq(1000,4000,1000), cex.axis=1.5, las=2)
title(xlab='Year', cex.lab=2.5, line=3.5)
title(ylab='Total Population Size', cex.lab=2.5, line=4.8)

points(x=5, y=3944, pch=16, cex=1, col='royalblue')
lines(x=c(4.4,5.6), y=c(3944,3944), lwd=1.2, col='royalblue')
vioplot(Nsum_post[8,]-median(Nsum_post[8,])+3944, at=10, add=T,
col='lightcoral', rectCol='grey36', lineCol='grey36', border=NA)
text(x= 5.6, y=3944, labels='True', pos=4, cex=1.8)
text(x=10.6, y=3944, labels='Estimated', pos=4, cex=1.8)
```

3.4 Spatial Dynamic N-Mixture Models with Population Growth

Even though the dynamic N-mixture model with population growth (Hostetler and Chandler 2015) contains a term for immigration, the spatial configuration of sites does not really influence the number of immigrants; thus, the model is non-spatial. To account for spatial dependency among sites, this model is further extended to a spatial version that explicitly considers movement among sites (Howell et al. 2020). That is to say, for our pond turtle, whether the number of turtles in a pond will increase or decrease is not only due to (1) the number of turtles in this pond, (2) the size of the pond, and (3) water temperature of the pond, but is also affected by (4) the number of turtles in other ponds and the chance that they move to this pond. Statistically speaking, the spatial dynamic N-mixture models with population growth describe initial population size in the same way as Eq. (3.3):

$$\left.\begin{array}{c} N_{i,1} \sim Poisson(\lambda_{i,1}) \\ \lambda_{i,1} = exp\left(\beta_0^{[0]} + \beta_1^{[0]} \times x_{i,1,1} + \ldots + \beta_{n_{covs}}^{[0]} \times x_{i,1,n_{covs}}\right) \end{array}\right\}, \tag{3.6}$$

in which $N_{i,\,1}$ is the abundance of the study species at site i in the first year, and $\lambda_{i,\,1}$ is the expectation of abundance, which is a function of covariates x's governed by intercept and slope parameters $\beta^{[0]}$'s. From the second year (i.e., $t \geq 2$), the process sub-model assumes that population size is due to local population growth from the previous year as well as movement between sites such that

$$\left.\begin{array}{c} N_{i_2,t} \sim Poisson(\lambda_{i_2,t}) \\ \lambda_{i_2,t} = \sum_{i_1=1}^{n_{site}} N_{i_1,t-1} \times \rho_{i_1,t-1} \times \theta_{i_1,i_2,t} \\ \rho_{i_1,t-1} = exp\left(\beta_0^{[\rho]} + \beta_{density}^{[\rho]} \times N_{i_1,t-1} + \beta_1^{[\rho]} \times x_{i_1,t,1} + \ldots + \beta_{n_{covs}}^{[\rho]} \times x_{i_1,t,n_{covs}}\right) \end{array}\right\}, \tag{3.7}$$

in which $\rho_{i_1,t-1}$ is the population growth rate at site i_1 and is a function of density and environmental covariates at that site, and $\theta_{i_1,i_2,t}$ is the standardized movement rate from site i_1 to site i_2. Movement rate $\theta_{i_1,i_2,t}$ can be constant over time (and thus $\theta_{i_1,i_2,t} = \theta_{i_1,i_2}$ for any t; see Sect. 3.4.1) or vary temporally (see Sect. 3.4.2), which we will specify later. The model basically says, to describe population size at any site i_2, we first need to describe potential population change at any site i_1 due to local conditions, including site i_2 itself (when $i_2 = i_1$); on top of that, we add up all individuals that either move from other sites to i_2 or stay at i_2. The model thus requires us to model all the sites, even though we normally would not be able to survey every site. We can model unsurveyed sites by giving missing values to the count data $y_{i,\,t,\,j}$ (see below), but locations and covariates are needed for all sites. An approach to dealing with missing values is introduced in Sect. 2.4.2, while model performance with missing values needs to be further investigated.

Finally, the observation sub-model links the observed counts to the true abundance in the same way as in Eq. (3.5):

$$\left.\begin{array}{r} y_{i,t,j} \sim Binomial\left(N_{i,t}, p_{i,t,j}\right) \\ p_{i,t,j} = logit^{-1}\left(\alpha_0 + \alpha_1 \times w_{i,t,j,1} + \ldots + \alpha_{n_{pcvs}} \times w_{i,t,j,n_{pcvs}}\right) \end{array}\right\}, \tag{3.8}$$

in which detection probability $p_{i,\,t,\,j}$ is a function of observation covariates $w_{i,t,j,n_{pcvs}}$ with intercept and slope parameters α's.

3.4.1 With Distance-Based Movement

In this model specification, movement rate between any pair of sites is assumed to only be related to the distance between these two sites and thus does not change over time (Howell et al. 2020). That is to say, our turtle has the same chance to move between any pair of ponds as long as the two ponds have the same distance between them. Here we use the same decay function as in Chap. 2 to have

$$\left.\begin{array}{r} \theta_{i_1,i_2} = \dfrac{\eta_{i_1,i_2}}{\sum_{i_2=1}^{n_{site}} \eta_{i_1,i_2}} \\ \eta_{i_1,i_2} = exp(-1 \times \kappa \times d_{i_1,i_2}) \end{array}\right\}, \tag{3.9}$$

in which θ_{i_1,i_2} is the standardized movement rate from site i_1 to site i_2, η_{i_1,i_2} is the unstandardized movement rate, and κ is a decay parameter governing the relationship between movement and distance, denoted d_{i_1,i_2}. A larger κ leads to a faster decrease in movement rate with increasing distance, meaning that long-distance movement is less likely to happen, and vice versa.

Now, let us use the code below to simulate data for this model specification. We assume positive effects of pond size on initial abundance ($\beta_{pond}^{[0]} = 1$) and population growth ($\beta_{pond}^{[\rho]} = 0.6$) and negative effects of water temperature on initial abundance ($\beta_{temperature}^{[0]} = -0.8$) and population growth ($\beta_{temperature}^{[\rho]} = -0.4$). We also assume a negative density dependence effect on population growth ($\beta_{density}^{[\rho]} = -0.8$) as well as a relatively strong rate of decay ($\kappa = 2$).

```
library(boot) # for using logit and inv.logit functions

set.seed(5)

# Basic values
nsite <- 360 # number of sites
nyear <- 20  # number of years
nreps <- 5  # number of within-season replicates
ncovs <- 2  # number of environmental covariates
npcvs <- 3  # number of observational covariates

beta0 <- c(1.8, 1, -0.8)      # intercept and slopes for initial abundance
beta_rho <- c(0, -0.8, 0.6, -0.4) # intercept and slopes for population
growth
kappa <- 2                # distance effect on movement
alpha <- c(0.3, -0.6, 0.3, -0.3) # intercept and slopes for detection
probability

# Read in environmental covariates
load('data/covariates.RData')
x1 <- (log(cov$size) - mean(log(cov$size))) / sd(log(cov$size)) #
standardized log habitat patch size
x2 <- (cov$temp - mean(cov$temp)) / sd(cov$temp) # standardized
temperature
x <- array(, dim=c(nsite, nyear, ncovs)) # environmental covariates
x[,,1] <- x1
x[,,2] <- x2
### Standardize covariates for each site
for (i in 1:nsite) {
 for (h in 1:ncovs) {
  x[i,2:nyear,h] <- (x[i,2:nyear,h] - mean(x[i,,h])) / sd(x[i,,h])
 } # h
} # i

# Simulate data
lambda <- matrix(, nsite, nyear) # expectation of abundance
N <- matrix(, nsite, nyear) # abundance
lambda[,1] <- exp(cbind(1,x[,1,]) %*% beta0) # expectation of initial
abundance
N[,1] <- rpois(nsite, lambda[,1]) # initial abundance

lon <- cov$lon # longitude or easting
lat <- cov$lat # latitude or northing
d <- matrix(, nsite, nsite) # distance matrix
for (i1 in 1:nsite) {
```

```
 for (i2 in 1:nsite) {
  d[i1,i2] <- sqrt((lon[i1] - lon[i2]) ^ 2 + (lat[i1] - lat[i2]) ^ 2)
 } # i2
} # i1
eta <- exp(-1 * kappa * d) # unstandardized colonization rate
theta <- eta / rowSums(eta) #  standardized colonization rate

rho <- matrix(0, nsite, nyear-1) # population growth rate
for (t in 2:nyear) {
 rho[,t-1] <- exp(cbind(1, (N[,t-1]-lambda[,1])/lambda[,1], x[,t,]) %
*% beta_rho)
 lambda[,t] <- (N[,t-1] * rho[,t-1]) %*% theta
 N[,t] <- rpois(nsite, lambda[,t])
} # t

w <- array(rnorm(nsite * nyear * nreps * npcvs, 0, 1), dim=c(nsite,
nyear, nreps, npcvs)) # observational covariates

p <- array(, dim=c(nsite, nyear, nreps)) # detection probability
for (i in 1:nsite) {
 for (t in 1:nyear) {
  p[i,t,] <- inv.logit(cbind(1,w[i,t,,]) %*% alpha)
 } # t
} # i

y <- array(, dim=c(nsite, nyear, nreps)) # count data
for (i in 1:nsite) {
 for (t in 1:nyear) {
  y[i,t,] <- rbinom(nreps, N[i,t], p[i,t,])
 } # t
} # i
```

Next, we will develop our own custom MCMC algorithm to estimate parameters $\beta^{[0]}$, $\beta^{[\rho]}$, κ, and α using data y, x, w, and d. Remember that we also use this model to illustrate how to calculate WAIC for model selection. To do so, a few lines are included to calculate and keep track of the posterior predictive density (*ppd*) of the data y given parameter estimates of population size N and detection probability p for site i, year t, visit j, and posterior sample k, i.e., $ppd_{i,t,j,k} = Pr(y_{i,t,j} | N_{i,j,k}, p_{i,t,j,k})$. These additional lines are highlighted in **bold**.

```
dist_growth_Nmix_mcmc <- function(y, x, w, d, nmcmc) {

 # Setup variables
 nsite <- dim(y)[1]
```

```
nyear <- dim(y)[2]
nreps <- dim(y)[3]
ncovs <- dim(x)[3]
npcvs <- dim(w)[4]
ymax <- apply(y, 1:2, max)

beta0_save <- matrix(0, ncovs+1, nmcmc)
beta_rho_save <- matrix(0, ncovs+2, nmcmc)
kappa_save <- rep(0, nmcmc)
alpha_save <- matrix(0, npcvs+1, nmcmc)
ppd_save <- array(0, dim=c(nsite, nyear, nreps, 2000)) # save posterior
predictive density for calculating WAIC

# Priors
beta0_mean <- rep(0, ncovs+1)
beta0_sd <- 2
beta_rho_mean <- rep(0, ncovs+2)
beta_rho_sd <- 2
log_kappa_mean <- 0
log_kappa_sd <- 2
alpha_mean <- rep(0, npcvs+1)
alpha_sd <- 2

# Starting values
beta0 <- rep(0, ncovs+1)
beta_rho <- rep(0, ncovs+2)
kappa <- 1
alpha <- rep(0, npcvs+1)
lambda0 <- exp(cbind(1,x[,1,]) %*% beta0) # expectation of initial
abundance
rho <- matrix(1, nsite, nyear-1)
eta <- exp(-1 * kappa * d)
theta <- eta / rowSums(eta)
N <- round((ymax + 1) / 0.5)
p <- array(0.5, dim=c(nsite, nyear, nreps))
# Tuning factors
beta0_tune <- 0.08
beta_rho_tune <- 0.15
kappa_tune <- 0.15
alpha_tune <- 0.02
N_tune <- 1

# Begin MCMC loop
for (k in 1:nmcmc) {
 ### Sample N
```

```
  N_star <- matrix(rpois(nsite*nyear, N + N_tune), nsite, nyear)
  mh1 <- apply(dbinom(y, N_star, p, log=T), 1:2, sum) +
     dpois(N_star, cbind(lambda0, t(t(N[,-nyear] * rho) %*% theta)),
log=T) +
     dpois(N, N_star+N_tune, log=T)
  mh2 <- apply(dbinom(y, N    , p, log=T), 1:2, sum) +
    dpois(N    , cbind(lambda0, t(t(N[,-nyear] * rho) %*% theta)), log=T)
+
     dpois(N_star, N+N_tune, log=T)
  mh <- exp(mh1 - mh2)
  Nkeep <- ((mh > runif(nsite*nyear)) & (N_star >= ymax))
  N[Nkeep] <- N_star[Nkeep]

  ### Sample beta0
  beta0_star <- rnorm(ncovs+1, beta0, beta0_tune)
  lambda0_star <- exp(cbind(1,x[,1,]) %*% beta0_star)
  mh1 <- sum(dpois(N[,1], lambda0_star, log=T)) +
     sum(dnorm(beta0_star, beta0_mean, beta0_sd, log=T))
  mh2 <- sum(dpois(N[,1], lambda0     , log=T)) +
     sum(dnorm(beta0     , beta0_mean, beta0_sd, log=T))
  mh <- exp(mh1 - mh2)
  if (mh > runif(1)) {
   beta0 <- beta0_star
   lambda0 <- lambda0_star
  }

  ### Sample beta_rho
  beta_rho_star <- rnorm(ncovs+2, beta_rho, beta_rho_tune)
  rho_star <- matrix(, nsite, nyear-1)
  for (t in 2:nyear) {
   rho_star[,t-1] <- exp(cbind(1, (N[,t-1]-lambda0)/lambda0, x[,t,]) %
*% beta_rho_star)
  } # t
  mh1 <- sum(dpois(N[,-1], t(t(N[,-nyear] * rho_star) %*% theta),
log=T)) +
     sum(dnorm(beta_rho_star, beta_rho_mean, beta_rho_sd, log=T))
  mh2 <- sum(dpois(N[,-1], t(t(N[,-nyear] * rho     ) %*% theta), log=T)) +
     sum(dnorm(beta_rho     , beta_rho_mean, beta_rho_sd, log=T))
  mh <- exp(mh1 - mh2)
  if (mh > runif(1)) {
   beta_rho <- beta_rho_star
   rho <- rho_star
  }
```

```
  ### Sample kappa
  kappa_star <- exp(rnorm(1, log(kappa), kappa_tune))
  eta_star <- exp(-1 * kappa_star * d)
  theta_star <- eta_star / rowSums(eta_star)
  mh1 <- sum(dpois(N[,-1], t(t(N[,-nyear] * rho) %*% theta_star),
log=T)) +
      dnorm(log(kappa_star), log_kappa_mean, log_kappa_sd, log=T)
  mh2 <- sum(dpois(N[,-1], t(t(N[,-nyear] * rho) %*% theta     ), log=T)) +
      dnorm(log(kappa     ), log_kappa_mean, log_kappa_sd, log=T)
  mh <- exp(mh1 - mh2)
  if (mh > runif(1)) {
   kappa <- kappa_star
   eta <- eta_star
   theta <- theta_star
  }

  ### Sample alpha
  alpha_star <- rnorm(npcvs+1, alpha, alpha_tune)
  p_star <- array(, dim=c(nsite, nyear, nreps))
  for (i in 1:nsite) {
   for (t in 1:nyear) {
    p_star[i,t,] <- inv.logit(cbind(1,w[i,t,,]) %*% alpha_star)
   } # t
  } # i
  mh1 <- sum(dbinom(y, N, p_star, log=TRUE), na.rm=T) +
      sum(dnorm(alpha_star, alpha_mean, alpha_sd, log=TRUE))
  mh2 <- sum(dbinom(y, N, p     , log=TRUE), na.rm=T) +
      sum(dnorm(alpha     , alpha_mean, alpha_sd, log=TRUE))
  mh <- exp(mh1 - mh2)
  if (mh > runif(1)) {
   alpha <- alpha_star
   p <- p_star
  }

  ### Calculate posterior predictive density for calculating WAIC
  ppd <- array(, dim=c(nsite, nyear, nreps))
  for (i in 1:nsite) {
   for (t in 1:nyear) {
    ppd[i,t,] <- dbinom(y[i,t,], N[i,t], p[i,t,], log=FALSE)
   } # t
  } # i

  ### Save samples
  beta0_save[,k] <- beta0
  beta_rho_save[,k] <- beta_rho
```

```
  kappa_save[k] <- kappa
  alpha_save[,k] <- alpha
  if (nmcmc > 2000) {
   if (k > nmcmc-2000) {
    ppd_save[,,,k-nmcmc+2000] <- ppd # save posterior predictive
density for calculating WAIC
   }
  }
 } # k

 # Write output
 list(beta0_save=beta0_save,
   beta_rho_save=beta_rho_save, kappa_save=kappa_save,
   alpha_save=alpha_save,
   ppd_save=ppd_save)

} # dist_growth_Nmix_mcmc
```

We then use the following code to implement the MCMC algorithm, which takes about 2.3 h to run. Using the code in Appendix 3.3 to create some traceplots, we can see good model convergence and agreement between parameter estimates of $\beta^{[0]}$, $\beta^{[\rho]}$, κ, and α and their true values (Fig. 3.5).

```
library(foreach)   # for parallel computing
library(doParallel) # for parallel computing

numCores <- round(detectCores() / 2) # only use half of the cores for
parallel computing
registerDoParallel(numCores) # setup parallel computing

nmcmc <- 50000 # number of iterations of MCMC computing
chain <- 3   # number of chains

start_time <- Sys.time() # start time of computing
out <- foreach (i=1:chain, .packages='boot') %dopar% {
 dist_growth_Nmix_mcmc(y=y, x=x, w=w, d=d, nmcmc=nmcmc)
} # i
stopImplicitCluster() # clean up the cluster
end_time <- Sys.time() # end time of computing
time_taken <- end_time - start_time # computing time
```

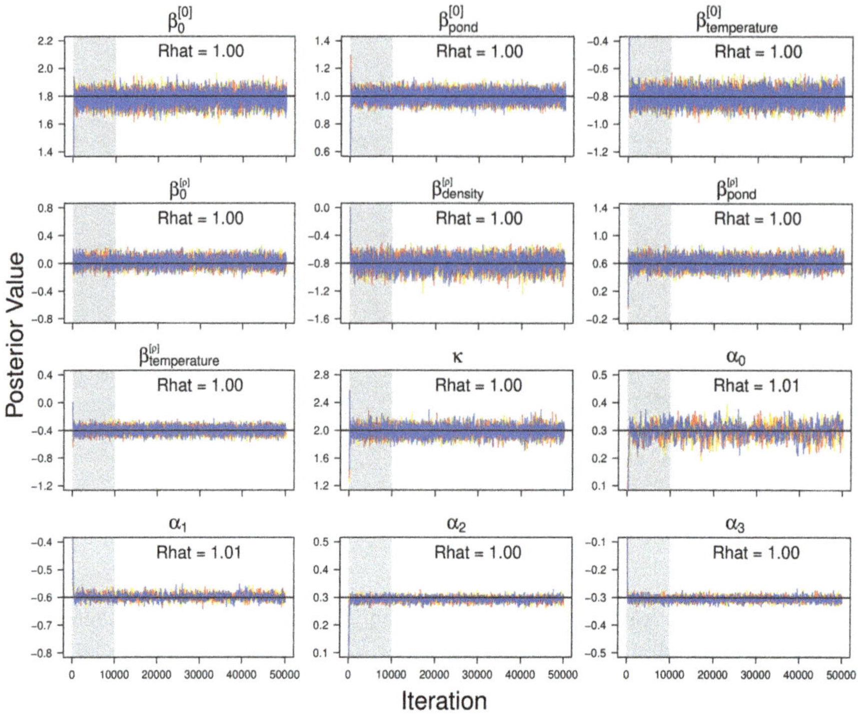

Fig. 3.5 Traceplots showing posterior chains for $\beta^{[0]}$, $\beta^{[\rho]}$, κ, and α in the spatial dynamic N-mixture model with population growth and distance-based movement. Rhat values are calculated after discarding the first 10,000 iterations. Horizontal lines represent the true values used in data simulation

Here, we can do a little experiment to predict population size with or without movement between sites, using the posterior samples. When predicting population size with movement, we use the data-generating model; when predicting population size without movement, we simply let $\lambda_{i,\,t} = N_{i,\,t\,-\,1} \times \rho_{i,\,t\,-\,1}$ by assuming that population size at a site is only driven by population growth at the same site. The code for predicting and plotting the results is provided below. We can see that, without movement, the total population size will be lower than when there is movement (except for the first few years), illustrating the importance of movement to sustain the population at a higher level (Fig. 3.6).

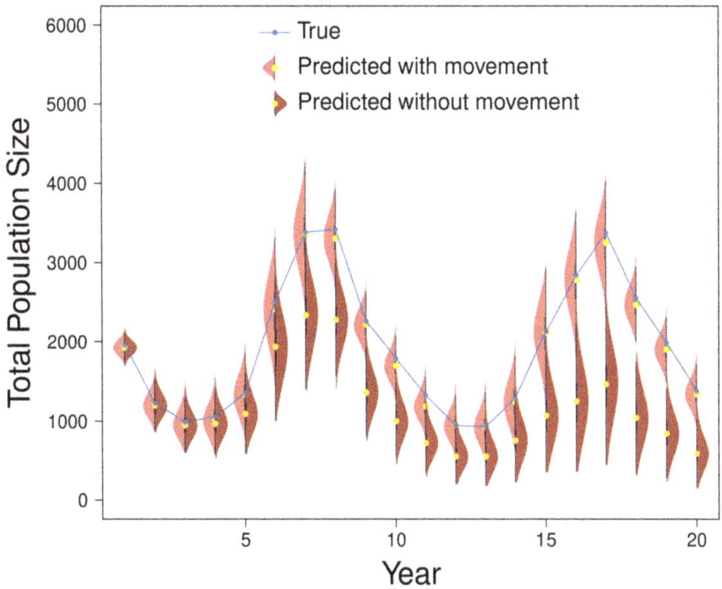

Fig. 3.6 Violin plots showing the predicted total population size with or without movement between sites, which is compared to the simulated true values

```
library(vioplot) # for making violin plots

for (i in 1:chain) {
 if (i == 1) {
  beta0_post <- out[[i]]$beta0_save[, (nmcmc-1999):nmcmc]
  beta_rho_post <- out[[i]]$beta_rho_save[, (nmcmc-1999):nmcmc]
  kappa_post <- out[[i]]$kappa_save[(nmcmc-1999):nmcmc]
 } else {
  beta0_post <- cbind(beta0_post, out[[i]]$beta0_save[, (nmcmc-1999):
nmcmc])
  beta_rho_post <- cbind(beta_rho_post, out[[i]]$beta_rho_save[,
(nmcmc-1999):nmcmc])
  kappa_post <- c(kappa_post, out[[i]]$kappa_save[(nmcmc-1999):
nmcmc])
 }
} # i
npost <- length(kappa_post)

lambda_pred_spatial <- lambda_pred_nonspatial <- array(, dim=c
(nsite, nyear, npost))
N_pred_spatial <- N_pred_nonspatial <- array(, dim=c(nsite, nyear,
```

```
npost))
lambda_pred_spatial[,1,] <- lambda_pred_nonspatial[,1,] <- exp(cbind
(1,x[,1,]) %*% beta0_post)
N_pred_spatial  [,1,] <- rpois(nsite*npost, lambda_pred_spatial
[,1,])
N_pred_nonspatial[,1,] <- rpois(nsite*npost, lambda_pred_nonspatial
[,1,])

eta_pred <- theta_pred <- array(, dim=c(nsite, nsite, npost))
for (k in 1:npost) {
 eta_pred[,,k] <- exp(-1 * kappa_post[k] * d)
 theta_pred[,,k] <- eta_pred[,,k] / rowSums(eta_pred[,,k])
} # k

for (k in 1:npost) {
 for (t in 2:nyear) {
  rho_pred_spatial  <- exp(cbind(1, (N_pred_spatial  [,t-1,k]-
lambda_pred_spatial  [,1,k])/lambda_pred_spatial  [,1,k], x[,t,]) %*%
beta_rho_post[,k])
  rho_pred_nonspatial <- exp(cbind(1, (N_pred_nonspatial[,t-1,k]-
lambda_pred_nonspatial[,1,k])/lambda_pred_nonspatial[,1,k], x[,
t,]) %*% beta_rho_post[,k])
  lambda_pred_spatial  [,t,k] <- t(N_pred_spatial  [,t-1,k] *
rho_pred_spatial  ) %*% theta_pred[,,k]
  lambda_pred_nonspatial[,t,k] <- t(N_pred_nonspatial[,t-1,k] *
rho_pred_nonspatial)
  N_pred_spatial  [,t,k] <- rpois(nsite, lambda_pred_spatial  [,t,k])
  N_pred_nonspatial[,t,k] <- rpois(nsite, lambda_pred_nonspatial[,t,
k])
 } # t
} # k

Nsum_pred_spatial  <- apply(N_pred_spatial  , 2:3, sum)
Nsum_pred_nonspatial <- apply(N_pred_nonspatial, 2:3, sum)

par(mfrow=c(1,1))
par(mar=c(5,7,1,1))

plot(1, xlim=c(1,nyear), ylim=c(0, 6000), type='n', axes=F, xlab='',
ylab='')
vioplot(t(Nsum_pred_spatial  ), col='lightcoral', rectCol='grey36',
lineCol='grey36', colMed='yellow', border=NA, side='left', add=T)
vioplot(t(Nsum_pred_nonspatial), col='firebrick', rectCol='grey16',
lineCol='grey16', colMed='gold',  border=NA, side='right', add=T)
lines(colSums(N), type='o', pch=16, cex=.8, lwd=1.2,
```

```
col='royalblue')
axis(1, at=seq(0,nyear,5), cex.axis=1.5)
axis(2, at=seq(0,6000,1000), cex.axis=1.5, las=2)
box()
title(xlab='Year', cex.lab=2.5, line=3.5)
title(ylab='Total Population Size', cex.lab=2.5, line=4.8)

points(x=6, y=5900, pch=16, cex=1, col='royalblue')
lines(x=c(5.5,6.5), y=c(5900,5900), lwd=1.2, col='royalblue')
text(x=6.5, y=5900, labels='True', pos=4, cex=1.8)
vioplot(rnorm(10000,5450,50), at=6, col='lightcoral',
rectCol='grey36', lineCol='grey36', colMed='yellow', border=NA,
side='left', add=T)
text(x=6.5, y=5450, labels='Predicted with movement', pos=4,
cex=1.8)
vioplot(rnorm(10000,5000,50), at=6, col='firebrick',
rectCol='grey16', lineCol='grey16', colMed='gold', border=NA,
side='right', add=T)
text(x=6.5, y=5000, labels='Predicted without movement', pos=4,
cex=1.8)
```

Eventually, we want to calculate *WAIC* using the posterior predictive density (*ppd*) values we saved earlier. Like other information-theoretic approaches, *WAIC* tries to deal with the trade-off between the goodness of fit and simplicity of the model. Thus, *WAIC* is calculated as $-2 \times lppd + 2 \times p_{waic}$, in which *lppd* equals $\sum_{i=1}^{n_{site}} \sum_{t=1}^{n_{year}} \sum_{j=1}^{n_{reps}} log\left[mean\left(ppd_{i,t,j,k}\right)\right]$ and quantifies the goodness of fit of the model to the data, and p_{waic} equals $\sum_{i=1}^{n_{site}} \sum_{t=1}^{n_{year}} \sum_{j=1}^{n_{reps}} Variance\left[log\left(ppd_{i,t,j,k}\right)\right]$ and represents model complexity. The mean and variance functions used in the calculation are applied over the posterior samples indexed by *k*. Alternatively, we can simply use the "WAIC" function in the R package "LaplacesDemon" to get the same values: $lppd = -41281$, $p_{waic} = 2407$ and $WAIC = 87376$. Imagine there is another model. We can calculate *WAIC* for that model as well and compare it with the *WAIC* value of the current model. The model with a lower *WAIC* value will be considered the better model.

```
ppd <- matrix(, nsite * nyear * nreps, 6000)
ppd[, 1:2000] <- out[[1]]$ppd_save
ppd[,2001:4000] <- out[[2]]$ppd_save
ppd[,4001:6000] <- out[[3]]$ppd_save
lppd <- sum(log(apply(ppd, 1, mean)))
pwaic <- sum(apply(log(ppd), 1, var))
waic <- -2 * lppd + 2 * pwaic
waic <- list(lppd=lppd, pwaic=pwaic, waic=waic)
waic
```

```
# Alternative approach
library(LaplacesDemon)
WAIC(log(ppd))
```

3.4.2 With Distance-, Density-, and Environment-Based Movement

More realistically, movement between two sites is not only related to the distance between the sites but also influenced by the population density and environmental conditions at these sites. That is to say, given the same distance, our turtle is more likely to move from a pond with more turtles to a pond with fewer turtles, or from a smaller pond to a larger pond, and vice versa. Using statistical language, we can embody such hypotheses as follows:

$$\left.\begin{array}{r}\theta_{i_1,i_2,t}=\frac{\eta_{i_1,i_2,t}}{\sum_{i_2=1}^{n_{site}}\eta_{i_1,i_2,t}}\\ \eta_{i_1,i_2,t}=exp\left[-1\times\kappa\times d_{i_1,i_2}+\beta_{density}^{[\eta]}\times f_{density}(N_{i_1,t-1},N_{i_2,t-1})+\beta_1^{[\eta]}\times f_1(x_{i_1,t,1},x_{i_2,t,1})+\ldots\right]\end{array}\right\}, \tag{3.10}$$

in which κ is a decay parameter governing the relationship between movement and distance, d_{i_1,i_2}, $\beta_{density}^{[\eta]}$ governs the effect of a density-dependence process on movement, and $\beta_1^{[\eta]}$, …, $\beta_{n_{covs}}^{[\eta]}$ govern the effects of environmental covariates on movement. This model specification allows movement rate $\theta_{i_1,i_2,t}$ to vary both spatially and temporally. Note that $f_{density}(N_{i,\ t\ -\ 1}, N_{j,\ t\ -\ 1})$ represents a function of $N_{i,\ t\ -\ 1}$ and $N_{j,\ t\ -\ 1}$ (e.g., $N_{j,\ t\ -\ 1} - N_{i,\ t\ -\ 1}$) that can be specific to the density-dependent process and thus different from f_1, etc.

Next, let us use the code below to simulate data for this model specification. We assume positive effects of pond size on initial abundance ($\beta_{pond}^{[0]}=1$), population growth ($\beta_{pond}^{[\rho]}=0.6$), and movement ($\beta_{pond}^{[\eta]}=0.6$), and negative effects of water temperature on initial abundance ($\beta_{temperature}^{[0]}=-0.8$) and population growth ($\beta_{temperature}^{[\rho]}=-0.4$). We also assume a negative density dependence effect on population growth ($\beta_{density}^{[\rho]}=-0.8$) and movement ($\beta_{density}^{[\eta]}=-0.8$). We assume a positive interaction between density and pond size on population growth ($\beta_{density\times pond}^{[\rho]}=0.2$) and movement ($\beta_{density\times pond}^{[\eta]}=0.2$) so that negative density dependence is weaker in larger ponds. Finally, we assume a relative strong rate of decay ($\kappa = 2$).

```
library(boot) # for using logit and inv.logit functions

set.seed(5)

# Basic values
nsite <- 360 # number of sites
```

```
nyear <- 20 # number of years
nreps <- 5  # number of within-season replicates
ncovs <- 2  # number of environmental covariates
npcvs <- 2  # number of observational covariates

beta0 <- c(1.8, 1, -0.8)         # intercept and slopes for initial
abundance
beta_rho <- c(0, -0.8, 0.6, -0.4, 0.2) # intercept and slopes for
population growth
kappa <- 2                  # distance effect on movement
beta_eta <- c(-0.8, 0.6, 0.2)     # slopes for movement
alpha <- c(0.3, -0.6, 0.4)       # intercept and slopes for detection
probability

# Read in environmental covariates
load('data/covariates.RData')
x1 <- (log(cov$size) - mean(log(cov$size))) / sd(log(cov$size)) #
standardized log habitat patch size
x2 <- (cov$temp - mean(cov$temp)) / sd(cov$temp) # standardized
temperature
x <- array(, dim=c(nsite, nyear, ncovs)) # environmental covariates
x[,,1] <- x1
x[,,2] <- x2
### Standardize covariates for each site
for (i in 1:nsite) {
 for (h in 1:ncovs) {
  x[i,2:nyear,h] <- (x[i,2:nyear,h] - mean(x[i,,h])) / sd(x[i,,h])
 } # h
} # i

# Simulate data
lambda <- matrix(, nsite, nyear) # expectation of abundance
N <- matrix(, nsite, nyear) # abundance
lambda[,1] <- exp(cbind(1,x[,1,]) %*% beta0) # expectation of initial
abundance
N[,1] <- rpois(nsite, lambda[,1]) # initial abundance

lon <- cov$lon # longitude or easting
lat <- cov$lat # latitude or northing
d <- matrix(, nsite, nsite) # distance matrix
for (i1 in 1:nsite) {
 for (i2 in 1:nsite) {
  d[i1,i2] <- sqrt((lon[i1] - lon[i2]) ^ 2 + (lat[i1] - lat[i2]) ^ 2)
 } # i2
} # i1
```

```
rho <- matrix(0, nsite, nyear-1) # population growth rate
eta <- array(, dim=c(nsite, nsite, nyear-1)) # unstandardized
colonization rate
theta <- array(, dim=c(nsite, nsite, nyear-1)) # standardized
colonization rate
for (t in 2:nyear) {
 rho[,t-1] <- exp(cbind(1, (N[,t-1]-lambda[,1])/lambda[,1], x[,t,],
(N[,t-1]-lambda[,1])/lambda[,1]*x[,t,1]) %*% beta_rho)
 for (i in 1:nsite) {
  eta[i,,t-1] <- exp(-1 * kappa * d[i,] + cbind((N[,t-1]-N[i,t-1])/exp
(beta0[1]), x[,t,1]-x[i,t,1], (N[,t-1]-N[i,t-1])/exp(beta0[1])*(x[,
t,1]-x[i,t,1])) %*% beta_eta)
 } # i
 theta[,,t-1] <- eta[,,t-1] / rowSums(eta[,,t-1])
 lambda[,t] <- (N[,t-1] * rho[,t-1]) %*% theta[,,t-1]
 N[,t] <- rpois(nsite, lambda[,t])
} # t

w <- array(rnorm(nsite * nyear * nreps * npcvs, 0, 1), dim=c(nsite,
nyear, nreps, npcvs)) # observational covariates

p <- array(, dim=c(nsite, nyear, nreps)) # detection probability
for (i in 1:nsite) {
 for (t in 1:nyear) {
  p[i,t,] <- inv.logit(cbind(1,w[i,t,,]) %*% alpha)
 } # t
} # i

y <- array(, dim=c(nsite, nyear, nreps)) # count data
for (i in 1:nsite) {
 for (t in 1:nyear) {
  y[i,t,] <- rbinom(nreps, N[i,t], p[i,t,])
 } # t
} # i
```

Next, we use the following code to define a custom MCMC algorithm to estimate parameters $\beta^{[0]}$, $\beta^{[\rho]}$, κ, $\beta^{[\eta]}$, and α using data y, x, w, and d:

```
dist_cov_growth_Nmix_mcmc <- function(y, x, w, d, nmcmc) {

 # Setup variables
 nsite <- dim(y)[1]
 nyear <- dim(y)[2]
 nreps <- dim(y)[3]
 ncovs <- dim(x)[3]
```

```
npcvs <- dim(w) [4]
ymax <- apply(y, 1:2, max)

beta0_save <- matrix(0, ncovs+1, nmcmc)
beta_rho_save <- matrix(0, ncovs+3, nmcmc)
kappa_save <- rep(0, nmcmc)
beta_eta_save <- matrix(0, ncovs+1, nmcmc)
alpha_save <- matrix(0, npcvs+1, nmcmc)
N_save <- array(0, dim=c(nsite, nyear, 2000))

# Priors
beta0_mean <- rep(0, ncovs+1)
beta0_sd <- 2
beta_rho_mean <- rep(0, ncovs+3)
beta_rho_sd <- 2
log_kappa_mean <- 0
log_kappa_sd <- 2
beta_eta_mean <- rep(0, ncovs+1)
beta_eta_sd <- 2
alpha_mean <- rep(0, npcvs+1)
alpha_sd <- 2

# Starting values
beta0 <- rep(0, ncovs+1)
beta_rho <- rep(0, ncovs+3)
kappa <- 1
beta_eta <- rep(0, ncovs+1)
alpha <- rep(0, npcvs+1)
lambda0 <- exp(cbind(1,x[,1,]) %*% beta0) # expectation of initial
abundance
rho <- matrix(1, nsite, nyear-1)
eta <- theta <- array(, dim=c(nsite, nsite, nyear-1))
for (t in 1:(nyear-1)) {
 for (i in 1:nsite) {
  eta[i,,t] <- exp(-1 * kappa * d[i,])
 } # i
 theta[,,t] <- eta[,,t] / rowSums(eta[,,t])
} # t
N <- round((ymax + 1) / 0.5)
p <- array(0.5, dim=c(nsite, nyear, nreps))

# Tuning factors
beta0_tune <- 0.08
beta_rho_tune <- 0.08
kappa_tune <- 0.15
```

```
beta_eta_tune <- 0.1
alpha_tune <- 0.02
N_tune <- 1

# Begin MCMC loop
for (k in 1:nmcmc) {
 ### Sample N
 N_star <- matrix(rpois(nsite*nyear, N + N_tune), nsite, nyear)
 prob1 <- prob2 <- matrix(, nsite, nyear)
 prob1[,1] <- dpois(N_star[,1], lambda0, log=T)
 prob2[,1] <- dpois(N     [,1], lambda0, log=T)
 for (t in 2:nyear) {
  prob1[,t] <- dpois(N_star[,t], (N[,t-1] * rho[,t-1]) %*% theta[,,
t-1], log=T)
  prob2[,t] <- dpois(N     [,t], (N[,t-1] * rho[,t-1]) %*% theta[,,t-1],
log=T)
 } # t
 mh1 <- apply(dbinom(y, N_star, p, log=T), 1:2, sum) +
     prob1 +
     dpois(N, N_star+N_tune, log=T)
 mh2 <- apply(dbinom(y, N     , p, log=T), 1:2, sum) +
     prob2 +
     dpois(N_star, N+N_tune, log=T)
 mh <- exp(mh1 - mh2)
 Nkeep <- ((mh > runif(nsite*nyear)) & (N_star >= ymax))
 N[Nkeep] <- N_star[Nkeep]

 ### Sample beta0
 beta0_star <- rnorm(ncovs+1, beta0, beta0_tune)
 lambda0_star <- exp(cbind(1,x[,1,]) %*% beta0_star)
 mh1 <- sum(dpois(N[,1], lambda0_star, log=T)) +
     sum(dnorm(beta0_star, beta0_mean, beta0_sd, log=T))
 mh2 <- sum(dpois(N[,1], lambda0     , log=T)) +
     sum(dnorm(beta0     , beta0_mean, beta0_sd, log=T))
 mh <- exp(mh1 - mh2)
 if (mh > runif(1)) {
  beta0 <- beta0_star
  lambda0 <- lambda0_star
 }

 ### Sample beta_rho
 beta_rho_star <- rnorm(ncovs+3, beta_rho, beta_rho_tune)
 rho_star <- matrix(, nsite, nyear-1)
 for (t in 2:nyear) {
  rho_star[,t-1] <- exp(cbind(1, (N[,t-1]-lambda0)/lambda0,x[,t,], (N
```

```
[,t-1]-lambda0)/lambda0*x[,t,1]) %*% beta_rho_star)
  } # t
  prob1 <- prob2 <- matrix(, nsite, nyear-1)
  for (t in 1:(nyear-1)) {
   prob1[,t] <- dpois(N[,t+1], (N[,t] * rho_star[,t]) %*% theta[,,t],
log=T)
   prob2[,t] <- dpois(N[,t+1], (N[,t] * rho     [,t]) %*% theta[,,t],
log=T)
  } # t
  mh1 <- sum(prob1) +
      sum(dnorm(beta_rho_star, beta_rho_mean, beta_rho_sd, log=T))
  mh2 <- sum(prob2) +
      sum(dnorm(beta_rho     , beta_rho_mean, beta_rho_sd, log=T))
  mh <- exp(mh1 - mh2)
  if (mh > runif(1)) {
   beta_rho <- beta_rho_star
   rho <- rho_star
  }

  ### Sample kappa
  kappa_star <- exp(rnorm(1, log(kappa), kappa_tune))
  eta_star <- theta_star <- array(, dim=c(nsite, nsite, nyear-1))
  for (t in 2:nyear) {
   for (i in 1:nsite) {
    eta_star[i,,t-1] <- exp(-1 * kappa_star * d[i,] + cbind((N[,t-1]-N
[i,t-1])/exp(beta0[1]),x[,t,1]-x[i,t,1],(N[,t-1]-N[i,t-1])/exp
(beta0[1])*(x[,t,1]-x[i,t,1])) %*% beta_eta)
   } # i
   theta_star[,,t-1] <- eta_star[,,t-1] / rowSums(eta_star[,,t-1])
  } # t
  prob1 <- prob2 <- matrix(, nsite, nyear-1)
  for (t in 1:(nyear-1)) {
   prob1[,t] <- dpois(N[,t+1], (N[,t] * rho[,t]) %*% theta_star[,,t],
log=T)
   prob2[,t] <- dpois(N[,t+1], (N[,t] * rho[,t]) %*% theta     [,,t],
log=T)
  } # t
  mh1 <- sum(prob1) +
      dnorm(log(kappa_star), log_kappa_mean, log_kappa_sd, log=T)
  mh2 <- sum(prob2) +
      dnorm(log(kappa     ), log_kappa_mean, log_kappa_sd, log=T)
  mh <- exp(mh1 - mh2)
  if (mh > runif(1)) {
   kappa <- kappa_star
```

```
    eta <- eta_star
    theta <- theta_star
  }

  ### Sample beta_eta
  beta_eta_star <- rnorm(ncovs+1, beta_eta, beta_eta_tune)
  eta_star <- theta_star <- array(, dim=c(nsite, nsite, nyear-1))
  for (t in 2:nyear) {
    for (i in 1:nsite) {
      eta_star[i,,t-1] <- exp(-1 * kappa * d[i,] + cbind((N[,t-1]-N[i,
t-1])/exp(beta0[1]),x[,t,1]-x[i,t,1],(N[,t-1]-N[i,t-1])/exp(beta0
[1])*(x[,t,1]-x[i,t,1])) %*% beta_eta_star)
    }
    theta_star[,,t-1] <- eta_star[,,t-1] / rowSums(eta_star[,,t-1])
  } # t
  prob1 <- prob2 <- matrix(, nsite, nyear-1)
  for (t in 1:(nyear-1)) {
    prob1[,t] <- dpois(N[,t+1], (N[,t] * rho[,t]) %*% theta_star[,,t],
log=T)
    prob2[,t] <- dpois(N[,t+1], (N[,t] * rho[,t]) %*% theta     [,,t],
log=T)
  } # t
  mh1 <- sum(prob1) +
      sum(dnorm(beta_eta_star, beta_eta_mean, beta_eta_sd, log=T))
  mh2 <- sum(prob2) +
      sum(dnorm(beta_eta     , beta_eta_mean, beta_eta_sd, log=T))
  mh <- exp(mh1 - mh2)
  if (mh > runif(1)) {
    beta_eta <- beta_eta_star
    eta <- eta_star
    theta <- theta_star
  }

  ### Sample alpha
  alpha_star <- rnorm(npcvs+1, alpha, alpha_tune)
  p_star <- array(, dim=c(nsite, nyear, nreps))
  for (i in 1:nsite) {
    for (t in 1:nyear) {
      p_star[i,t,] <- inv.logit(cbind(1,w[i,t,,]) %*% alpha_star)
    } # t
  } # i
  mh1 <- sum(dbinom(y, N, p_star, log=TRUE), na.rm=T) +
      sum(dnorm(alpha_star, alpha_mean, alpha_sd, log=TRUE))
  mh2 <- sum(dbinom(y, N, p     , log=TRUE), na.rm=T) +
      sum(dnorm(alpha     , alpha_mean, alpha_sd, log=TRUE))
```

```
  mh <- exp(mh1 - mh2)
  if (mh > runif(1)) {
   alpha <- alpha_star
   p <- p_star
  }

  ### Save samples
  beta0_save[,k] <- beta0
  beta_rho_save[,k] <- beta_rho
  kappa_save[k] <- kappa
  beta_eta_save[,k] <- beta_eta
  alpha_save[,k] <- alpha
  if (nmcmc > 2000) {
   if (k > nmcmc-2000) {
    N_save[,,k-nmcmc+2000] <- N
   }
  }
 } # k

 # Write output
 list(beta0_save=beta0_save,
    beta_rho_save=beta_rho_save, kappa_save=kappa_save,
beta_eta_save=beta_eta_save,
    alpha_save=alpha_save,
    N_save=N_save)

} # dist_cov_growth_Nmix_mcmc
```

Using the following code to implement the MCMC algorithm, the model takes more than 14 h to run. Using code in Appendix 3.4, we can create traceplots that show good model convergence and agreement between the estimates of parameters $\beta^{[0]}$, $\beta^{[\rho]}$, κ, $\beta^{[\eta]}$, and α and their true values used in data simulation (Fig. 3.7).

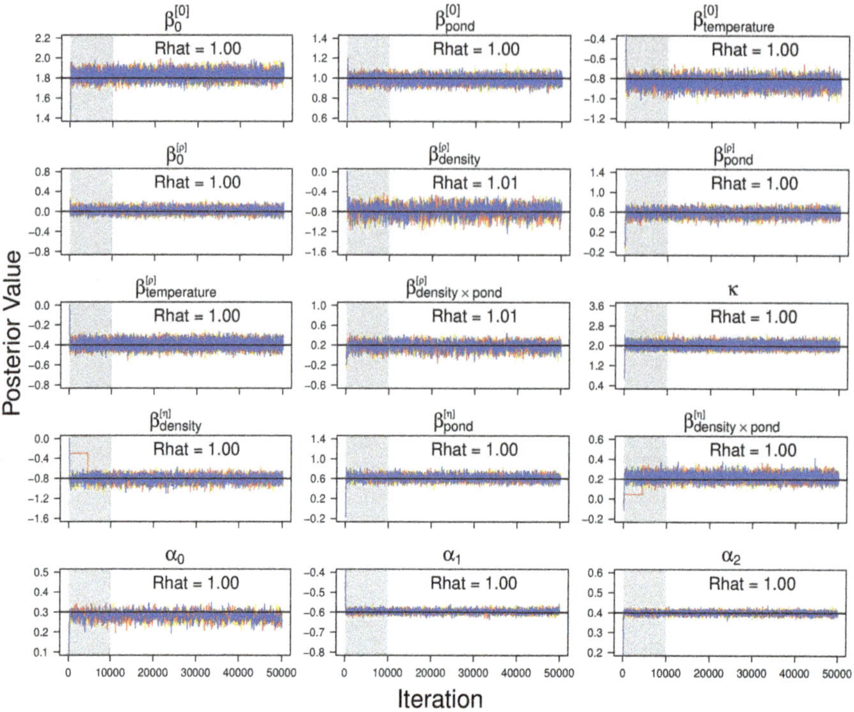

Fig. 3.7 Traceplots showing posterior chains for $\beta^{[0]}$, $\beta^{[\rho]}$, κ, $\beta^{[\eta]}$, and α in the spatial dynamic N-mixture model with population growth and distance-, density-, and environment-based movements. Rhat values are calculated after discarding the first 10,000 iterations. Horizontal lines represent the true values used in data simulation

```
library(foreach)  # for parallel computing
library(doParallel) # for parallel computing

numCores <- round(detectCores() / 2) # only use half of the cores for
parallel computing
registerDoParallel(numCores) # setup parallel computing

nmcmc <- 50000 # number of iterations of MCMC computing
chain <- 3   # number of chains

start_time <- Sys.time() # start time of computing
out <- foreach (i=1:chain, .packages='boot') %dopar% {
 dist_cov_growth_Nmix_mcmc(y=y, x=x, w=w, d=d, nmcmc=nmcmc)
} # i
stopImplicitCluster() # clean up the cluster
```

```
end_time <- Sys.time() # end time of computing
time_taken <- end_time - start_time # computing time
```

We can also use the following code to calculate the standardized movement rate, θ, for different years given the specific density and environmental conditions in that year. Note that θ is asymmetric (i.e., $\theta_{i_1,i_2,t} \neq \theta_{i_2,i_1,t}$) due to the fact that it is a function of density and environmental differences between each pair of sites. We can see evident spatial and temporal variation in θ, using years 7 and 14 as examples, as well as satisfying agreement between model estimates and the simulated truth (Fig. 3.8).

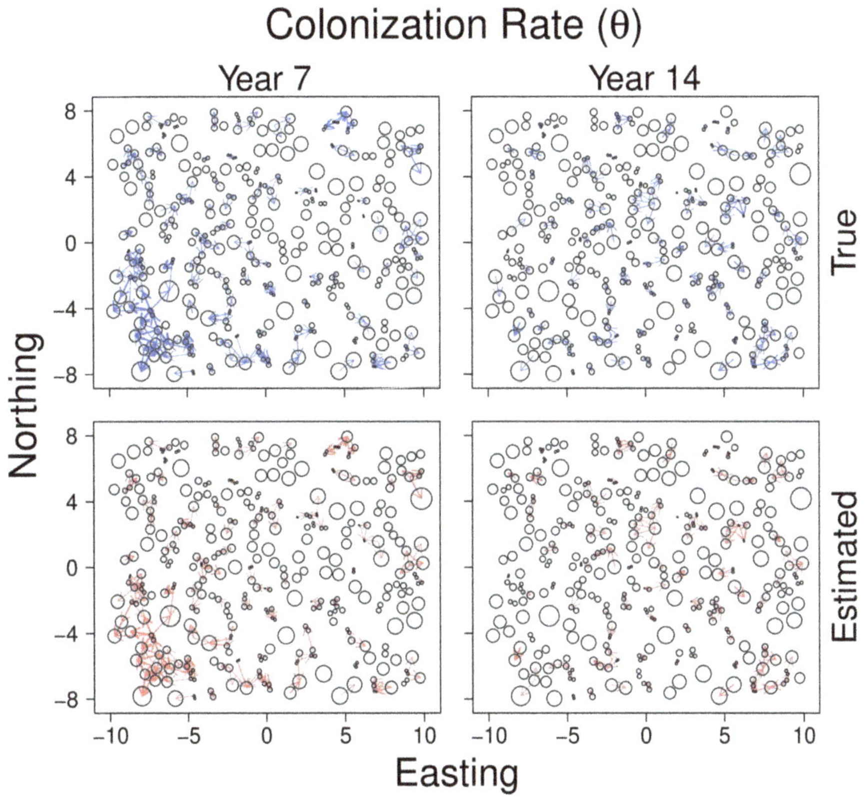

Fig. 3.8 Simulated true (upper row) and estimated (lower row) standardized movement rates (θ) in year 7 and year 14. The arrow points to the direction of net movement, and a thicker line represents a higher movement rate. Only θ's that are greater than 0.1 are plotted

```
kappa_post <- c(out[[1]]$kappa[(nmcmc/2+1):nmcmc], out[[2]]$kappa
[(nmcmc/2+1):nmcmc], out[[3]]$kappa[(nmcmc/2+1):nmcmc])
beta0_post <- cbind(out[[1]]$beta0[,(nmcmc/2+1):nmcmc], out[[2]]$
beta0[,(nmcmc/2+1):nmcmc], out[[3]]$beta0[,(nmcmc/2+1):nmcmc])
beta_eta_post <- cbind(out[[1]]$beta_eta[,(nmcmc/2+1):nmcmc], out
[[2]]$beta_eta[,(nmcmc/2+1):nmcmc], out[[3]]$beta_eta[,(nmcmc/
2+1):nmcmc])
N_post <- array(, dim=c(nsite, nyear, 6000))
N_post[,, 1:2000] <- out[[1]]$N
N_post[,,2001:4000] <- out[[2]]$N
N_post[,,4001:6000] <- out[[3]]$N
kappa_med <- median(kappa_post)
beta0_med <- apply(beta0_post, 1, median)
beta_eta_med <- apply(beta_eta_post, 1, median)
N_med <- apply(N_post, 1:2, median)

years_plot <- c(7, 14)
eta_true <- theta_true <- eta_pred <- theta_pred <- array(, dim=c
(nsite, nsite, length(years_plot)))
for (t in 1:length(years_plot)) {
 yr <- years_plot[t]
 for (i in 1:nsite) {
  eta_true[i,,t] <- exp(-1 * kappa   * d[i,] + cbind((N   [,yr]-N   [i,
yr])/exp(beta0   [1]),x[,yr,1]-x[i,yr,1],(N   [,yr]-N   [i,yr])/exp
(beta0   [1])*(x[,yr,1]-x[i,yr,1])) %*% beta_eta   )
  eta_pred[i,,t] <- exp(-1 * kappa_med * d[i,] + cbind((N_med[,yr]-
N_med[i,yr])/exp(beta0_med[1]),x[,yr,1]-x[i,yr,1],(N_med[,yr]-
N_med[i,yr])/exp(beta0_med[1])*(x[,yr,1]-x[i,yr,1])) %*%
beta_eta_med)
 } # i
 theta_true[,,t] <- eta_true[,,t] / rowSums(eta_true[,,t])
 theta_pred[,,t] <- eta_pred[,,t] / rowSums(eta_pred[,,t])
} # t

plot_cut <- 0.1

par(mfrow=c(2,2))
par(mar=c(1,1,1,1))
par(oma=c(4,5,5,2))

for (t in 1:length(years_plot)) {
 plot(cov$lon, cov$lat, xlim=c(-10.2,10.2), ylim=c(-8.2,8.2),
axes=F, xlab='', ylab='', type='n')
 axis(1, at=seq(-10,10,5), labels=rep('',5))
 if (t == 1) {
```

```
   axis(2, at=seq(-08,08,4), cex.axis=1.6, las=2)
  } else {
   axis(2, at=seq(-08,08,4), labels=rep('',5))
   axis(4, at=0, labels='True', cex.axis=2.5, tick=F, line=.5)
  }
 axis(3, at=0, labels=paste('Year ', years_plot[t], sep=''), cex.
axis=2.5, line=-.8, tick=F)
 box()
 for (i in 1:(nsite-1)) {
  for (j in (i+1):nsite) {
   length <- ifelse(d[i,j] / 8 > 0.1, 0.1, d[i,j] / 8)
   if (theta_true[i,j,t] > theta_true[j,i,t]) {
    if ((theta_true[i,j,t] - theta_true[j,i,t]) > plot_cut) {
     arrows(x0=cov$lon[j], y0=cov$lat[j], x1=cov$lon[i], y1=cov$lat
[i],
       lwd=(theta_true[i,j,t] - theta_true[j,i,t])*5, length=length,
col='royalblue')
    }
   } else {
    if ((theta_true[j,i,t] - theta_true[i,j,t]) > plot_cut) {
     arrows(x0=cov$lon[i], y0=cov$lat[i], x1=cov$lon[j], y1=cov$lat
[j],
       lwd=(theta_true[j,i,t] - theta_true[i,j,t])*5, length=length,
col='royalblue')
    }
   }
  } # j
 } # i
 points(cov$lon, cov$lat, pch=21, cex=rowMeans(cov$size)*4.6,
col='grey18', bg=NA)
} # t

for (t in 1:length(years_plot)) {
 plot(cov$lon, cov$lat, xlim=c(-10.2,10.2), ylim=c(-8.2,8.2),
axes=F, xlab='', ylab='', type='n')
 axis(1, at=seq(-10,10,5), cex.axis=1.6)
 if (t == 1) {
  axis(2, at=seq(-08,08,4), cex.axis=1.6, las=2)
 } else {
  axis(2, at=seq(-08,08,4), labels=rep('',5))
  axis(4, at=0, labels='Estimated', cex.axis=2.5, tick=F, line=.5)
 }
 box()
 for (i in 1:(nsite-1)) {
  for (j in (i+1):nsite) {
```

```
   length <- ifelse(d[i,j] / 8 > 0.1, 0.1, d[i,j] / 8)
   if (theta_pred[i,j,t] > theta_pred[j,i,t]) {
    if ((theta_pred[i,j,t] - theta_pred[j,i,t]) > plot_cut) {
     arrows(x0=cov$lon[j], y0=cov$lat[j], x1=cov$lon[i], y1=cov$lat
[i],
       lwd=(theta_pred[i,j,t] - theta_pred[j,i,t])*5, length=length,
col='lightcoral')
    }
   } else {
    if ((theta_pred[j,i,t] - theta_pred[i,j,t]) > plot_cut) {
     arrows(x0=cov$lon[i], y0=cov$lat[i], x1=cov$lon[j], y1=cov$lat
[j],
       lwd=(theta_pred[j,i,t] - theta_pred[i,j,t])*5, length=length,
col='lightcoral')
    }
   }
  } # j
 } # i
 points(cov$lon, cov$lat, pch=21, cex=rowMeans(cov$size)*4.6,
col='grey18', bg=NA)
} # t

title(xlab='Easting' , line=2.6, cex.lab=2.8, outer=T)
title(ylab='Northing', line=2.4, cex.lab=2.8, outer=T)
title(main=expression(paste('Colonization Rate (', theta, ')',
sep='')), line=3.2, cex.main=3, outer=T)
```

3.5 Spatial Dynamic N-Mixture Models with Population Growth for Interacting Species

The single-species spatial dynamic N-mixture model with population growth (Howell et al. 2020) is further extended to allow for inter-specific interactions such as predator–prey interaction or competition (Zhao et al. 2022). To illustrate this model with a system of two competitive species, we can image that our little turtle is in fact the Blanding's turtle and the ponds it lives in are, rather unfortunately, invaded by another turtle species, the red-eared slider. Because the invasive red-eared slider holds advantages in competition over the native Blanding's turtle, the number of the Blanding's turtles in a pond is highly negatively affected by the number of red-eared slider in that pond. By contrast, the number of red-eared slider in a pond is mainly affected by its own number and abiotic environment but not the number of the Blanding's turtles. Finally, both turtle species can move between ponds, although the Blanding's turtle tends to move a longer distance to find suitable ponds than does red-eared slider. Using statistical language, the process model first specifies the first-year abundance for each species as

$$\left.\begin{array}{c} N_{i,1,s} \sim Poisson(\lambda_{i,1,s}) \\ \lambda_{i,1,s} = exp\left(\beta_{0,s}^{[0]} + \beta_{1,s}^{[0]} \times x_{i,1,1} + \ldots + \beta_{n_{covs},s}^{[0]} \times x_{i,1,n_{covs}}\right) \end{array}\right\}, \tag{3.11}$$

in which $N_{i,\ 1,\ s}$ is the abundance of species s at site i in the first year, $\lambda_{i,\ 1,\ s}$ is the corresponding expectation of abundance which is a function of covariates x's with species-specific intercept and slope parameters $\beta^{[0]}$'s.

From the second year (i.e., $t \geq 2$), the process sub-model specifies population size as

$$\left.\begin{array}{c} N_{i_2,t,s} \sim Poisson(\lambda_{i_2,t,s}) \\ \lambda_{i_2,t,s} = \sum_{i_1=1}^{n_{site}} N_{i_1,t-1,s} \times \rho_{i_1,t-1,s} \times \theta_{i_1,i_2,s} \\ \rho_{i_1,t-1,s} = exp\left(\beta_{0,s}^{[\rho]} + \beta_{density,s}^{[\rho]} \times N_{i_1,t-1,s} + \beta_{competition,s}^{[\rho]} \times N_{i_1,t-1,-s} + \beta_{1,s}^{[\rho]} \times x_{i_1,t,1} + \ldots\right) \\ \theta_{i_1,i_2,s} = \frac{\eta_{i_1,i_2,s}}{\sum_{i_2=1}^{n_{site}} \eta_{i_1,i_2,s}} \\ \eta_{i_1,i_2,s} = exp(-1 \times \kappa_s \times d_{i_1,i_2}) \end{array}\right\}, \tag{3.12}$$

in which $\rho_{i_1,t-1,s}$ is population growth rate of species s at the site i_1 in year t-1, $\theta_{i_1,i_2,s}$ is standardized movement rate for species s from sites i_1 and i_2, and $\eta_{i_1,i_2,s}$ is unstandardized movement rate. Population growth rate $\rho_{i_1,t-1,s}$ is a function of density and environmental covariates, with $\beta_{0,s}^{[\rho]}$ being the intercept, $\beta_{density,s}^{[\rho]}$ being a slope for intra-specific density-dependent process, $\beta_{compitition,s}^{[\rho]}$ being another slope for inter-specific competition, and $\beta_{1,s}^{[\rho]}, \ldots, \beta_{n_{covs},s}^{[\rho]}$ being slopes for environmental effects on population growth. The index $-s$ represents the species other than species s. For example, if we consider a system with two species, we will have $-s = Blanding's$ for the native Blanding's turtle when $s = red - eared$ for the invasive red-eared slider, and vice versa.

Finally, we have an observation sub-model that links the observed count data to the true abundance:

$$\left.\begin{array}{c} y_{i,t,j,s} \sim Binomial\left(N_{i,t,s}, p_{i,t,j}\right) \\ p_{i,t,j} = logit^{-1}\left(\alpha_0 + \alpha_1 \times w_{i,t,j,1} + \ldots + \alpha_{n_{pcvs}} \times w_{i,t,j,n_{pcvs}}\right) \end{array}\right\}. \tag{3.13}$$

in which detection probability $p_{i,\ t,\ j}$ is a function of observation covariates $w_{i,t,j,n_{pcvs}}$ with intercept and slope parameters α's. We can further extend α's to be species-specific.

Using the code below, we can simulate some data under this model. We assume that pond size has positive effects on the initial abundance ($\beta_{pond,red-eared}^{[0]} = 0.3$, $\beta_{pond,Blanding's}^{[0]} = 0.6$) and population growth ($\beta_{pond,red-eared}^{[\rho]} = 0.1$, $\beta_{pond,Blanding's}^{[\rho]} = 0.2$) of both the invasive red-eared slider and native Blanding's turtle. We also assume that water temperature has positive effects on the initial abundance ($\beta_{temperature,red-eared}^{[0]} = 0.3$) and population growth ($\beta_{temperature,red-eared}^{[\rho]} = 0.1$) of red-eared slider, but negative effects on the initial abundance ($\beta_{temperature,Blanding's}^{[0]} =$

– 0.6) and population growth ($\beta^{[\rho]}_{temperature, Blanding's} = -0.2$) of Blanding's turtle. We assume that population growth of both species is driven by negative intra-specific density dependence ($\beta^{[\rho]}_{density, red-eared} = -0.4$, $\beta^{[\rho]}_{density, Blanding's} = -0.2$). While population growth of Blanding's turtle is negatively affected by the number of red-eared slider ($\beta^{[\rho]}_{competition, Blanding's} = -0.8$), population growth of red-eared slider is not affected by the number of Blanding's turtles ($\beta^{[\rho]}_{competition, red-eared} = 0$). Finally, we assume a stronger decay for the red-eared slider ($\kappa_{red-eared} = 3$) than for the Blanding's turtle ($\kappa_{Blanding's} = 1$) in the dispersal function, meaning that the latter has a higher rate of long-distance movement than the former.

```
library(boot) # for using logit and inv.logit functions

set.seed(6)

# Basic values
nsite <- 360 # number of sites
nyear <- 20  # number of years
nreps <- 5   # number of within-season replicates
nspec <- 2   # number of species
ncovs <- 2   # number of environmental covariates
npcvs <- 2   # number of observational covariates

### For beta0 and beta_lambda, the first column is for the invasive
red-eared slider, and the second column is for the native Blanding's
turtle
beta0 <- matrix(c(3, 0.3, 0.3, 1.6, 0.6, -0.6), ncovs+1, nspec)
# intercept and slopes for initial population size
beta_rho <- matrix(c(0, -0.4, 0, 0.1, 0.1, 0, -0.8, -0.2, 0.2, -0.2),
nspec+ncovs+1, nspec) # intercept and slopes for local population
growth rate
kappa <- c(3, 1)                                  # distance effect on
movement
alpha <- c(0.4, -0.6, 0.3)                        # intercept and slopes
for detection probability

# Read in environmental covariates
load('data/covariates.RData')
x1 <- (log(cov$size) - mean(log(cov$size))) / sd(log(cov$size)) #
standardized log food patch size
x2 <- (cov$temp - mean(cov$temp)) / sd(cov$temp) # standardized
temperature
x <- array(, dim=c(nsite, nyear, ncovs)) # environmental covariates
x[,,1] <- x1
x[,,2] <- x2
```

```
### Standardize covariates for each site
for (i in 1:nsite) {
 for (h in 1:ncovs) {
  x[i,2:nyear,h] <- (x[i,2:nyear,h] - mean(x[i,,h])) / sd(x[i,,h])
 } # h
} # i

# Simulate data
lambda <- array(, dim=c(nsite, nyear, nspec)) # expectation of
abundance
N <- array(, dim=c(nsite, nyear, nspec)) # abundance
for (s in 1:nspec) {
 lambda[,1,s] <- exp(cbind(1,x[,1,]) %*% beta0[,s]) # expectation of
initial abundance
 N[,1,s] <- rpois(nsite, lambda[,1,s]) # initial abundance
} # s

lon <- cov$lon # longitude or easting
lat <- cov$lat # latitude or northing
d <- matrix(, nsite, nsite) # distance matrix
for (i1 in 1:nsite) {
 for (i2 in 1:nsite) {
  d[i1,i2] <- sqrt((lon[i1] - lon[i2]) ^ 2 + (lat[i1] - lat[i2]) ^ 2)
 } # i2
} # i1

eta  <- array(, dim=c(nsite, nsite, nspec)) # unstandardized
colonization rate
theta <- array(, dim=c(nsite, nsite, nspec)) #  standardized
colonization rate
for (s in 1:nspec) {
 eta[,,s] <- exp(-1 * kappa[s] * d)
 theta[,,s] <- eta[,,s] / rowSums(eta[,,s])
} # s

rho <- array(, dim=c(nsite, nyear-1, nspec)) # population growth rate
for (t in 2:nyear) {
 for (s in 1:nspec) {
  rho[,t-1,s] <- exp(cbind(1, (N[,t-1,]-lambda[,1,])/lambda[,1,], x[,
t,]) %*% beta_rho[,s])
   lambda[,t,s] <- (N[,t-1,s] * rho[,t-1,s]) %*% theta[,,s]
  N[,t,s] <- rpois(nsite, lambda[,t,s])
 } # s
} # t

w <- array(rnorm(nsite * nyear * nreps * npcvs, 0, 1), dim=c(nsite,
nyear, nreps, npcvs)) # observational covariates
```

```
p <- array(, dim=c(nsite, nyear, nreps)) # detection probability
for (t in 1:nyear) {
 for (j in 1:nreps) {
  p[,t,j] <- inv.logit(cbind(1,w[,t,j,]) %*% alpha)
 } # j
} # t

y <- array(, dim=c(nsite, nyear, nreps, nspec)) # count data
for (t in 1:nyear) {
 for (j in 1:nreps) {
  for (s in 1:nspec) {
   y[,t,j,s] <- rbinom(nsite, N[,t,s], p[,t,j])
  } # s
 } # j
} # t
```

We then develop a custom MCMC algorithm to estimate parameters $\beta^{[0]}$, $\beta^{[\rho]}$, κ, and α using data y, x, w, and d, using the code below:

```
spatial_growth_Nmix_ii_mcmc <- function(y, x, w, d, nmcmc) {

 # Setup variables
 nsite <- dim(y)[1]
 nyear <- dim(y)[2]
 nreps <- dim(y)[3]
 nspec <- dim(y)[4]
 ncovs <- dim(x)[3]
 npcvs <- dim(w)[4]
 ymax <- apply(y, c(1,2,4), max)

 beta0_save <- array(0, dim=c(ncovs+1, nspec, nmcmc))
 beta_rho_save <- array(0, dim=c(nspec+ncovs+1, nspec, nmcmc))
 kappa_save <- matrix(0, nspec, nmcmc)
 alpha_save <- matrix(0, npcvs+1, nmcmc)
 N_save <- array(0, dim=c(nsite, nyear, nspec, 2000))

 # Priors
 beta0_mean <- matrix(0, ncovs+1, nspec)
 beta0_sd <- 2
 beta_rho_mean <- matrix(0, nspec+ncovs+1, nspec)
 beta_rho_sd <- 2
 log_kappa_mean <- rep(0, nspec)
 log_kappa_sd <- 2
 alpha_mean <- rep(0, npcvs+1)
 alpha_sd <- 2
```

```
# Starting values
beta0 <- matrix(0, ncovs+1, nspec)
beta_rho <- matrix(0, nspec+ncovs+1, nspec)
kappa <- rep(1, nspec)
alpha <- rep(0, npcvs+1)
eta <- theta <- array(, dim=c(nsite, nsite, nspec))
for (s in 1:nspec) {
 eta[,,s] <- exp(-1 * kappa[s] * d)
 theta[,,s] <- eta[,,s] / rowSums(eta[,,s])
} # s
rho <- array(1, dim=c(nsite, nyear-1, nspec))
p <- array(0.5, dim=c(nsite, nyear, nreps))
N <- round((ymax + 1) / 0.5)

# Tuning factors
beta0_tune <- matrix(0.035, ncovs+1, nspec)
beta_rho_tune <- matrix(c(0.05, 0.07), nspec+ncovs+1, nspec,
byrow=T)
kappa_tune <- 0.15
alpha_tune <- c(0.02, 0.02, 0.02)
N_tune <- 1

# Begin MCMC loop
for (k in 1:nmcmc) {
 ### Sample N
 N_star <- array(rpois(nsite * nyear * nspec, N + N_tune), dim=c(nsite,
nyear, nspec))
  lambda0 <- matrix(, nsite, nspec) # lambda for the first year
  lambdar <- array(, dim=c(nsite, nyear-1, nspec)) # lambda except for
the first year
  mh1 <- mh2 <- array(, dim=c(nsite, nyear, nspec))
  for (s in 1:nspec) {
   lambda0[,s] <- exp(cbind(1, x[,1,]) %*% beta0[,s])
   for (t in 2:nyear) {
    lambdar[,t-1,s] <- (N[,t-1,s] * rho[,t-1,s]) %*% theta[,,s]
   } # t
  mh1[,,s] <- apply(dbinom(y[,,,s], N_star[,,s], p, log=T), 1:2, sum)
+
        dpois(N_star[,,s], cbind(lambda0[,s], lambdar[,,s]), log=T) +
         dpois(N[,,s], N_star[,,s] + N_tune, log=T)
   mh2[,,s] <- apply(dbinom(y[,,,s], N   [,,s], p, log=T), 1:2, sum) +
         dpois(N   [,,s], cbind(lambda0[,s], lambdar[,,s]), log=T) +
         dpois(N_star[,,s], N[,,s] + N_tune, log=T)
  } # s
  mh <- exp(mh1 - mh2)
  Nkeep <- ((mh > runif(nsite*nyear*nspec)) & (N_star >= ymax))
  N[Nkeep] <- N_star[Nkeep]
```

```
  ### Sample beta0
  beta0_star <- matrix(rnorm((ncovs+1)*nspec, beta0, beta0_tune),
ncovs+1, nspec)
  lambda0_star <- matrix(, nsite, nspec)
  for (s in 1:nspec) {
   lambda0_star[,s] <- exp(cbind(1, x[,1,]) %*% beta0_star[,s])
   mh1 <- sum(dpois(N[,1,s], exp(cbind(1, x[,1,]) %*% beta0_star[,s]),
log=T)) +
       sum(dnorm(beta0_star[,s], beta0_mean[,s], beta0_sd, log=T))
   mh2 <- sum(dpois(N[,1,s], exp(cbind(1, x[,1,]) %*% beta0    [,s]),
log=T)) +
       sum(dnorm(beta0    [,s], beta0_mean[,s], beta0_sd, log=T))
   mh <- exp(mh1 - mh2)
   if (mh > runif(1)) {
    beta0[,s] <- beta0_star[,s]
    lambda0[,s] <- lambda0_star[,s]
   }
  } # s

  ### Sample beta_rho
  beta_rho_star <- matrix(rnorm((nspec+ncovs+1)*nspec, beta_rho,
beta_rho_tune), nspec+ncovs+1, nspec)
  rho_star <- lambdar <- lambdar_star <- array(, dim=c(nsite, nyear-1,
nspec))
  for (s in 1:nspec) {
   for (t in 2:nyear) {
    rho_star[,t-1,s] <- exp(cbind(1,(N[,t-1,]-lambda0)/lambda0,x[,
t,]) %*% beta_rho_star[,s])
   } # t
   lambdar     [,,s] <- t(t(N[,-nyear,s] * rho     [,,s]) %*% theta[,,s])
   lambdar_star[,,s] <- t(t(N[,-nyear,s] * rho_star[,,s]) %*% theta[,,
s])
   mh1 <- sum(dpois(N[,-1,s], lambdar_star[,,s], log=T)) +
       sum(dnorm(beta_rho_star[,s], beta_rho_mean[,s], beta_rho_sd,
log=T))
   mh2 <- sum(dpois(N[,-1,s], lambdar     [,,s], log=T)) +
       sum(dnorm(beta_rho     [,s], beta_rho_mean[,s], beta_rho_sd,
log=T))
   mh <- exp(mh1 - mh2)
   if (mh > runif(1)) {
    beta_rho[,s] <- beta_rho_star[,s]
    rho[,,s] <- rho_star[,,s]
   }
  } # s
```

```
### Sample kappa
kappa_star <- exp(rnorm(nspec, log(kappa), kappa_tune))
eta_star <- theta_star <- array(, dim=c(nsite, nsite, nspec))
lambdar <- lambdar_star <- array(, dim=c(nsite, nyear-1, nspec))
for (s in 1:nspec) {
 eta_star[,,s] <- exp(-1 * kappa_star[s] * d)
 theta_star[,,s] <- eta_star[,,s] / rowSums(eta_star[,,s])
 lambdar   [,,s] <- t(t(N[,-nyear,s] * rho[,,s]) %*% theta   [,,s])
 lambdar_star[,,s] <- t(t(N[,-nyear,s] * rho[,,s]) %*% theta_star[,,
s])
 mh1 <- sum(dpois(N[,-1,s], lambdar_star[,,s], log=T)) +
   sum(dnorm(log(kappa_star[s]), log_kappa_mean[s], log_kappa_sd,
log=T))
 mh2 <- sum(dpois(N[,-1,s], lambdar   [,,s], log=T)) +
    sum(dnorm(log(kappa   [s]), log_kappa_mean[s], log_kappa_sd,
log=T))
 mh <- exp(mh1 - mh2)
 if (mh > runif(1)) {
  kappa[s] <- kappa_star[s]
  eta[,,s] <- eta_star[,,s]
  theta[,,s] <- theta_star[,,s]
 }
} # s

### Sample alpha
alpha_star <- rnorm(npcvs+1, alpha, alpha_tune)
p_star <- array(0, dim=c(nsite, nyear, nreps))
for (t in 1:nyear) {
 for (j in 1:nreps) {
  p_star[,t,j] <- inv.logit(cbind(1,w[,t,j,]) %*% alpha_star)
 } # j
} # t
mh1t <- mh2t <- rep(0, nspec)
for (s in 1:nspec) {
 mh1t[s] <- sum(dbinom(y[,,,s], N[,,s], p_star, log=T))
 mh2t[s] <- sum(dbinom(y[,,,s], N[,,s], p    , log=T))
} # s
mh1 <- sum(mh1t) +
   sum(dnorm(alpha_star, alpha_mean, alpha_sd, log=T))
mh2 <- sum(mh2t) +
   sum(dnorm(alpha    , alpha_mean, alpha_sd, log=T))
mh <- exp(mh1 - mh2)
if (mh > runif(1)) {
 alpha <- alpha_star
 p <- p_star
}
```

```
    ### Save samples
    beta0_save[,,k] <- beta0
    beta_rho_save[,,k] <- beta_rho
    kappa_save[,k] <- kappa
    alpha_save[,k] <- alpha
    if (nmcmc > 2000) {
     if (k > nmcmc-2000) {
      N_save[,,,k-nmcmc+2000] <- N
     }
    }
  } # k

  # Write output
  list(beta0_save=beta0_save,
     beta_rho_save=beta_rho_save, kappa_save=kappa_save,
     alpha_save=alpha_save,
     N_save=N_save)

} # spatial_growth_Nmix_ii_mcmc
```

Using the following code to implement the algorithm, we can see that the model takes about 3 h to run. Using the code in Appendix 3.5, we can create some traceplots that show good model convergence and agreement between parameter estimates of $\beta^{[0]}$, $\beta^{[\rho]}$, κ, and α and their true values used in data simulation (Fig. 3.9).

```
library(foreach)  # for parallel computing
library(doParallel) # for parallel computing

numCores <- round(detectCores() / 2) # only use half of the cores for
parallel computing
registerDoParallel(numCores) # setup parallel computing

nmcmc <- 50000 # number of iterations of MCMC computing
chain <- 3    # number of chains

start_time <- Sys.time() # start time of computing
out <- foreach (i=1:chain, .packages='boot') %dopar% {
 spatial_growth_Nmix_ii_mcmc(y=y, x=x, w=w, d=d, nmcmc=nmcmc)
} # i
stopImplicitCluster() # clean up the cluster
end_time <- Sys.time() # end time of computing
time_taken <- end_time - start_time # computing time
```

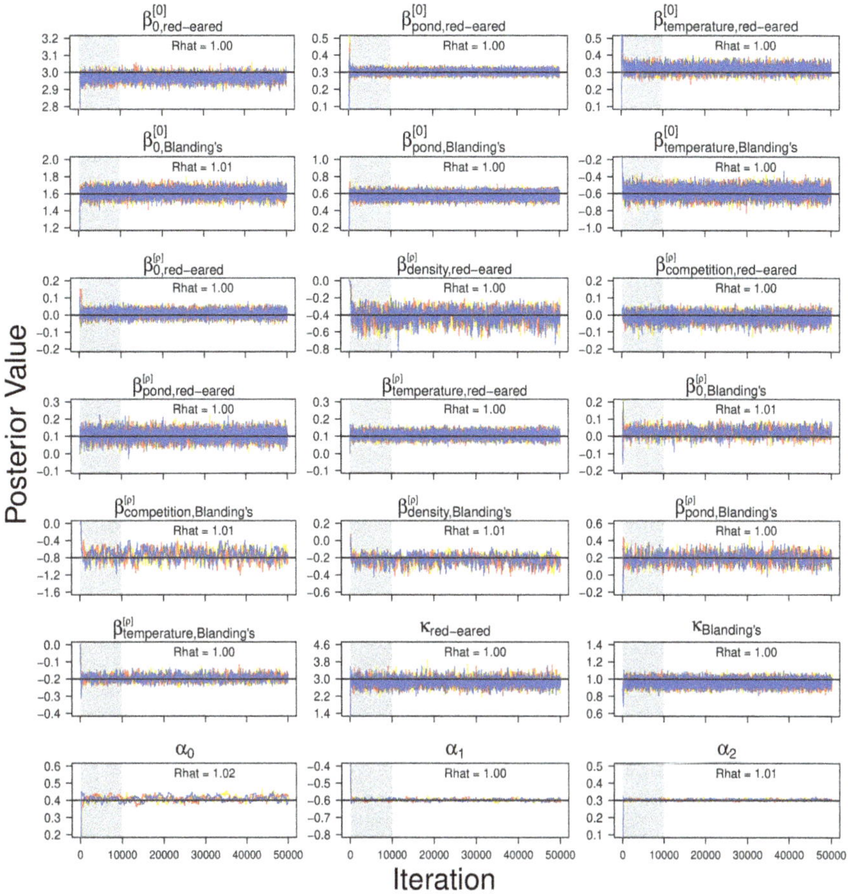

Fig. 3.9 Traceplots showing posterior chains for $\beta^{[0]}$, $\beta^{[\rho]}$, κ, and α in the spatial dynamic N-mixture model with population growth for two competitive species, the invasive red-eared slider and native Blanding's turtle. Rhat values are calculated after discarding the first 10,000 iterations. Horizontal lines represent the true values used in data simulation

Next, let us also use the following code to plot the estimated total abundance for each species, which shows excellent agreement with the simulated true values (Fig. 3.10). The temporal variation of the total abundance of the invasive red-eared slider is mainly driven by water temperature. The temporal variation of the total abundance of the native Blanding's turtle, on the other hand, is mainly driven by the number of red-eared sliders. Therefore, we can see that the population of Blanding's turtle goes down when the population of red-eared sliders goes up, and vice versa.

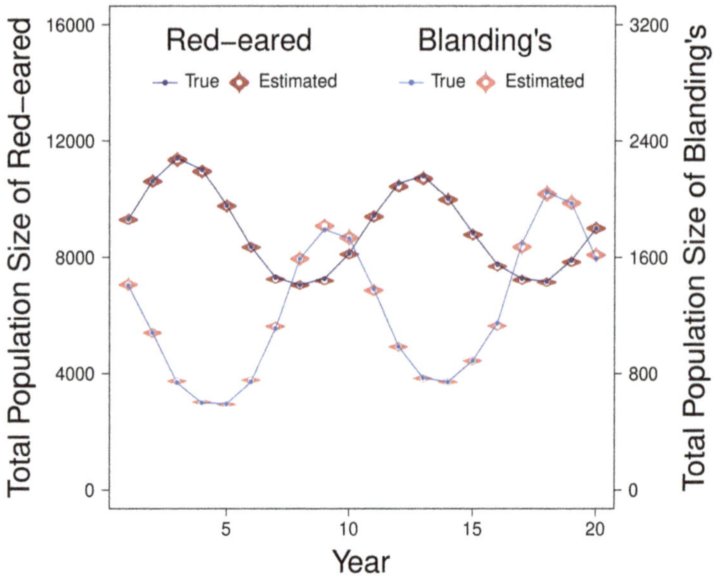

Fig. 3.10 Violin plots showing agreement between the estimated total abundance ($\sum_{i=1}^{n_{site}} N_{i,t}$; violin) and their simulated true values (blue points) for the invasive red-eared slider (deeper colors) and native Blanding's turtle (lighter colors)

```
library(vioplot) # for making violin plots

N_post <- array(, dim=c(nsite, nyear, nspec, 2000*chain))
for (i in 1:chain) {
 N_post[,,,(i-1)*2000+1:2000] <- out[[i]]$N_save
} # i
Nsum <- apply(N, 2:3, sum)
Nsum_post <- apply(N_post, 2:4, sum)

par(mfrow=c(1,1))
par(mar=c(5,7.5,1,7))

plot(1, xlim=c(1,nyear), ylim=c(0, 16000), type='n', axes=F,
xlab='', ylab='')
vioplot(t(Nsum_post[,1,]*1), col='firebrick' , rectCol=NA,
lineCol=NA, border=NA, add=T)
vioplot(t(Nsum_post[,2,]*5), col='lightcoral', rectCol=NA,
lineCol=NA, border=NA, add=T)
lines(Nsum[,1]*1, type='o', pch=16, cex=.8, lwd=1.2, col='navy'    )
lines(Nsum[,2]*5, type='o', pch=16, cex=.8, lwd=1.2,
col='royalblue')
```

```
axis(1, at=seq(0,nyear,5), cex.axis=1.5)
axis(2, at=seq(0,16000,4000), cex.axis=1.5, las=2)
axis(4, at=seq(0,16000,4000), labels=seq(0,16000,4000)/5, cex.
axis=1.5, las=2)
box()
title(xlab='Year', cex.lab=2.5, line=3.5)
axis(2, at=8000, labels="Total Population Size of Red-eared" , cex.
axis=2.5, line=4.4, tick=F)
axis(4, at=8000, labels="Total Population Size of Blanding's", cex.
axis=2.5, line=4.6, tick=F)

text(x= 5.5, y=16000, labels="Red-eared", pos=1, cex=2.5)
text(x=15.5, y=16000, labels="Blanding's", pos=1, cex=2.5)
points(x=2.5, y=14000, pch=16, cex=1, col='navy')
lines(x=c(2,3), y=c(14000,14000), lwd=1.2, col='navy')
text(x=3, y=14000, labels='True', pos=4, cex=1.5)
vioplot(rnorm(1e4,14000,100), at=5.5, col='firebrick', rectCol=NA,
lineCol=NA, border=NA, add=T)
text(x=6, y=14000, labels='Estimated', pos=4, cex=1.5)
points(x=12.5, y=14000, pch=16, cex=1, col='royalblue')
lines(x=c(12,13), y=c(14000,14000), lwd=1.2, col='royalblue')
text(x=13, y=14000, labels='True', pos=4, cex=1.5)
vioplot(rnorm(1e4,14000,100), at=15.5, col='lightcoral',
rectCol=NA, lineCol=NA, border=NA, add=T)
text(x=16, y=14000, labels='Estimated', pos=4, cex=1.5)
```

We can also use the following code to plot the proportion of the abundance of red-eared slider at each pond and year. Using years 3 and 18 as examples, we can see the spatial variation of this proportion as well as how it changes over time (Fig. 3.11). In year 3 when the total abundance of red-eared slider reaches its apex, it is the majority species in most of the ponds. In year 18 when the total abundance of red-eared slider is relatively low and the total abundance of the Blanding's turtle is relatively high, we can see a substantial proportion of the Blanding's turtle in a large number of ponds. The estimates show excellent agreement with the simulated true values.

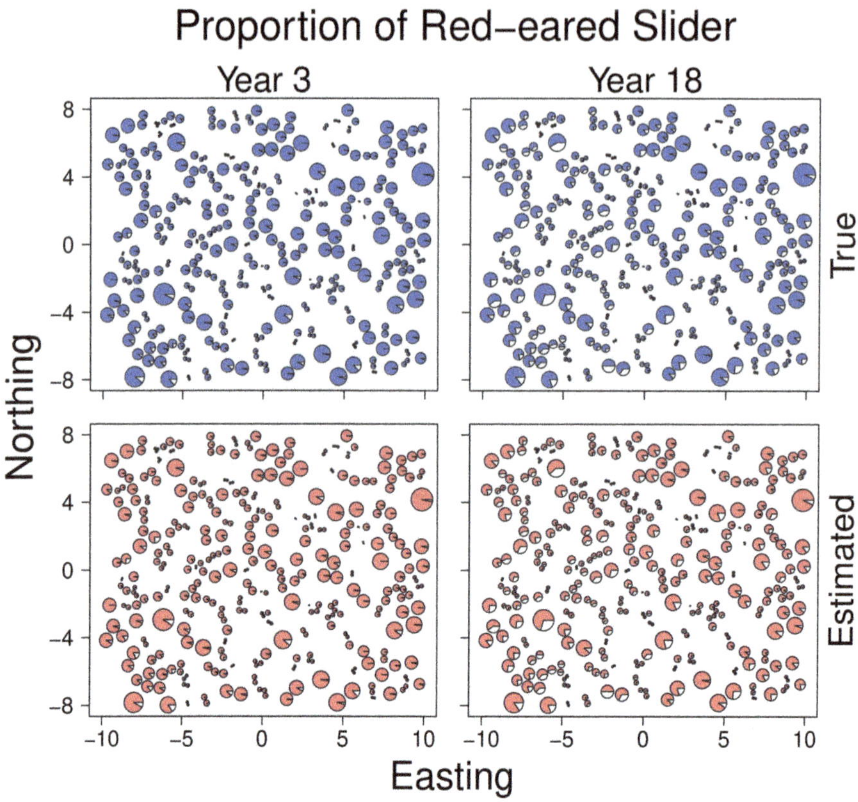

Fig. 3.11 The simulated true (blue) and estimated (red) proportion of the abundance of the invasive red-eared slider in each pond in years 3 and 18, against the proportion of the abundance of the native Blanding's turtle (white)

```
library(plotrix) # for making pie charts at selected positions

years_plot <- c(3, 18)
N_prop <- N[,years_plot,1] / apply(N[,years_plot,], 1:2, sum)
N_post_med <- apply(N_post, 1:3, median)
N_prop_post <- N_post_med[,years_plot,1] / apply(N_post_med[,
years_plot,], 1:2, sum)
size <- rowMeans(cov$size)

par(mfrow=c(2,2))
par(mar=c(1,1,1,1))
par(oma=c(4,5,5,2))

for (t in 1:2) {
 plot(1, xlim=c(-10, 10), ylim=c(-8, 8), type='n', axes=F, xlab='',
```

```
ylab='')
 for (i in 1:nsite) {
  floating.pie(xpos=cov$lon[i], ypos=cov$lat[i], x=c(N_prop[i,t],1-
N_prop[i,t]),
        radius=size[i]*0.9, col=c('royalblue','white'),
border='grey18')
 } # i
 axis(1, at=seq(-10,10,5), labels=rep('',5))
 if (t == 1) {
  axis(2, at=seq(-8,8,4), las=2, cex.axis=1.6)
 } else {
  axis(2, at=seq(-8,8,4), labels=rep('',5))
 }
 axis(3, at=0, labels=paste('Year', years_plot[t], sep=' '), cex.
axis=2.5, line=-.8, tick=F)
 if (t == 2) {
  axis(4, at=0, labels='True', cex.axis=2.5, line=0.4, tick=F)
 }
 box()
} # t

for (t in 1:2) {
 plot(1, xlim=c(-10, 10), ylim=c(-8, 8), type='n', axes=F, xlab='',
ylab='')
 for (i in 1:nsite) {
  floating.pie(xpos=cov$lon[i], ypos=cov$lat[i], x=c(N_prop_post[i,
t],1-N_prop_post[i,t]),
        radius=size[i]*0.9, col=c('lightcoral','white'),
border='grey18')
 } # i
 axis(1, at=seq(-10,10,5), cex.axis=1.6)
 if (t == 1) {
  axis(2, at=seq(-8,8,4), las=2, cex.axis=1.6)
 } else {
  axis(2, at=seq(-8,8,4), labels=rep('',5))
 }
 if (t == 2) {
  axis(4, at=0, labels='Estimated', cex.axis=2.5, line=0.4, tick=F)
 }
 box()
} # t

title(xlab='Easting' , line=2.6, cex.lab=2.8, outer=T)
title(ylab='Northing', line=2.4, cex.lab=2.8, outer=T)
```

```
title(main=expression("Proportion of Red-eared Slider"), line=3.2,
cex.main=3, outer=T)
```

3.6 Dail-Madsen Models

The dynamic N-mixture models we introduced so far describe temporal changes in abundance with population growth. The Dail and Madsen (2011) model, on the other hand, describe temporal changes in abundance with apparent survival (true survival minus emigration) and recruitment (reproduction plus immigration). While apparent survival can be considered the opposite of loss (death plus emigration), recruitment can also be called gain, providing a different perspective on population changes than population growth. These demographic parameters can then be expressed as functions of density and environmental covariates (Bellier et al. 2016). While the original Dail and Madsen (2011) model does not consider age structure, it was later extended to an age-structured, or generally speaking, multistate version (Zipkin et al. 2014). We will introduce these versions in detail below.

3.6.1 Without Age Structure

The Dail and Madsen (2011) model describes the changes in population size with apparent survival and recruitment, which are in turn functions of density and environmental covariates. Using our pond turtle as an example, the model basically says there will be more turtles that survive and stay in a pond if there are fewer turtles in that pond, the pond is larger, or water temperature is lower; meanwhile, there will also be more turtles that either are born or move into that pond if the above-mentioned conditions are satisfied. Using statistical language, the model first specifies the first-year abundance in the same way as previous dynamic N-mixture models:

$$\left.\begin{array}{c} N_{i,1} \sim Poisson(\lambda_{i,1}) \\ \lambda_{i,1} = exp\left(\beta_0^{[0]} + \beta_1^{[0]} \times x_{i,1,1} + \ldots + \beta_{n_{covs}}^{[0]} \times x_{i,1,n_{covs}}\right) \end{array}\right\}, \tag{3.14}$$

in which $N_{i,\ 1}$ is the first-year abundance of the study species at site i, $\lambda_{i,\ 1}$ is the expectation of the abundance and is a function of covariates x's governed by intercept and slope parameters $\beta^{[0]}$'s.

From the second year (i.e., $t \geq 2$), the process sub-model assumes that population size is due to apparent survival and recruitment from the previous year such that

$$\left.\begin{array}{r}N_{i,t} \sim Poisson(\lambda_{i,t}) \\ \lambda_{i,t} = N_{i,t-1} \times \phi_{i,t-1} + \gamma_{i,t-1} \\ \phi_{i,t-1} = logit^{-1}\left(\beta_0^{[\phi]} + \beta_{density}^{[\phi]} \times N_{i,t-1} + \beta_1^{[\phi]} \times x_{i,t,1} + \ldots + \beta_{n_{covs}}^{[\phi]} \times x_{i,t,n_{covs}}\right) \\ \gamma_{i,t-1} = exp\left(\beta_0^{[\gamma]} + \beta_{density}^{[\gamma]} \times N_{i,t-1} + \beta_1^{[\gamma]} \times x_{i,t,1} + \ldots + \beta_{n_{covs}}^{[\gamma]} \times x_{i,t,n_{covs}}\right)\end{array}\right\}, \tag{3.15}$$

in which $\phi_{i,\ t\ -\ 1}$ is the probability of apparent survival and is governed by density-dependent processes ($\beta_{density}^{[\phi]}$) and environmental conditions ($\beta_1^{[\phi]}, \ldots, \beta_{n_{covs}}^{[\phi]}$), and $\gamma_{i,\ t\ -\ 1}$ is the expected number of recruits and is also governed by density-dependent processes ($\beta_{density}^{[\gamma]}$) and the environment ($\beta_1^{[\gamma]}, \ldots, \beta_{n_{covs}}^{[\gamma]}$). Note that $\gamma_{i,\ t\ -\ 1}$ is not multiplied by $N_{i,\ t\ -\ 1}$; in this way, $\lambda_{i,\ t}$, and thus $N_{i,\ t}$, is allowed to be greater than 0 when $N_{i,\ t\ -\ 1}$ is 0, representing a potential of colonization through immigration after a site goes extinct. Note that the statistical form introduced here is a modified version of the original Dail-Madsen (2011) model as the latter specifies two latent state variables, the number of apparent survivors, $A_{i,\ t\ -\ 1} \sim Binomial(N_{i,\ t\ -\ 1}, \phi_{i,\ t\ -\ 1})$, and the number of new recruits, $G_{i,\ t\ -\ 1} \sim Poisson(\gamma_{i,\ t\ -\ 1})$. The population size in the next year is then calculated as the sum of these two latent state variables such that $N_{i,\ t} = A_{i,\ t\ -\ 1} + G_{i,\ t\ -\ 1}$. Our version achieves more efficient sampling by dropping these latent state variables.

Eventually, we have an observation sub-model that links the observed count data to the true abundance:

$$\left.\begin{array}{r}y_{i,t,j} \sim Binomial\left(N_{i,t}, p_{i,t,j}\right) \\ p_{i,t,j} = logit^{-1}\left(\alpha_0 + \alpha_1 \times w_{i,t,j,1} + \ldots + \alpha_{n_{pcvs}} \times w_{i,t,j,n_{pcvs}}\right)\end{array}\right\}, \tag{3.16}$$

in which detection probability $p_{i,\ t,\ j}$ is a function of observation covariates $w_{i,t,j,n_{pcvs}}$ with intercept and slope parameters α's.

Using the code below, we can simulate some data under the Dail-Madsen model. We assume positive effects of pond size on initial abundance ($\beta_{pond}^{[0]} = 1$), apparent survival ($\beta_{pond}^{[\phi]} = 0.4$), and recruitment ($\beta_{pond}^{[\gamma]} = 0.6$); negative effects of water temperature on initial abundance ($\beta_{temperature}^{[0]} = -0.8$), apparent survival ($\beta_{temperature}^{[\phi]} = -0.2$), and recruitment ($\beta_{temperature}^{[\gamma]} = -0.3$); and negative density dependence on apparent survival ($\beta_{density}^{[\phi]} = -0.4$) and recruitment ($\beta_{density}^{[\gamma]} = -0.6$).

```
library(boot) # for using logit and inv.logit functions

set.seed(6)

# Basic values
nsite <- 360 # number of sites
nyear <- 20  # number of years
```

```
nreps <- 5  # number of within-season replicates
ncovs <- 2  # number of environmental covariates
npcvs <- 3  # number of observational covariates

beta0 <- c(1.8, 1, -0.8)        # intercept and slopes for the initial
abundance
beta_phi  <- c(0.8, -0.4, 0.4, -0.2) # intercept and slopes for apparent
survival
beta_gamma <- c(0.4, -0.6, 0.6, -0.3) # intercept and slopes for gain
(recruitment)
alpha <- c(0.3, -0.6, 0.3, -0.3)    # intercept and slopes for detection

# Read in environmental covariates
load('data/covariates.RData')
x1 <- (log(cov$size) - mean(log(cov$size))) / sd(log(cov$size)) #
standardized log habitat patch size
x2 <- (cov$temp - mean(cov$temp)) / sd(cov$temp) # standardized
temperature
x <- array(, dim=c(nsite, nyear, ncovs)) # environmental covariates
x[,,1] <- x1
x[,,2] <- x2
### Standardize covariates for each site
for (i in 1:nsite) {
 for (h in 1:ncovs) {
  x[i,2:nyear,h] <- (x[i,2:nyear,h] - mean(x[i,,h])) / sd(x[i,,h])
 } # h
} # i

# Simulate data
lambda <- matrix(, nsite, nyear) # expectation of abundance
N <- matrix(, nsite, nyear) # abundance
lambda[,1] <- exp(cbind(1,x[,1,]) %*% beta0) # expectation of initial
abundance
N[,1] <- rpois(nsite, lambda[,1]) # initial abundance

phi  <- matrix(, nsite, nyear-1) # apparent survival
gamma <- matrix(, nsite, nyear-1) # recrutiment
for (t in 2:nyear) {
 phi [,t-1] <- inv.logit(cbind(1, (N[,t-1]-lambda[,1])/lambda[,1], x
[,t,]) %*% beta_phi )
 gamma[,t-1] <- exp   (cbind(1, (N[,t-1]-lambda[,1])/lambda[,1], x[,
t,]) %*% beta_gamma)
 lambda[,t] <- N[,t-1] * phi[,t-1] + gamma[,t-1]
 N[,t] <- rpois(nsite, lambda[,t])
} # t
```

```
w <- array(rnorm(nsite * nyear * nreps * npcvs, 0, 1), dim=c(nsite,
nyear, nreps, npcvs)) # observational covariates

p <- array(, dim=c(nsite, nyear, nreps)) # detection probability
for (i in 1:nsite) {
 for (t in 1:nyear) {
  p[i,t,] <- inv.logit(cbind(1,w[i,t,,]) %*% alpha)
 } # t
} # i

y <- array(, dim=c(nsite, nyear, nreps)) # count data
for (i in 1:nsite) {
 for (t in 1:nyear) {
  y[i,t,] <- rbinom(nreps, N[i,t], p[i,t,])
 } # t
} # i
```

We then can use the following code to develop a custom MCMC algorithm to estimate $\beta^{[0]}$, $\beta^{[\phi]}$, $\beta^{[\gamma]}$, and α using data y, x, and w, using the code below:

```
Dail_Madsen_mcmc <- function(y, x, w, nmcmc) {

 # Setup variables
 nsite <- dim(y)[1]
 nyear <- dim(y)[2]
 nreps <- dim(y)[3]
 ncovs <- dim(x)[3]
 npcvs <- dim(w)[4]
 ymax <- apply(y, 1:2, max)

 beta0_save <- matrix(0, ncovs+1, nmcmc)
 beta_phi_save  <- matrix(0, ncovs+2, nmcmc)
 beta_gamma_save <- matrix(0, ncovs+2, nmcmc)
 alpha_save <- matrix(0, npcvs+1, nmcmc)
 N_save <- matrix(0, nyear, nmcmc)
 phi_save  <- matrix(0, nyear-1, nmcmc)
 gamma_save <- matrix(0, nyear-1, nmcmc)

 # Priors
 beta0_mean <- rep(0, ncovs+1)
 beta0_sd <- 2
 beta_phi_mean <- rep(0, ncovs+2)
 beta_phi_sd <- 2
 beta_gamma_mean <- rep(0, ncovs+2)
 beta_gamma_sd <- 2
```

```
alpha_mean <- rep(0, npcvs+1)
alpha_sd <- 2

# Starting values
beta0 <- rep(0, ncovs+1)
beta_phi  <- rep(0, ncovs+2)
beta_gamma <- rep(0, ncovs+2)
alpha <- rep(0, npcvs+1)
lambda0 <- rep(1, nsite) # expectation of initial abundance
phi  <- matrix(0.5, nsite, nyear-1)
gamma <- matrix(1 , nsite, nyear-1)
N <- round((ymax + 1) / 0.5)
p <- array(0.5, dim=c(nsite, nyear, nreps))

# Tuning factors
beta0_tune <- 0.05
beta_phi_tune  <- c(0.15, 0.10, 0.10, 0.10)
beta_gamma_tune <- c(0.08, 0.05, 0.05, 0.05)
alpha_tune <- 0.01
N_tune <- 1

# Begin MCMC loop
for (k in 1:nmcmc) {
 ### Sample N
 N_star <- matrix(rpois(nsite*nyear, N + N_tune), nsite, nyear)
 mh1 <- apply(dbinom(y, N_star, p, log=T), 1:2, sum) +
     dpois(N_star, cbind(lambda0, N[,-nyear] * phi + gamma), log=T) +
     dpois(N, N_star + N_tune, log=T)
 mh2 <- apply(dbinom(y, N    , p, log=T), 1:2, sum) +
     dpois(N    , cbind(lambda0, N[,-nyear] * phi + gamma), log=T) +
     dpois(N_star, N + N_tune, log=T)
 mhN <- exp(mh1 - mh2)
 Nkeep <- ((mhN > runif(nsite*nyear)) & (N_star >= ymax))
 N[Nkeep] <- N_star[Nkeep]

 ### Sample beta0
 beta0_star <- rnorm(ncovs+1, beta0, beta0_tune)
 lambda0_star <- exp(cbind(1,x[,1,]) %*% beta0_star)
 mh1 <- sum(dpois(N[,1], lambda0_star, log=T)) +
     sum(dnorm(beta0_star, beta0_mean, beta0_sd, log=T))
 mh2 <- sum(dpois(N[,1], lambda0    , log=T)) +
     sum(dnorm(beta0    , beta0_mean, beta0_sd, log=T))
 mh <- exp(mh1 - mh2)
 if (mh > runif(1)) {
  beta0 <- beta0_star
```

```
    lambda0 <- lambda0_star
  }

  ### Sample beta_phi
  beta_phi_star <- rnorm(ncovs+2, beta_phi, beta_phi_tune)
  phi <- phi_star <- matrix(, nsite, nyear-1)
  for (t in 2:nyear) {
    phi     [,t-1] <- inv.logit(cbind(1,(N[,t-1]-lambda0)/lambda0,x[,
t,]) %*% beta_phi     )
    phi_star[,t-1] <- inv.logit(cbind(1,(N[,t-1]-lambda0)/lambda0,x[,
t,]) %*% beta_phi_star)
  } # t
  mh1 <- sum(dpois(N[,-1], N[,-nyear] * phi_star + gamma, log=T)) +
      sum(dnorm(beta_phi_star, beta_phi_mean, beta_phi_sd, log=T))
  mh2 <- sum(dpois(N[,-1], N[,-nyear] * phi      + gamma, log=T)) +
      sum(dnorm(beta_phi     , beta_phi_mean, beta_phi_sd, log=T))
  mh <- exp(mh1 - mh2)
  if (mh > runif(1)) {
    beta_phi <- beta_phi_star
  }

  ### Sample beta_gamma
  beta_gamma_star <- rnorm(ncovs+2, beta_gamma, beta_gamma_tune)
  gamma <- gamma_star <- matrix(, nsite, nyear-1)
  for (t in 2:nyear) {
    gamma     [,t-1] <- exp(cbind(1,(N[,t-1]-lambda0)/lambda0,x[,t,]) %*%
beta_gamma     )
    gamma_star[,t-1] <- exp(cbind(1,(N[,t-1]-lambda0)/lambda0,x[,t,])
%*% beta_gamma_star)
  } # t
  mh1 <- sum(dpois(N[,-1], N[,-nyear] * phi + gamma_star, log=T)) +
      sum(dnorm(beta_gamma_star, beta_gamma_mean, beta_gamma_sd,
log=T))
  mh2 <- sum(dpois(N[,-1], N[,-nyear] * phi + gamma     , log=T)) +
     sum(dnorm(beta_gamma     , beta_gamma_mean, beta_gamma_sd, log=T))
  mh <- exp(mh1 - mh2)
  if (mh > runif(1)) {
    beta_gamma <- beta_gamma_star
  }

  ### Sample alpha
  alpha_star <- rnorm(npcvs+1, alpha, alpha_tune)
  p_star <- array(, dim=c(nsite, nyear, nreps))
  for (i in 1:nsite) {
    for (t in 1:nyear) {
      p_star[i,t,] <- inv.logit(cbind(1,w[i,t,,]) %*% alpha_star)
```

```
    } # t
   } # i
   mh1 <- sum(dbinom(y, N, p_star, log=TRUE), na.rm=T) +
       sum(dnorm(alpha_star, alpha_mean, alpha_sd, log=TRUE))
   mh2 <- sum(dbinom(y, N, p    , log=TRUE), na.rm=T) +
       sum(dnorm(alpha    , alpha_mean, alpha_sd, log=TRUE))
   mh <- exp(mh1 - mh2)
   if (mh > runif(1)) {
    alpha <- alpha_star
    p <- p_star
   }

   ### Save samples
   beta0_save[,k] <- beta0
   beta_phi_save[,k] <- beta_phi
   beta_gamma_save[,k] <- beta_gamma
   alpha_save[,k] <- alpha
   N_save[,k] <- colMeans(N)
   phi_save[,k] <- colMeans(phi)
   gamma_save[,k] <- colMeans(gamma)
  } # k

  # Write output
  list(beta0_save=beta0_save,
    beta_phi_save=beta_phi_save, beta_gamma_save=beta_gamma_save,
    alpha_save=alpha_save,
    N_save=N_save, phi_save=phi_save, gamma_save=gamma_save)

} # Dail_Madsen_mcmc
```

Now, let us use the following code to implement the MCMC algorithm, which takes about 1.6 h to run. We then use the code in Appendix 3.6 to create some traceplots; we see good model convergence and agreement between parameter estimates of $\beta^{[0]}$, $\beta^{[\phi]}$, $\beta^{[\gamma]}$, and α and their true values used in data simulation (Fig. 3.12).

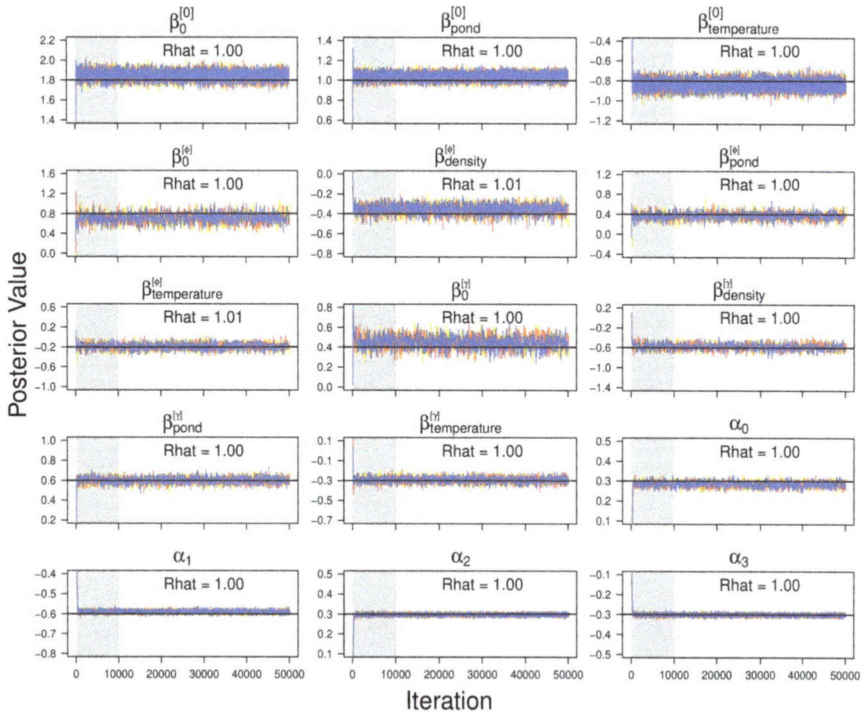

Fig. 3.12 Traceplots showing posterior chains for $\beta^{[0]}$, $\beta^{[\phi]}$, $\beta^{[\gamma]}$, and α in the Dail-Madsen model. Rhat values are calculated after discarding the first 10,000 iterations. Horizontal lines represent the true values used in data simulation

```
library(foreach)  # for parallel computing
library(doParallel) # for parallel computing

numCores <- round(detectCores() / 2) # only use half of the cores for
parallel computing
registerDoParallel(numCores) # setup parallel computing

nmcmc <- 50000 # number of iterations of MCMC computing
chain <- 3    # number of chains

start_time <- Sys.time() # start time of computing
out <- foreach (i=1:chain, .packages='boot') %dopar% {
 Dail_Madsen_mcmc(y=y, x=x, w=w, nmcmc=nmcmc)
} # i
stopImplicitCluster() # clean up the cluster
end_time <- Sys.time() # end time of computing
time_taken <- end_time - start_time # computing time
```

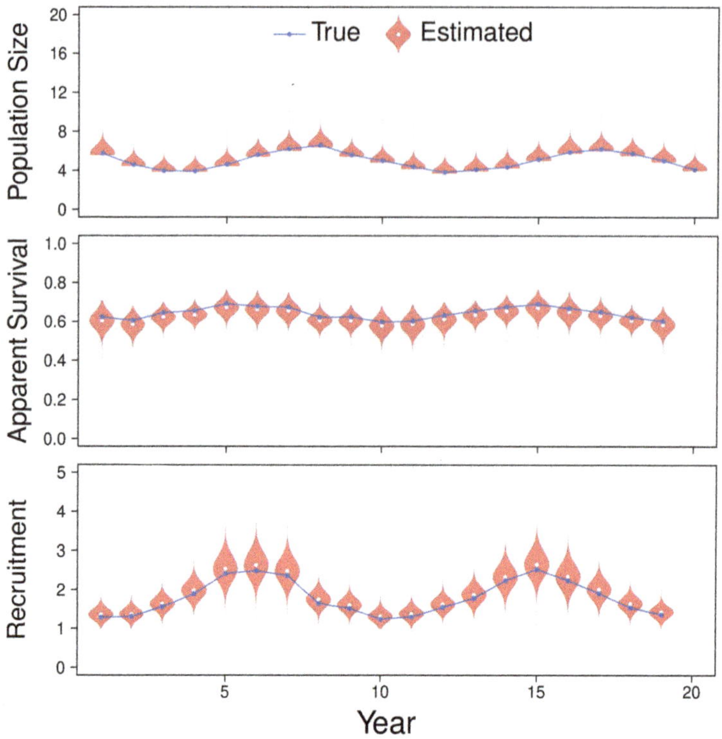

Fig. 3.13 Violin plots showing the consistency between the estimated mean population size, apparent survival, and recruitment (violins) and the simulated true values (blue points)

We can also plot the annual means of abundance, apparent survival, and recruitment across all sites, using the code below. We can see that the estimated values adequately capture the temporal variation of these state variables (Fig. 3.13).

```
library(vioplot) # for making violin plots

for (i in 1:chain) {
 if (i == 1) {
  N_mean <- out[[1]]$N_save
  phi_mean <- out[[1]]$phi_save
  gamma_mean <- out[[1]]$gamma_save
 } else {
  N_mean <- cbind(N_mean, out[[1]]$N_save)
  phi_mean <- cbind(phi_mean, out[[1]]$phi_save)
  gamma_mean <- cbind(gamma_mean, out[[1]]$gamma_save)
 }
} # i
```

```
par(mfrow=c(3,1))
par(mar=c(1.5,1.5,0,0))
par(oma=c(4,5,1,1))

plot(1, xlim=c(1,nyear), ylim=c(0,20), type='l', axes=F, xlab='',
ylab='')
vioplot(t(N_mean), add=T, col='lightcoral', rectCol=NA, lineCol=NA,
border=NA)
lines(colMeans(N), type='o', pch=16, col='royalblue')
axis(1, at=seq(0,nyear,5), labels=rep('',5))
axis(2, at=seq(0,20,4), cex.axis=1.6, las=2)
axis(2, at=10, labels='Population Size', cex.axis=2.5, line=3.2,
tick=F)

points(x=7, y=18, pch=16, cex=1, col='royalblue')
lines(x=c(6.5,7.5), y=c(18,18), lwd=1.2, col='royalblue')
vioplot(rnorm(10000,18,0.5), at=10.5, add=T, col='lightcoral',
rectCol=NA, lineCol=NA, border=NA)
text(x=7.5, y=18, labels='True', pos=4, cex=2.5)
text(x=11, y=18, labels='Estimated', pos=4, cex=2.5)

plot(1, xlim=c(1,nyear), ylim=c(0,1), type='l', axes=F, xlab='',
ylab='')
vioplot(t(phi_mean), add=T, col='lightcoral', rectCol=NA,
lineCol=NA, border=NA)
lines(colMeans(phi), type='o', pch=16, col='royalblue')
axis(1, at=seq(0,nyear,5), labels=rep('',5))
axis(2, at=seq(0,1,0.2), cex.axis=1.6, las=2)
axis(2, at=0.5, labels='Apparent Survival', cex.axis=2.5, line=3.2,
tick=F)

plot(1, xlim=c(1,nyear), ylim=c(0,5), type='l', axes=F, xlab='',
ylab='')
vioplot(t(gamma_mean), add=T, col='lightcoral', rectCol=NA,
lineCol=NA, border=NA)
lines(colMeans(gamma), type='o', pch=16, col='royalblue')
axis(1, at=seq(0,nyear,5), cex.axis=1.6)
axis(2, at=seq(0,5,1), cex.axis=1.6, las=2)
axis(2, at=2.5, labels='Recruitment', cex.axis=2.5, line=3.2,
tick=F)

title(xlab='Year', cex.lab=3, line=2.4, outer=T)
```

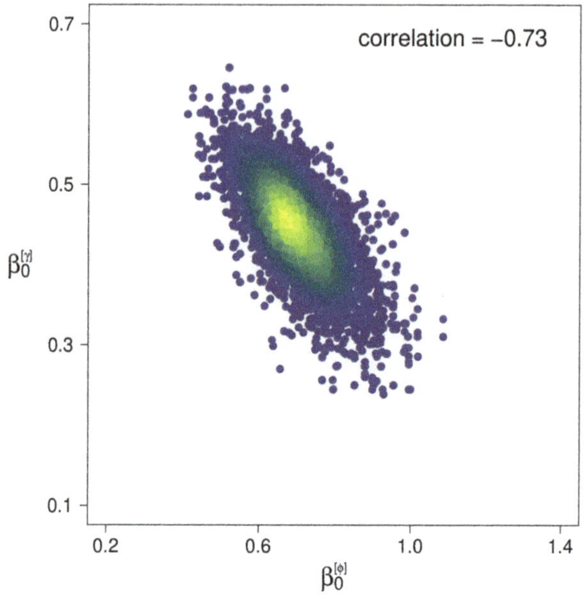

Fig. 3.14 A scatterplot showing the negative correlation between two parameters, $\beta_0^{[\phi]}$ and $\beta_0^{[\gamma]}$, in the Dail-Madsen model. Each point represents a pair of draws of the two parameters during one iteration of the MCMC algorithm

Also note that, even though the model converges well, it is relatively difficult to identify some parameters, in particular $\beta_0^{[\phi]}$ (mean apparent survival probability on the logit scale) and $\beta_0^{[\gamma]}$ (mean expected number of recruits on the log scale). Using the code below, we can see a strong negative correlation (−0.73) between these parameters (Fig. 3.14). Such a challenge in parameter identification is likely the reason to constrain further extension of the model into a spatial version (Zhao et al. 2017).

```
for (i in 1:chain) {
 if (i == 1) {
  beta_phi0_post  <- out[[i]]$beta_phi  [1,(nmcmc/2+1):nmcmc]
  beta_gamma0_post <- out[[i]]$beta_gamma[1,(nmcmc/2+1):nmcmc]
 } else {
  beta_phi0_post  <- c(beta_phi0_post , out[[i]]$beta_phi  [1,(nmcmc/
2+1):nmcmc])
  beta_gamma0_post <- c(beta_gamma0_post, out[[i]]$beta_gamma[1,
(nmcmc/2+1):nmcmc])
 }
} # i

nplot <- 10000
sel <- sort(sample(1:(nmcmc/2*chain), nplot, replace=F))
a <- beta_phi0_post[sel]
b <- beta_gamma0_post[sel]
a2 <- (a - min(a)) / (max(a) - min(a))
```

```
b2 <- (b - min(b)) / (max(b) - min(b))
ab <- matrix(, nplot, nplot)
for (i in 1:nplot) {
 for (j in 1:nplot) {
  ab[i,j] <- sqrt((a2[i] - a2[j]) ^ 2 + (b2[i] - b2[j]) ^ 2)
 } # j
} # i

ab01 <- ifelse(ab < 0.08, 1, 0)
ab_scale <- rowSums(ab01) - min(rowSums(ab01)) + 1
col2 <- c(colorRampPalette(c('navy','seagreen'))(floor(max(ab_scale)
*0.6)),
    colorRampPalette(c('seagreen','yellow'))(ceiling(max(ab_scale)
*0.4)))

par(mfrow=c(1,1))
par(mar=c(5,6,1,1))

plot(a, b, col=col2[ab_scale], pch=16, cex=1.4, xlim=c(0.2,1.4),
ylim=c(0.1,0.7), axes=F, xlab='', ylab='')
axis(1, at=seq(0.2,1.6,0.4), cex.axis=1.5)
axis(2, at=seq(0.1,0.8,0.2), cex.axis=1.5, las=2)
axis(1, at=0.8, labels=ylab_beta_phi [1], cex.axis=2, line=2.5,
tick=F)
axis(2, at=0.4, labels=ylab_beta_gamma[1], cex.axis=2, line=2.8,
las=2, tick=F)
box()

text(x=0.84, y=0.68, labels=paste('correlation = ', round(cor(a, b),
digits=2), sep=''), pos=4, cex=1.8)
```

3.6.2 With Age Structure

When age-structured count data are available, we can extend the Dail and Madsen (2011) model to account for age structure, in particular to accommodate transitions between age classes, in a population (Zipkin et al. 2014). As a matter of fact, the multistate versions of the Dail and Madsen (2011) model can be applied to any population structure (e.g., disease structure; DiRenzo et al. 2018, 2019), as long as appropriate data are available. Using our pond turtle as an example, the age-structured Dail-Madsen model assumes there are adult turtles that lay eggs and juvenile turtles that do not. There is a chance that juvenile turtles will become mature and start laying eggs in a year, after which they will continue doing so. The number of adults in a pond is given by the number of adults from the previous year that survive until this year and stay in the same pond, plus the number of juveniles that survive and stay from the previous year and start laying eggs this year. The number of juveniles in a pond is given by the number of juveniles that survive and stay from the previous year but have not yet become mature to lay eggs, as well as turtles that are newly recruited. As for new recruits, they include newly hatched turtles plus turtles that move in (which have to be juveniles according to our assumption), but we do not further distinguish them as we do not consider movement in this model. In this model, the chances of surviving, recruiting, and starting laying eggs are all affected by the number of turtles in a pond (adults and juveniles combined), the size of the pond, and water temperature. Statistically speaking, the age-structured Dail-Madsen model first specifies initial abundance in a way that is not that different from the version without age structure, except that we have separate abundance and parameters for adult and juvenile animals now:

$$\left.\begin{array}{r}N_{i,1,adu} \sim Poisson(\lambda_{i,1,adu}) \\ N_{i,1,juv} \sim Poisson(\lambda_{i,1,juv}) \\ \lambda_{i,1,adu} = exp\left(\beta^{[0]}_{0,adu} + \beta^{[0]}_{1,adu} \times x_{i,1,1} + \ldots + \beta^{[0]}_{n_{covs},adu} \times x_{i,1,n_{covs}}\right) \\ \lambda_{i,1,juv} = exp\left(\beta^{[0]}_{0,juv} + \beta^{[0]}_{1,juv} \times x_{i,1,1} + \ldots + \beta^{[0]}_{n_{covs},juv} \times x_{i,1,n_{covs}}\right)\end{array}\right\}. \tag{3.17}$$

Here, $N_{i,\ 1,\ adu}$ is the first-year abundance for adult animals at site i, $N_{i,\ 1,\ juv}$ is the first-year abundance for juvenile animals, $\lambda_{i,\ 1,\ adu}$ and $\lambda_{i,\ 1,\ juv}$ are expectations of the first-year adult and juvenile abundance, respectively, and are functions of covariates x's governed by age-class-specific intercept and slope parameters $\beta^{[0]}$'s.

From the second year (i.e., $t \geq 2$), the process sub-model assumes that adult abundance is due to the apparent survival of adult animals from the previous year as well as transition of juveniles to adults; juvenile abundance, on the other hand, is due to newly recruited juveniles as well as juveniles from the previous year that haven't transitioned to the adult stage. Using statistical language, we have

$$\left.\begin{array}{r}N_{i,t,adu} \sim Poisson(\lambda_{i,t,adu}) \\ N_{i,t,juv} \sim Poisson(\lambda_{i,t,juv}) \\ \lambda_{i,t,adu} = N_{i,t-1,adu} \times \phi_{i,t-1,adu} + N_{i,t-1,juv} \times \phi_{i,t-1,juv} \times \omega_{i,t-1} \\ \lambda_{i,t,juv} = N_{i,t-1,juv} \times \phi_{i,t-1,juv} \times (1 - \omega_{i,t-1}) + \gamma_{i,t-1} \\ \phi_{i,t-1,adu} = logit^{-1}\left[\beta^{[\phi]}_{0,adu} + \beta^{[\phi]}_{density,adu} \times (N_{i,t-1,adu} + N_{i,t-1,juv}) + \beta^{[\phi]}_{1,adu} \times x_{i,t,1} + \ldots\right] \\ \phi_{i,t-1,juv} = logit^{-1}\left[\beta^{[\phi]}_{0,juv} + \beta^{[\phi]}_{density,juv} \times (N_{i,t-1,adu} + N_{i,t-1,juv}) + \beta^{[\phi]}_{1,juv} \times x_{i,t,1} + \ldots\right] \\ \omega_{i,t-1} = logit^{-1}\left[\beta^{[\omega]}_{0} + \beta^{[\omega]}_{density} \times (N_{i,t-1,adu} + N_{i,t-1,juv}) + \beta^{[\omega]}_{1} \times x_{i,t,1} + \ldots\right] \\ \gamma_{i,t-1} = exp\left[\beta^{[\gamma]}_{0} + \beta^{[\gamma]}_{density} \times (N_{i,t-1,adu} + N_{i,t-1,juv}) + \beta^{[\gamma]}_{1} \times x_{i,t,1} + \ldots\right]\end{array}\right\}, \tag{3.18}$$

in which $\phi_{i,\ t\ -\ 1,\ adu}$ is the probability of adult apparent survival, $\phi_{i,\ t\ -\ 1,\ juv}$ is the probability of juvenile apparent survival, $\omega_{i,\ t\ -\ 1}$ is the probability of transitioning from juvenile to adult, and $\gamma_{i,\ t\ -\ 1}$ is the expected number of recruits, all are driven by density-dependent processes ($\beta^{[\phi]}_{density,adu}$, $\beta^{[\phi]}_{density,juv}$, $\beta^{[\omega]}_{density}$, $\beta^{[\gamma]}_{density}$) and the environment ($\beta_{1,adu}^{[\phi]}, \ldots, \beta_{n_{covs},adu}^{[\phi]}, \beta_{1,juv}^{[\phi]}, \ldots, \beta_{n_{covs},juv}^{[\phi]}, \beta_{1}^{[\omega]}, \ldots, \beta_{n_{covs}}^{[\omega]}, \beta_{1}^{[\gamma]}, \ldots, \beta_{n_{covs}}^{[\gamma]}$). Note that this model is slightly different from the formulation specified by Zipkin et al. (2014): we again do not include any state variables (e.g., number of adult apparent survivors) in order to improve sampling efficiency.

Last, we have an observation sub-model that links the age-specific count data to the true abundance:

$$\left.\begin{array}{r} y_{i,t,j,adu} \sim Binomial\left(N_{i,t,adu}, p_{i,t,j}\right) \\ y_{i,t,j,juv} \sim Binomial\left(N_{i,t,juv}, p_{i,t,j}\right) \\ p_{i,t,j} = logit^{-1}\left(\alpha_0 + \alpha_1 \times w_{i,t,j,1} + \ldots + \alpha_{n_{pcvs}} \times w_{i,t,j,n_{pcvs}}\right) \end{array}\right\}, \quad (3.19)$$

in which detection probability $p_{i,\ t,\ j}$ is a function of observation covariates $w_{i,t,j,n_{pcvs}}$ with intercept and slope parameters α's. Here we assume the same detection $p_{i,\ t,\ j}$ for adults and juveniles, which can be further extended to allow age-dependent detection (i.e., $p_{i,\ t,\ j,\ adu}$ and $p_{i,\ t,\ j,\ juv}$).

Using the code below, we can simulate some data under the age-structured Dail-Madsen model. We assume positive effects of pond size on initial adult abundance ($\beta_{pond,adu}^{[0]} = 0.5$), initial juvenile abundance ($\beta_{pond,juv}^{[0]} = 1$), adult apparent survival ($\beta_{pond,adu}^{[\phi]} = 0.4$), juvenile apparent survival ($\beta_{pond,juv}^{[\phi]} = 0.6$), transition ($\beta_{pond}^{[\omega]} = 0.5$), and recruitment ($\beta_{pond}^{[\gamma]} = 0.6$); negative effects of temperature on initial adult abundance ($\beta_{temperature,adu}^{[0]} = -0.4$), initial juvenile abundance ($\beta_{temperature,juv}^{[0]} = -0.8$), adult apparent survival ($\beta_{temperature,adu}^{[\phi]} = -0.2$), juvenile apparent survival ($\beta_{temperature,juv}^{[\phi]} = -0.3$), transition ($\beta_{temperature}^{[\omega]} = -0.3$), and recruitment ($\beta_{temperature}^{[\gamma]} = -0.3$); and negative density dependence on adult apparent survival ($\beta_{density,adu}^{[\phi]} = -0.3$), juvenile apparent survival ($\beta_{density,juv}^{[\phi]} = -0.5$), transition ($\beta_{density}^{[\omega]} = -0.1$), and recruitment ($\beta_{density}^{[\gamma]} = -0.6$).

```
library(boot) # for using logit and inv.logit functions

set.seed(6)

# Basic values
nsite <- 360 # number of sites
nyear <- 20 # number of years
nreps <- 5  # number of within-season replicates
ncovs <- 2  # number of environmental covariates
npcvs <- 1  # number of observational covariates

beta0_adu <- c(1.8, 0.5, -0.4)      # intercept and slopes for adult
initial abundance
beta0_juv <- c(1.4,  1, -0.8)      # intercept and slopes for juvenile
initial abundance
beta_phi_adu <- c(0.8, -0.3, 0.4, -0.2) # intercept and slopes for adult
apparent survival
```

```
beta_phi_juv <- c(0.4, -0.5, 0.6, -0.3) # intercept and slopes for
juvenile apparent survival
beta_omega  <- c(0.5, -0.1, 0.5, -0.3) # intercept and slopes for
transition probability from juvenile to adult
beta_gamma  <- c(0.4, -0.6, 0.6, -0.3) # intercept and slopes for gain
(recruitment)
alpha <- c(0.3, -0.6)            # intercept and slopes for detection

# Read in environmental covariates
load('data/covariates.RData')
x1 <- (log(cov$size) - mean(log(cov$size))) / sd(log(cov$size)) #
standardized log food patch size
x2 <- (cov$temp - mean(cov$temp)) / sd(cov$temp) # standardized
temperature
x <- array(, dim=c(nsite, nyear, ncovs)) # environmental covariates
x[,,1] <- x1
x[,,2] <- x2
### Standardize covariates for each site
for (i in 1:nsite) {
 for (h in 1:ncovs) {
  x[i,2:nyear,h] <- (x[i,2:nyear,h] - mean(x[i,,h])) / sd(x[i,,h])
 } # h
} # i

# Simulate data
lambda_adu <- matrix(, nsite, nyear) # expectation of adult abundance
lambda_juv <- matrix(, nsite, nyear) # expectation of juvenile
abundance
N_adu <- matrix(, nsite, nyear) # adult abundance
N_juv <- matrix(, nsite, nyear) # juvenile abundance
N <- matrix(, nsite, nyear) # total abundance

lambda_adu[,1] <- exp(cbind(1,x[,1,]) %*% beta0_adu) # expectation of
adult initial abundance
lambda_juv[,1] <- exp(cbind(1,x[,1,]) %*% beta0_juv) # expectation of
juvenile initial abundance
lambda0 <- lambda_adu[,1] + lambda_juv[,1] # expectation of total
initial abundance
N_adu[,1] <- rpois(nsite, lambda_adu[,1]) # adult initial abundance
N_juv[,1] <- rpois(nsite, lambda_juv[,1]) # juvenile initial abundance
N[,1] <- N_adu[,1] + N_juv[,1] # total initial abundance

phi_adu <- matrix(, nsite, nyear-1) # adult apparent survival
phi_juv <- matrix(, nsite, nyear-1) # juvenile apparent survival
omega  <- matrix(, nsite, nyear-1) # transition from juvenile to adult
gamma  <- matrix(, nsite, nyear-1) # recrutiment
```

```
for (t in 2:nyear) {
 phi_adu[,t-1] <- inv.logit(cbind(1, (N[,t-1]-lambda0)/lambda0, x[,
t,]) %*% beta_phi_adu)
 phi_juv[,t-1] <- inv.logit(cbind(1, (N[,t-1]-lambda0)/lambda0, x[,
t,]) %*% beta_phi_juv)
 omega[,t-1] <- inv.logit(cbind(1, (N[,t-1]-lambda0)/lambda0, x[,
t,]) %*% beta_omega)
 gamma[,t-1] <- exp   (cbind(1, (N[,t-1]-lambda0)/lambda0, x[,t,]) %*%
beta_gamma)
 lambda_adu[,t] <- N_adu[,t-1] * phi_adu[,t-1] + N_juv[,t-1] * phi_juv
[,t-1] * omega[,t-1]
 lambda_juv[,t] <- N_juv[,t-1] * phi_juv[,t-1] * (1 - omega[,t-1]) +
gamma[,t-1]
 N_adu[,t] <- rpois(nsite, lambda_adu[,t])
 N_juv[,t] <- rpois(nsite, lambda_juv[,t])
 N[,t] <- N_adu[,t] + N_juv[,t]
} # t

w <- array(rnorm(nsite * nyear * nreps * npcvs, 0, 1), dim=c(nsite,
nyear, nreps, npcvs)) # observational covariates

p <- array(, dim=c(nsite, nyear, nreps)) # detection probability
for (i in 1:nsite) {
 for (t in 1:nyear) {
  p[i,t,] <- inv.logit(cbind(1,w[i,t,,]) %*% alpha)
 } # t
} # i

y_adu <- array(, dim=c(nsite, nyear, nreps)) # count data of adult
y_juv <- array(, dim=c(nsite, nyear, nreps)) # count data of juvenile
for (i in 1:nsite) {
 for (t in 1:nyear) {
  y_adu[i,t,] <- rbinom(nreps, N_adu[i,t], p[i,t,])
  y_juv[i,t,] <- rbinom(nreps, N_juv[i,t], p[i,t,])
 } # t
} # i
```

We then can use the following code to develop a custom MCMC algorithm to estimate $\beta^{[0]}$, $\beta^{[\phi]}$, $\beta^{[\omega]}$, $\beta^{[\gamma]}$, and α using data y, x, and w, using the code below:

```
age_Dail_Madsen_mcmc <- function(y_adu, y_juv, x, w, nmcmc) {

 # Setup variables
 nsite <- dim(y_adu)[1]
 nyear <- dim(y_adu)[2]
 nreps <- dim(y_adu)[3]
```

```
ncovs <- dim(x)[3]
npcvs <- dim(w)[4]
y_adu_max <- apply(y_adu, 1:2, max)
y_juv_max <- apply(y_juv, 1:2, max)

beta0_adu_save <- matrix(0, ncovs+1, nmcmc)
beta0_juv_save <- matrix(0, ncovs+1, nmcmc)
beta_phi_adu_save <- matrix(0, ncovs+2, nmcmc)
beta_phi_juv_save <- matrix(0, ncovs+2, nmcmc)
beta_omega_save  <- matrix(0, ncovs+2, nmcmc)
beta_gamma_save  <- matrix(0, ncovs+2, nmcmc)
alpha_save <- matrix(0, npcvs+1, nmcmc)
N_adu_save <- matrix(0, nyear, nmcmc)
N_juv_save <- matrix(0, nyear, nmcmc)
phi_adu_save <- matrix(0, nyear-1, nmcmc)
phi_juv_save <- matrix(0, nyear-1, nmcmc)
omega_save  <- matrix(0, nyear-1, nmcmc)
gamma_save  <- matrix(0, nyear-1, nmcmc)

# Priors
beta0_adu_mean <- rep(0, ncovs+1)
beta0_adu_sd <- 2
beta0_juv_mean <- rep(0, ncovs+1)
beta0_juv_sd <- 2
beta_phi_adu_mean <- rep(0, ncovs+2)
beta_phi_adu_sd <- 2
beta_phi_juv_mean <- rep(0, ncovs+2)
beta_phi_juv_sd <- 2
beta_omega_mean <- rep(0, ncovs+2)
beta_omega_sd <- 2
beta_gamma_mean <- rep(0, ncovs+2)
beta_gamma_sd <- 2
alpha_mean <- rep(0, npcvs+1)
alpha_sd <- 2

# Starting values
beta0_adu <- rep(0, ncovs+1)
beta0_juv <- rep(0, ncovs+1)
beta_phi_adu <- rep(0, ncovs+2)
beta_phi_juv <- rep(0, ncovs+2)
beta_omega <- rep(0, ncovs+2)
beta_gamma <- rep(0, ncovs+2)
alpha <- rep(0, npcvs+1)
lambda0_adu <- exp(cbind(1,x[,1,]) %*% beta0_adu)
lambda0_juv <- exp(cbind(1,x[,1,]) %*% beta0_juv)
lambda0 <- lambda0_adu + lambda0_juv
```

```
phi_adu <- matrix(0.5, nsite, nyear-1)
phi_juv <- matrix(0.5, nsite, nyear-1)
omega <- matrix(0.5, nsite, nyear-1)
gamma <- matrix(1, nsite, nyear-1)
N_adu <- round((y_adu_max + 1) / 0.5)
N_juv <- round((y_juv_max + 1) / 0.5)
N <- N_adu + N_juv
p <- array(0.5, dim=c(nsite, nyear, nreps))

# Tuning factors
beta0_adu_tune <- 0.05
beta0_juv_tune <- 0.05
beta_phi_adu_tune <- c(0.12, 0.100, 0.100, 0.100)
beta_phi_juv_tune <- c(0.15, 0.100, 0.100, 0.100)
beta_omega_tune <- c(0.08, 0.050, 0.050, 0.050)
beta_gamma_tune <- c(0.05, 0.035, 0.035, 0.035)
alpha_tune <- 0.01
N_adu_tune <- 1
N_juv_tune <- 1

# Begin MCMC loop
for (k in 1:nmcmc) {
  ### Sample N_adu
  N_adu_star <- matrix(rpois(nsite*nyear, N_adu + N_adu_tune), nsite,
nyear)
  mh1 <- apply(dbinom(y_adu, N_adu_star, p, log=T), 1:2, sum) +
      dpois(N_adu_star, cbind(lambda0_adu, N_adu[,-nyear] * phi_adu +
N_juv[,-nyear] * phi_juv * omega  + 1e-6), log=T) +
      dpois(N_adu, N_adu_star + N_adu_tune, log=T)
  mh2 <- apply(dbinom(y_adu, N_adu     , p, log=T), 1:2, sum) +
     dpois(N_adu    , cbind(lambda0_adu, N_adu[,-nyear] * phi_adu + N_juv
[,-nyear] * phi_juv * omega + 1e-6), log=T) +
      dpois(N_adu_star, N_adu + N_adu_tune, log=T)
  mhNa <- exp(mh1 - mh2)
  Nakeep <- ((mhNa > runif(nsite*nyear)) & (N_adu_star >= y_adu_max))
  N_adu[Nakeep] <- N_adu_star[Nakeep]
  N <- N_adu + N_juv

  ### Sample N_juv
  N_juv_star <- matrix(rpois(nsite*nyear, N_juv + N_juv_tune), nsite,
nyear)
  mh1 <- apply(dbinom(y_juv, N_juv_star, p, log=T), 1:2, sum) +
      dpois(N_juv_star, cbind(lambda0_juv, N_juv[,-nyear] * phi_juv *
(1 - omega) + gamma), log=T) +
      dpois(N_juv, N_juv_star + N_juv_tune, log=T)
  mh2 <- apply(dbinom(y_juv, N_juv     , p, log=T), 1:2, sum) +
```

```
    dpois(N_juv    , cbind(lambda0_juv, N_juv[,-nyear] * phi_juv * (1 -
omega) + gamma), log=T) +
     dpois(N_juv_star, N_juv + N_juv_tune, log=T)
  mhNj <- exp(mh1 - mh2)
  Njkeep <- ((mhNj > runif(nsite*nyear)) & (N_juv_star >= y_juv_max))
  N_juv[Njkeep] <- N_juv_star[Njkeep]
  N <- N_adu + N_juv

  ### Sample beta0_adu
  beta0_adu_star <- rnorm(ncovs+1, beta0_adu, beta0_adu_tune)
  lambda0_adu_star <- exp(cbind(1,x[,1,]) %*% beta0_adu_star)
  mh1 <- sum(dpois(N_adu[,1], lambda0_adu_star, log=T)) +
     sum(dnorm(beta0_adu_star, beta0_adu_mean, beta0_adu_sd, log=T))
  mh2 <- sum(dpois(N_adu[,1], lambda0_adu    , log=T)) +
     sum(dnorm(beta0_adu    , beta0_adu_mean, beta0_adu_sd, log=T))
  mh0a <- exp(mh1 - mh2)
  if (mh0a > runif(1)) {
   beta0_adu <- beta0_adu_star
   lambda0_adu<- lambda0_adu_star
  }

  ### Sample beta0_juv
  beta0_juv_star <- rnorm(ncovs+1, beta0_juv, beta0_juv_tune)
  lambda0_juv_star <- exp(cbind(1,x[,1,]) %*% beta0_juv_star)
  mh1 <- sum(dpois(N_juv[,1], lambda0_juv_star, log=T)) +
     sum(dnorm(beta0_juv_star, beta0_juv_mean, beta0_juv_sd, log=T))
  mh2 <- sum(dpois(N_juv[,1], lambda0_juv    , log=T)) +
     sum(dnorm(beta0_juv    , beta0_juv_mean, beta0_juv_sd, log=T))
  mh0j <- exp(mh1 - mh2)
  if (mh0j > runif(1)) {
   beta0_juv <- beta0_juv_star
   lambda0_juv <- lambda0_juv_star
  }
  lambda0 <- lambda0_adu + lambda0_juv

  ### Sample beta_phi_adu
  beta_phi_adu_star <- rnorm(ncovs+2, beta_phi_adu,
beta_phi_adu_tune)
  phi_adu <- phi_adu_star <- matrix(, nsite, nyear-1)
  for (t in 2:nyear) {
   phi_adu    [,t-1] <- inv.logit(cbind(1,(N[,t-1]-lambda0)/lambda0,x
[,t,]) %*% beta_phi_adu     )
   phi_adu_star[,t-1] <- inv.logit(cbind(1,(N[,t-1]-lambda0)/
lambda0,x[,t,]) %*% beta_phi_adu_star)
  } # t
 mh1 <- sum(dpois(N_adu[,-1], N_adu[,-nyear] * phi_adu_star + N_juv[,-
```

```
nyear] * phi_juv * omega + 1e-6, log=T)) +
      sum(dnorm(beta_phi_adu_star, beta_phi_adu_mean,
beta_phi_adu_sd, log=T))
  mh2 <- sum(dpois(N_adu[,-1], N_adu[,-nyear] * phi_adu     + N_juv[,-
nyear] * phi_juv * omega + 1e-6, log=T)) +
      sum(dnorm(beta_phi_adu    , beta_phi_adu_mean, beta_phi_adu_sd,
log=T))
  mhSa <- exp(mh1 - mh2)
  if (mhSa > runif(1)) {
   beta_phi_adu <- beta_phi_adu_star
  }

  ### Sample beta_phi_juv
  beta_phi_juv_star <- rnorm(ncovs+2, beta_phi_juv,
beta_phi_juv_tune)
  phi_juv <- phi_juv_star <- matrix(, nsite, nyear-1)
  for (t in 2:nyear) {
   phi_juv    [,t-1] <- inv.logit(cbind(1,(N[,t-1]-lambda0)/lambda0,x
[,t,]) %*% beta_phi_juv    )
   phi_juv_star[,t-1] <- inv.logit(cbind(1,(N[,t-1]-lambda0)/
lambda0,x[,t,]) %*% beta_phi_juv_star)
  } # t
  mh1 <- sum(dpois(N_adu[,-1], N_adu[,-nyear] * phi_adu + N_juv[,-
nyear] * phi_juv_star * omega + 1e-6, log=T)) +
     sum(dpois(N_juv[,-1], N_juv[,-nyear] * phi_juv_star * (1 - omega) +
gamma, log=T)) +
      sum(dnorm(beta_phi_juv_star, beta_phi_juv_mean,
beta_phi_juv_sd, log=T))
  mh2 <- sum(dpois(N_adu[,-1], N_adu[,-nyear] * phi_adu + N_juv[,-
nyear] * phi_juv     * omega + 1e-6, log=T)) +
      sum(dpois(N_juv[,-1], N_juv[,-nyear] * phi_juv     * (1 - omega) +
gamma, log=T)) +
      sum(dnorm(beta_phi_juv    , beta_phi_juv_mean, beta_phi_juv_sd,
log=T))
  mhSj <- exp(mh1 - mh2)
  if (mhSj > runif(1)) {
   beta_phi_juv <- beta_phi_juv_star
  }

  ### Sample beta_omega
  beta_omega_star <- rnorm(ncovs+2, beta_omega, beta_omega_tune)
  omega <- omega_star <- matrix(, nsite, nyear-1)
  for (t in 2:nyear) {
   omega    [,t-1] <- inv.logit(cbind(1,(N[,t-1]-lambda0)/lambda0,x[,
t,]) %*% beta_omega    )
   omega_star[,t-1] <- inv.logit(cbind(1,(N[,t-1]-lambda0)/lambda0,x
```

```
[,t,]) %*% beta_omega_star)
  } # t
  mh1 <- sum(dpois(N_adu[,-1], N_adu[,-nyear] * phi_adu + N_juv[,-
nyear] * phi_juv * omega_star + 1e-6, log=T)) +
     sum(dpois(N_juv[,-1], N_juv[,-nyear] * phi_juv * (1 - omega_star) +
gamma, log=T)) +
      sum(dnorm(beta_omega_star, beta_omega_mean, beta_omega_sd,
log=T))
  mh2 <- sum(dpois(N_adu[,-1], N_adu[,-nyear] * phi_adu + N_juv[,-
nyear] * phi_juv * omega     + 1e-6, log=T)) +
      sum(dpois(N_juv[,-1], N_juv[,-nyear] * phi_juv * (1 - omega    ) +
gamma, log=T)) +
     sum(dnorm(beta_omega    , beta_omega_mean, beta_omega_sd, log=T))
  mhD <- exp(mh1 - mh2)
  if (mhD > runif(1)) {
   beta_omega <- beta_omega_star
  }

  ### Sample beta_gamma
  beta_gamma_star <- rnorm(ncovs+2, beta_gamma, beta_gamma_tune)
  gamma <- gamma_star <- matrix(, nsite, nyear-1)
  for (t in 2:nyear) {
  gamma    [,t-1] <- exp(cbind(1,(N[,t-1]-lambda0)/lambda0,x[,t,]) %*%
beta_gamma     )
  gamma_star[,t-1] <- exp(cbind(1,(N[,t-1]-lambda0)/lambda0,x[,t,])
%*% beta_gamma_star)
  } # t
  mh1 <- sum(dpois(N_juv[,-1], N_juv[,-nyear] * phi_juv * (1 - omega) +
gamma_star, log=T)) +
      sum(dnorm(beta_gamma_star, beta_gamma_mean, beta_gamma_sd,
log=T))
  mh2 <- sum(dpois(N_juv[,-1], N_juv[,-nyear] * phi_juv * (1 - omega) +
gamma     , log=T)) +
     sum(dnorm(beta_gamma    , beta_gamma_mean, beta_gamma_sd, log=T))
  mhG <- exp(mh1 - mh2)
  if (mhG > runif(1)) {
   beta_gamma <- beta_gamma_star
  }

  ### Sample alpha
  alpha_star <- rnorm(npcvs+1, alpha, alpha_tune)
  p_star <- array(, dim=c(nsite, nyear, nreps))
  for (i in 1:nsite) {
   for (t in 1:nyear) {
    p_star[i,t,] <- inv.logit(cbind(1,w[i,t,,]) %*% alpha_star)
   } # t
```

```
  } # i
  mh1 <- sum(dbinom(y_adu, N_adu, p_star, log=TRUE), na.rm=T) +
      sum(dbinom(y_juv, N_juv, p_star, log=TRUE), na.rm=T) +
      sum(dnorm(alpha_star, alpha_mean, alpha_sd, log=TRUE))
  mh2 <- sum(dbinom(y_adu, N_adu, p    , log=TRUE), na.rm=T) +
      sum(dbinom(y_juv, N_juv, p    , log=TRUE), na.rm=T) +
      sum(dnorm(alpha     , alpha_mean, alpha_sd, log=TRUE))
  mhA <- exp(mh1 - mh2)
  if (mhA > runif(1)) {
   alpha <- alpha_star
   p <- p_star
  }

  ### Save samples
  beta0_adu_save[,k] <- beta0_adu
  beta0_juv_save[,k] <- beta0_juv
  beta_phi_adu_save[,k] <- beta_phi_adu
  beta_phi_juv_save[,k] <- beta_phi_juv
  beta_omega_save[,k] <- beta_omega
  beta_gamma_save[,k] <- beta_gamma
  alpha_save[,k] <- alpha
  N_adu_save[,k] <- colMeans(N_adu)
  N_juv_save[,k] <- colMeans(N_juv)
  phi_adu_save[,k] <- colMeans(phi_adu)
  phi_juv_save[,k] <- colMeans(phi_juv)
  omega_save[,k] <- colMeans(omega)
  gamma_save[,k] <- colMeans(gamma)
 } # k

 # Write output
 list(beta0_adu_save=beta0_adu_save,
beta0_juv_save=beta0_juv_save,
    beta_phi_adu_save=beta_phi_adu_save,
beta_phi_juv_save=beta_phi_juv_save,
    beta_omega_save=beta_omega_save,
beta_gamma_save=beta_gamma_save,
    alpha_save=alpha_save,
    N_adu_save=N_adu_save, N_juv_save=N_juv_save,
    phi_adu_save=phi_adu_save, phi_juv_save=phi_juv_save,
    omega_save=omega_save, gamma_save=gamma_save)

} # age_Dail_Madsen_mcmc
```

Now, let us use the following code to implement the MCMC algorithm, which takes about 2.5 h to run. Then, we can use the code in Appendix 3.7 to create some

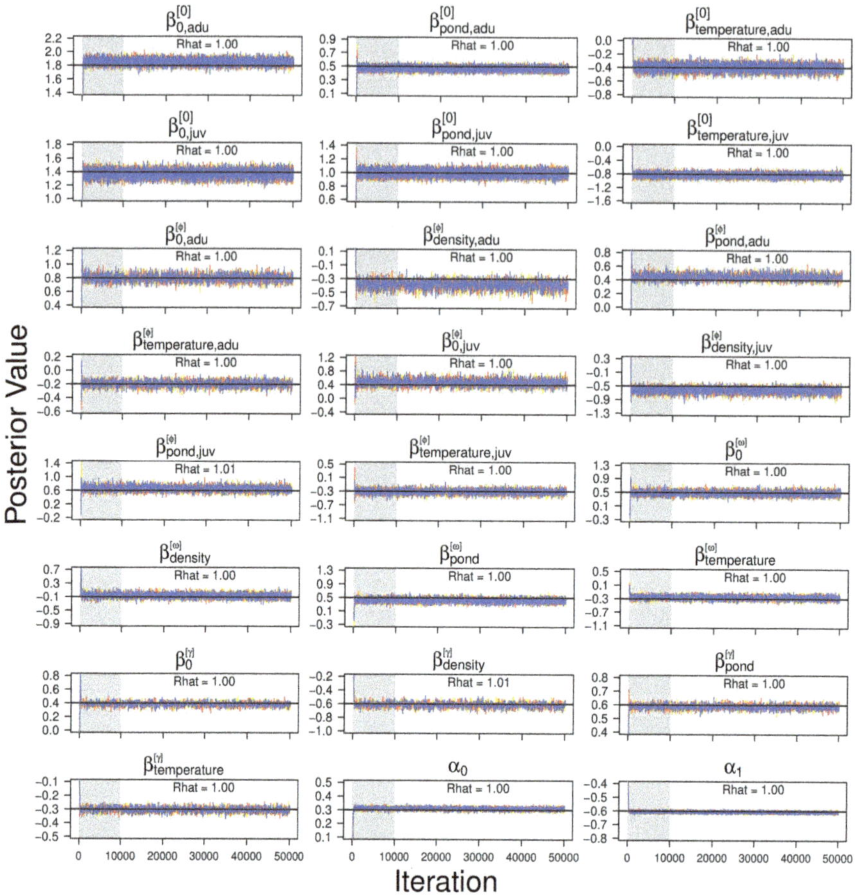

Fig. 3.15 Traceplots showing posterior chains for $\beta^{[0]}$, $\beta^{[\phi]}$, $\beta^{[\omega]}$, $\beta^{[\gamma]}$, and α in the age-structured Dail-Madsen model. Rhat values are calculated after discarding the first 10,000 iterations. Horizontal lines represent the true values used in data simulation

traceplots; we see good model convergence and consistency between parameter estimates of $\beta^{[0]}$, $\beta^{[\phi]}$, $\beta^{[\omega]}$, $\beta^{[\gamma]}$, and α and their true values used in data simulation (Fig. 3.15).

```
library(foreach)  # for parallel computing
library(doParallel) # for parallel computing

numCores <- round(detectCores() / 2) # only use half of the cores for
parallel computing
registerDoParallel(numCores) # setup parallel computing

nmcmc <- 50000 # number of iterations of MCMC computing
chain <- 3    # number of chains
```

```
start_time <- Sys.time() # start time of computing
out <- foreach (i=1:chain, .packages='boot') %dopar% {
  age_Dail_Madsen_mcmc(y_adu=y_adu, y_juv=y_juv, x=x, w=w,
nmcmc=nmcmc)
} # i
stopImplicitCluster() # clean up the cluster
end_time <- Sys.time() # end time of computing
time_taken <- end_time - start_time # computing time
```

We can also plot yearly means of adult abundance, juvenile abundance, adult apparent survival, juvenile apparent survival, transition, and recruitment across all sites, using the code below. We can see that the estimated values capture the temporal variation of these state variables well (Fig. 3.16).

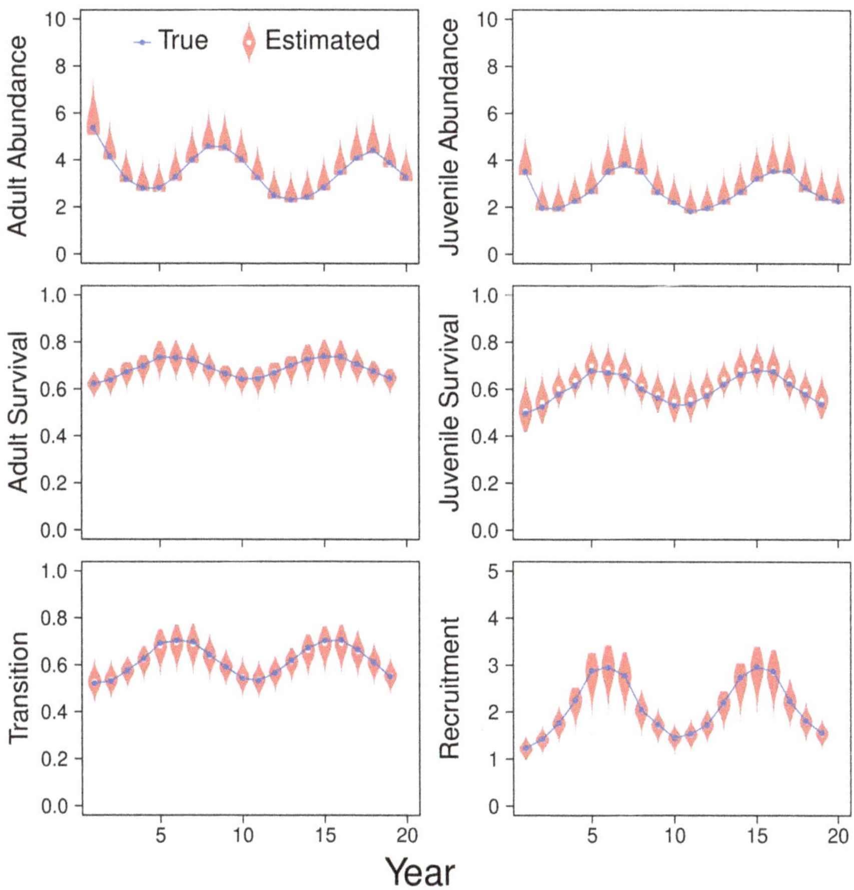

Fig. 3.16 Violin plots showing the consistency between the estimated mean adult abundance, juvenile abundance, adult apparent survival, juvenile apparent survival, transition, and recruitment (violins) and the simulated true values (blue points)

```
library(vioplot) # for making violin plots

for (i in 1:chain) {
 if (i == 1) {
  N_adu_mean <- out[[1]]$N_adu_save
  N_juv_mean <- out[[1]]$N_juv_save
  phi_adu_mean <- out[[1]]$phi_adu_save
  phi_juv_mean <- out[[1]]$phi_juv_save
  omega_mean <- out[[1]]$omega_save
  gamma_mean <- out[[1]]$gamma_save
 } else {
  N_adu_mean <- cbind(N_adu_mean, out[[1]]$N_adu_save)
  N_juv_mean <- cbind(N_juv_mean, out[[1]]$N_juv_save)
  phi_adu_mean <- cbind(phi_adu_mean, out[[1]]$phi_adu_save)
  phi_juv_mean <- cbind(phi_juv_mean, out[[1]]$phi_juv_save)
  omega_mean <- cbind(omega_mean, out[[1]]$omega_save)
  gamma_mean <- cbind(gamma_mean, out[[1]]$gamma_save)
 }
} # i

par(mfrow=c(3,2))
par(mar=c(1.5,6.5,0,0))
par(oma=c(4,0,1,1))

plot(1, xlim=c(1,nyear), ylim=c(0,10), type='l', axes=F, xlab='',
ylab='')
vioplot(t(N_adu_mean), add=T, col='lightcoral', rectCol=NA,
lineCol=NA, border=NA)
lines(colMeans(N_adu), type='o', pch=16, col='royalblue')
axis(1, at=seq(0,nyear,5), labels=rep('',5))
axis(2, at=seq(0,10,2), cex.axis=1.6, las=2)
axis(2, at=5, labels='Adult Abundance', cex.axis=2.2, line=2.6,
tick=F)

points(x=4, y=9, pch=16, cex=1, col='royalblue')
lines(x=c(3.5,4.5), y=c(9,9), lwd=1.2, col='royalblue')
vioplot(rnorm(10000,9,0.25), at=10.5, add=T, col='lightcoral',
rectCol=NA, lineCol=NA, border=NA)
text(x=4.5, y=9, labels='True', pos=4, cex=2.2)
text(x=11, y=9, labels='Estimated', pos=4, cex=2.2)

plot(1, xlim=c(1,nyear), ylim=c(0,10), type='l', axes=F, xlab='',
ylab='')
vioplot(t(N_juv_mean), add=T, col='lightcoral', rectCol=NA,
lineCol=NA, border=NA)
lines(colMeans(N_juv), type='o', pch=16, col='royalblue')
```

```
axis(1, at=seq(0,nyear,5), labels=rep('',5))
axis(2, at=seq(0,10,2), cex.axis=1.6, las=2)
axis(2, at=5, labels='Juvenile Abundance', cex.axis=2.2, line=2.6,
tick=F)

plot(1, xlim=c(1,nyear), ylim=c(0,1), type='l', axes=F, xlab='',
ylab='')
vioplot(t(phi_adu_mean), add=T, col='lightcoral', rectCol=NA,
lineCol=NA, border=NA)
lines(colMeans(phi_adu), type='o', pch=16, col='royalblue')
axis(1, at=seq(0,nyear,5), labels=rep('',5))
axis(2, at=seq(0,1,0.2), cex.axis=1.6, las=2)
axis(2, at=0.5, labels='Adult Survival', cex.axis=2.2, line=2.6,
tick=F)

plot(1, xlim=c(1,nyear), ylim=c(0,1), type='l', axes=F, xlab='',
ylab='')
vioplot(t(phi_juv_mean), add=T, col='lightcoral', rectCol=NA,
lineCol=NA, border=NA)
lines(colMeans(phi_juv), type='o', pch=16, col='royalblue')
axis(1, at=seq(0,nyear,5), labels=rep('',5))
axis(2, at=seq(0,1,0.2), cex.axis=1.6, las=2)
axis(2, at=0.5, labels='Juvenile Survival', cex.axis=2.2, line=2.6,
tick=F)

plot(1, xlim=c(1,nyear), ylim=c(0,1), type='l', axes=F, xlab='',
ylab='')
vioplot(t(omega_mean), add=T, col='lightcoral', rectCol=NA,
lineCol=NA, border=NA)
lines(colMeans(omega), type='o', pch=16, col='royalblue')
axis(1, at=seq(0,nyear,5), cex.axis=1.6)
axis(2, at=seq(0,1,0.2), cex.axis=1.6, las=2)
axis(2, at=0.5, labels='Transition', cex.axis=2.2, line=2.6,
tick=F)

plot(1, xlim=c(1,nyear), ylim=c(0,5), type='l', axes=F, xlab='',
ylab='')
vioplot(t(gamma_mean), add=T, col='lightcoral', rectCol=NA,
lineCol=NA, border=NA)
lines(colMeans(gamma), type='o', pch=16, col='royalblue')
axis(1, at=seq(0,nyear,5), cex.axis=1.6)
axis(2, at=seq(0,5,1), cex.axis=1.6, las=2)
axis(2, at=2.5, labels='Recruitment', cex.axis=2.2, line=2.6,
tick=F)

title(xlab='Year', cex.lab=3, line=2.4, outer=T)
```

3.7 Summary and Additional Reading

In this chapter, we talked about static N-mixture models and their two main routes toward considering population dynamics, one describes population growth without separating apparent survival and recruitment, whereas the other breaks down population growth into components of apparent survival and recruitment. Dynamic N-mixture models with population growth are further extended to spatial versions for single and multiple species, and dynamic N-mixture models with apparent survival and recruitment, the Dail-Madsen (2011) model, is further extended to allow age structure. The Dail-Madsen (2011) model has also been extended to allow disease structure (DiRenzo et al. 2018, 2019) and disease transmission among sites (Brintz et al. 2021). In this chapter, we have focused on models that use naïve count data, while other types of count data, e.g., distance sampling data, can also be used to estimate the spatial distributions (Royle et al. 2004) and dynamics of abundance (Sollmann et al. 2015). Magically, abundance can also be estimated with detection/non-detection data, using the Royle-Nichols model (Royle and Nichols 2003) and its dynamic extension (Rossman et al. 2016). Joint analysis of count and detection/non-detection data can lead to improved inferential performance in comparison to analyzing count data only (Zipkin et al. 2017). Other extensions of N-mixture models can be found in synthesis book chapters (Kéry and Royle 2016, 2021). Thus, we conclude the chapter on N-mixture models here. In the next chapter, we will introduce spatial integrated population models that provide further insights into the demographic foundations of spatially structured population dynamics by jointly analyzing count and capture-recapture data.

References

Bellier E et al (2016) Simulation-based assessment of dynamic N-mixture models in the presence of density dependence and environmental stochasticity. Methods Ecol Evol 7:1029–1040

Brintz BJ et al (2021) A spatially explicit N-mixture model for the estimation of disease prevalence. Stat Model 23:31. 1471082X211020872

Chandler RB et al (2011) Inference about density and temporary emigration in unmarked populations. Ecology 92:1429–1435

Dail D, Madsen L (2011) Models for estimating abundance from repeated counts of an open metapopulation. Biometrics 67:577–587

DiRenzo GV et al (2018) Eco-evolutionary rescue promotes host–pathogen coexistence. Ecol Appl 28:1948–1962

DiRenzo GV et al (2019) Disease-structured N-mixture models: a practical guide to model disease dynamics using count data. Ecol Evol 9:899–909

Hostetler JA, Chandler RB (2015) Improved state-space models for inference about spatial and temporal variation in abundance from count data. Ecology 96:1713–1723

Howell PE et al (2020) Informing amphibian conservation efforts with abundance-based metapopulation models. Herpetologica 76:240–250

Kéry M, Royle JA (2016) Applied hierarchical modeling in ecology: analysis of distribution, abundance and species richness in R and BUGS, Volume 1: Prelude and static models. Elsevier

Kéry M, Royle JA (2021) Applied hierarchical modeling in ecology: analysis of distribution, abundance and species richness in R and BUGS, Volume 2: Dynamic and advanced models. Elsevier

Rossman S et al (2016) Dynamic N-occupancy models: estimating demographic rates and local abundance from detection-nondetection data. Ecology 97:3300

Royle JA (2004) N-mixture models for estimating population size from spatially replicated counts. Biometrics 60:108–115

Royle JA, Nichols JD (2003) Estimating abundance from repeated presence-absence data or point counts. Ecology 84:777–790

Royle JA et al (2004) Modeling abundance effects in distance sampling. Ecology 85:1951–1957

Sollmann R et al (2015) An open-population hierarchical distance sampling model. Ecology 96: 325–331

Stanley TR, Royle JA (2005) Estimating site occupancy and abundance using indirect detection indices. J Wildl Manag 69:874–883

Wyatt RJ (2002) Estimating riverine fish population size from single- and multiple-pass removal sampling using a hierarchical model. Can J Fish Aquat Sci 59:695–706

Zhao Q et al (2017) Spatially explicit dynamic N-mixture models. Popul Ecol 59:293–300

Zhao Q et al (2022) Spatial dynamic N-mixture models with interspecific interactions. Methods Ecol Evol 13:2209–2221

Zipkin EF et al (2014) Modeling structured population dynamics using data from unmarked individuals. Ecology 95:22–29

Zipkin EF et al (2017) Integrating count and detection–nondetection data to model population dynamics. Ecology 98:1640–1650

Integrated Population Models

4

4.1 Introduction

In this chapter, we will introduce integrated population models (IPMs; Fournier and Archibald 1982; Besbeas et al. 2002; Brooks et al. 2004; Schaub and Abadi 2011; Schaub and Kéry 2022; Frost et al. 2023) with a focus on their spatial versions (Borysiewicz et al. 2009; Péron et al. 2010; McCrea et al. 2010; Zhao 2020; Schaub and Kéry 2022; Chapter 20). The idea of data integration in IPMs is to leverage abundance and demographic information by jointly analyzing multiple data sets. A "classic" IPM contains three sub-models, a population sub-model for counts of adult individuals or breeders, a reproduction sub-model for counts of juvenile individuals or offspring, and a survival sub-model (e.g., Cormack-Jolly-Seber/CJS model; Cormack 1964; Jolly 1965; Seber 1965) for capture-recapture data. By analyzing all these data in a joint likelihood, IPMs often achieve improved precision in comparison to the analysis of each data set separately (Schaub and Abadi 2011). Also, IPMs allow for the estimation of latent state variables (e.g., reproduction) without specific data (e.g., counts of juveniles; Besbeas et al. 2002; Tavecchia et al. 2009; Zhao et al. 2019).

In addition to inheriting all the above-mentioned characteristics of classic IPMs, spatial IPMs also explicitly consider spatial locations of, and thus movement between local populations. More specifically, a spatial IPM also contains three sub-models, a population sub-model for counts of adults, a reproduction sub-model for counts of juveniles, and a survival-movement sub-model (e.g., a multistate model; Arnason 1972, 1973) for capture-recapture data, but now spatial locations of counts and capture-recaptures are explicitly considered in all three sub-models. In this way, spatial IPMs provide inference for the entire BIDE process, an advantage over dynamic N-mixture models (see Chap. 3). Considering that count data often cover larger spatiotemporal extents than capture-recapture data, but the

Supplementary Information The online version contains supplementary material available at https://doi.org/10.1007/978-3-031-64518-1_4.

Q. Zhao, *Bayesian Analysis of Spatially Structured Population Dynamics*, Analysis and Synthesis 253, https://doi.org/10.1007/978-3-031-64518-1_4

latter provide stronger information for the BIDE processes than the former, leveraging the strength of these data in spatial IPMs provides opportunities to understand spatial structures and dynamics of populations at spatiotemporal scales that cannot be achieved by using each data alone.

Next, we will first introduce the CJS model for estimating apparent survival as well as its spatial, or more generally, multistate extension, the multistate model which also allows inference for movement. These capture-recapture models are introduced because they are often used as components of IPMs, as mentioned above (also see Schaub and Kéry 2022). We then will introduce four spatial IPMs with increasing complexity, all using a combination of count and capture-recapture data. We will start with a spatial IPM without age-specific survival and movement (Zhao 2020), and then extend it to allow age-specific survival and movement, body-class-specific reproduction, survival, and movement, and eventually gene-class-specific reproduction, survival, and movement. We will end this chapter with a brief summary and suggestions for additional reading.

4.2 Cormack-Jolly-Seber Models

The Cormack-Jolly-Seber (CJS) model is a discrete-time model for animal apparent survival from capture-recapture data collected on marked or otherwise individually recognizable animals (Cormack 1964; Jolly 1965; Seber 1965; Lebreton et al. 1992). It assumes that individual animals can survive over a time period of one to several years (or other primary occasions) while the probability of survival is driven by some covariates. On top of that, researchers have a chance of capturing an individual given it is alive. When captured for the first time, an individual must have some natural mark (e.g., a color pattern, DNA) or is given an artificial mark (e.g., a tag) to allow individual identification in later captures. Such individual identification is assumed to be perfect so that we will know for sure that an individual is alive when it is captured, but we will not be able to tell if an individual is alive or dead if it is not captured. In other words, we only allow false-negative error (alive but not captured), but not false-positive error (not alive but captured). However, for instance, if we captured an individual in the first year and recaptured it in the third year, we would be able to tell that it was alive in the second year, even though we did not capture it in that year. Also note that, when an individual disappears, there is no way to tell if it is dead or it has left the study area with capture-recapture data only. Therefore, the CJS model estimates the probability of apparent survival, i.e., the situation when an individual survives and stays in the study area. That means, when we said "alive" above, we actually meant "alive in the study area." Consequently, the apparent survival probability being estimated is the product of two components, the probability that an individual survives from one year to the next and the probability that it remains in the study area.

Imagine some fish species living in patchy coral reefs, their probability of surviving and staying in the area is driven by the average size of reefs and average temperature across the entire area (as we are not considering the spatial structure of the population yet). The population is formed by individuals, some barely survive the first year while some others survive (and stay) for a much longer period of time (e.g., 5 years). When researchers go to the area over a number of years, they may capture some individuals and uniquely identify them from their natural color patterns. Eventually, researchers have a record of the so-called capture history for each individual. For example, an individual with code 9527 was first captured in the first year and was captured again in the fourth and fifth years of a 5-year study period, but was missed in the second and third years. On the other hand, another individual with code 8848 was first captured in the second year but was never captured again in later years of the study. The captured histories of these two individuals will be coded as 10011 and 01000, respectively, in which 1 stands for "captured" and 0 for "not captured." Researchers will then use these data to estimate the probability of apparent survival in relation to average reef size and temperature.

Using statistical language, we first have the following for the underlying apparent survival state:

$$\left.\begin{array}{c} z_{v,t=f_v} = 1 \\ z_{v,t>f_v} \sim Bernoulli(z_{v,t-1} \times \phi_{t-1}) \\ \phi_{t-1} = logit^{-1}\left(\beta_0 + \beta_1 \times x_{t-1,1} + \ldots + \beta_{n_{covs}} \times x_{t-1,n_{covs}}\right) \end{array}\right\}, \tag{4.1}$$

in which $z_{v,\ t}$ represents the alive/dead state such that $z_{v,\ t} = 1$ when individual v is alive and stays in the study area in year t, and 0 otherwise. We always have $z_{v,t=f_v} = 1$ for the year in which individual v is captured for the first time, meaning that every individual is known to be alive in that year. In the years after the first capture, $z_{v,t>f_v}$ depends on $z_{v,\ t\ -\ 1}$ as well as the probability of apparent survival, $\phi_{t\ -\ 1}$, which is a function of environmental covariates x's with intercept and slopes parameters β's. More specifically, individual v has a probability of $\phi_{t\ -\ 1}$ to survive from year $t - 1$ to year t if it was alive in year $t - 1$ (i.e., $z_{v,\ t\ -\ 1} = 1$), while it will always stay dead (or more precisely, disappeared) if it was already dead in the previous year (i.e., $z_{v,\ t\ -\ 1} = 0$). Here, we do not consider the effect of density-dependent processes on apparent survival, but we will do so later in IPMs (Sect. 4.4).

Given the survival history, $z_{v,\ t}$, the model specifies the observed capture history as follows:

$$\left.\begin{array}{c} c_{v,t=f_v} = z_{v,t=f_v} = 1 \\ c_{v,t>f_v} \sim Bernoulli(z_{v,t} \times p_t) \\ p_t = logit^{-1}\left(\alpha_0 + \alpha_1 \times w_{t,1} + \ldots + \alpha_{n_{pcvs}} \times w_{t,n_{pcvs}}\right) \end{array}\right\}, \tag{4.2}$$

in which $c_{v,\ t}$ is the capture history of individual v in year t, which equals 1 when individual v is captured in year t, and 0 when it is not captured. We have $c_{v,t=f_v} = z_{v,t=f_v} = 1$ for the year of first capture; in later years, $c_{v,t>f_v}$ depends on $z_{v,\ t}$ as well as a capture probability, p_t, which is a function of capture covariates w's with intercept and slope parameters α's. This means that an individual has a probability of p_t to be captured if it is alive (i.e., $z_{v,\ t} = 1$), but it will never by captured when it is dead (i.e., $z_{v,\ t} = 0$). This means we only allow false-negative error, but not false-positive error. Studies have extended the model to also allow false-positive errors due to, for example, misidentification (Schwarz and Stobo 1999; Rakhimberdiev et al. 2022). For CJS model, we do not need to consider the years before the first capture. Fascinatingly, CJS model allows the inference of the frequency of an event, death (or emigration), even if we never observe a single dead animal. For a related type of model based on recovered dead animals, see Brownie et al. (1978, 1985).

Our main interest is to estimate ϕ_t and β for understanding the ecology of the study species. We sometimes are also interested in p_t and α so that we can learn more about our data collection process. Next, in Sect. 4.2.1, we will use the individual capture history data, $c_{v,\ t}$, to estimate the above-mentioned parameters, which also involves sampling $z_{v,\ t}$. However, sampling $z_{v,\ t}$ can be time consuming, and we want to circumvent the trouble of sampling $z_{v,\ t}$ when $z_{v,\ t}$ is not of interest. Therefore, in Sect. 4.2.2, we will introduce another approach in which we convert $c_{v,\ t}$ into a format called m-array with a multinomial likelihood to achieve much faster computing.

4.2.1 With Data in Individual Capture History

To illustrate how to use individual capture history data in a CJS model, we will first use the following code to set up some basic information and simulate the ecological process (i.e., individual survival history) for the model. We assume that the average reef size has a positive effect ($\beta_{reef} = 0.1$) and average temperature has a negative effect ($\beta_{temperature} = -0.8$) on apparent survival at a logit scale.

```
library(boot) # for using logit and inv.logit functions

set.seed(6)

# Basic values
nindv <- 1800 # number of captured individuals
nyear <- 10  # number of years
ncovs <- 2   # number of environmental covariates
npcvs <- 2   # number of capture covariates

beta <- c( 0.4,  0.1, -0.8) # intercept and slopes for apparent survival
probability
alpha <- c(-1.4, -0.8,  0.4) # intercept and slopes for capture
probability

# Read in environmental covariates
load('data/covariates.RData')
x1 <- colMeans(cov$size)[1:(nyear-1)] # mean habitat patch size
x2 <- colMeans(cov$temp)[1:(nyear-1)] # mean temperature
x1 <- (log(x1) - mean(log(x1))) / sd(log(x1))
x2 <- (x2 - mean(x2)) / sd(x2)
x <- cbind(x1, x2) # environmental covariates
```

```
# Simulate data
### Process
phi <- inv.logit(cbind(1,x) %*% beta) # apparent survival probability

f <- rep(1:(nyear-1), each=nindv/(nyear-1)) # year of first capture

z <- matrix(0, nindv, nyear) # individual survival history
for (v in 1:nindv) {
 z[v,f[v]] <- 1
 for (t in (f[v]+1):nyear) {
  z[v,t] <- rbinom(1, 1, z[v,t-1] * phi[t-1])
 } # t
} # v
```

The simulated survival histories of the first three individuals during the 10-year study period look like the following. These values mean that individual 1 survives from year 1 through year 3, individual 2 is alive only in year 1, and individual 3 survives from year 1 through year 8.

	Survival history									
	Year 1	Year 2	Year 3	Year 4	Year 5	Year 6	Year 7	Year 8	Year 9	Year 10
Individual 1	1	1	1	0	0	0	0	0	0	0
Individual 2	1	0	0	0	0	0	0	0	0	0
Individual 3	1	1	1	1	1	1	1	1	0	0

```
### Capture
w <- matrix(rnorm((nyear-1)*npcvs, 0, 1), nyear-1, npcvs) # capture
covariates

p <- inv.logit(cbind(1,w) %*% alpha) # capture probability

c <- matrix(0, nindv, nyear) # individual capture history
for (v in 1:nindv) {
 c[v,f[v]] <- z[v,f[v]]
 for (t in (f[v]+1):nyear) {
  c[v,t] <- rbinom(1, 1, z[v,t] * p[t-1])
 } # t
} # v
```

We then can use the following code to simulate the individual capture history data.

The capture histories of the first three individuals look like the following. These values mean that individual 1 is captured in years 1 and 3, individual 2 is captured in year 1, and individual 3 is captured in years 1, 5, and 8. From the data, we can tell that individual 1 is alive but missed in year 2, and individual 3 is alive but missed in years 2, 3, 4, 6, and 7. However, we will not be able to tell if these individuals are dead or simply missed after their last capture (e.g., year 4 onward for individual 1).

	Capture history									
	Year 1	Year 2	Year 3	Year 4	Year 5	Year 6	Year 7	Year 8	Year 9	Year 10
Individual 1	1	0	1	0	0	0	0	0	0	0
Individual 2	1	0	0	0	0	0	0	0	0	0
Individual 3	1	0	0	0	1	0	0	1	0	0

We then develop the MCMC algorithm to estimate parameters β, α, and ϕ given data c, x, and w.

```
cjs_indv_mcmc <- function(c, x, w, nmcmc) {

 # Setup variables
 nindv <- dim(c)[1]
 nyear <- dim(c)[2]
 ncovs <- dim(x)[2]
 npcvs <- dim(w)[2]

 beta_save <- matrix(0, ncovs+1, nmcmc)
 alpha_save <- matrix(0, npcvs+1, nmcmc)
 phi_save <- matrix(0, nyear-1, nmcmc)

 # Prior
 beta_mean <- rep(0, ncovs+1)
 beta_sd <- 2
 alpha_mean <- rep(0, npcvs+1)
 alpha_sd <- 2

 # Starting values
 beta <- rep(0, ncovs+1)
 alpha <- rep(0, npcvs+1)
 phi <- rep(0.5, nyear-1)
 p <- rep(0.5, nyear-1)
 zi <- c
 f <- l <- rep(0, nindv)
 for (v in 1:nindv) {
  f[v] <- min(which(c[v,]==1))
  l[v] <- max(which(c[v,]==1))
  zi[v, f[v]:l[v]] <- 1
 } # v
 z <- zi
```

```
# Tuning factor
beta_tune <- 0.05
alpha_tune <- 0.05

# Begin MCMC loop
for (k in 1:nmcmc) {
  # Sample beta
  beta_star <- rnorm(ncovs+1, beta, beta_tune)
  phi_star <- inv.logit(cbind(1,x) %*% beta_star)
  prob1 <- prob2 <- numeric(nyear-1)
  for (t in 1:(nyear-1)) {
   prob1[t] <- sum(dbinom(z[which(f <= t),t+1], 1, z[which(f <= t),t] *
phi_star[t], log=TRUE))
   prob2[t] <- sum(dbinom(z[which(f <= t),t+1], 1, z[which(f <= t),t] *
phi    [t], log=TRUE))
  } # t
  mh1 <- sum(prob1) +
      sum(dnorm(beta_star, beta_mean, beta_sd, log=T))
  mh2 <- sum(prob2) +
      sum(dnorm(beta    , beta_mean, beta_sd, log=T))
  mh <- exp(mh1 - mh2)
  if (mh > runif(1)) {
   beta <- beta_star
   phi <- phi_star
  }

  # Sample alpha
  alpha_star <- rnorm(npcvs+1, alpha, alpha_tune)
  p_star <- inv.logit(cbind(1,w) %*% alpha_star)
  prob1 <- prob2 <- numeric(nyear-1)
  for (t in 1:(nyear-1)) {
   prob1[t] <- sum(dbinom(c[which(f <= t),t+1], 1, z[which(f <= t),t+1]
* p_star[t], log=TRUE))
   prob2[t] <- sum(dbinom(c[which(f <= t),t+1], 1, z[which(f <= t),t+1]
* p    [t], log=TRUE))
  } # t
  mh1 <- sum(prob1) +
      sum(dnorm(alpha_star, alpha_mean, alpha_sd))
  mh2 <- sum(prob2) +
      sum(dnorm(alpha    , alpha_mean, alpha_sd))
  mh <- exp(mh1 - mh2)
  if (mh > runif(1)) {
   alpha <- alpha_star
   p <- p_star
  }
```

```
  # Sample z
  for (v in 1:nindv) {
   if (l[v] <= nyear-2) {
    for (t in (l[v]+1):(nyear-1)) {
     if (z[v,t-1] == 1 & z[v,t+1] == 0) {
      num_temp <- phi[t-1] * (1 - phi[t]) * (1 - p[t-1])
      psi_temp <- num_temp / (num_temp + 1 - phi[t-1])
      z[v,t] <- rbinom(1, 1, psi_temp)
     }
     if (z[v,t-1] == 0) {
      z[v,t] <- 0
     }
     if (z[v,t+1] == 1) {
      z[v,t] <- 1
     }
    } # t

    if (z[v,nyear-1] == 1) {
     num_temp <- phi[nyear-1] * (1 - p[nyear-1])
     psi_temp <- num_temp / (num_temp + 1 - phi[nyear-1])
     z[v,nyear] <- rbinom(1, 1, psi_temp)
    }
    if (z[v,nyear-1] == 0) {
     z[v,nyear] <- 0
    }
   }

   if (l[v] == nyear-1) {
    num_temp <- phi[nyear-1] * (1 - p[nyear-1])
    psi_temp <- num_temp / (num_temp + 1 - phi[nyear-1])
    z[v,nyear] <- rbinom(1, 1, psi_temp)
   }
  } # v

  # Save samples
  beta_save[,k] <- beta
  alpha_save[,k] <- alpha
  phi_save[,k] <- phi
 } # k

 list(beta_save=beta_save, alpha_save=alpha_save,
phi_save=phi_save)
} # cjs_indv_mcmc
```

When using the following code to implement the algorithm, it takes about 38 min to run. Then using code in Appendix 4.1, we can create traceplots for the monitored parameters, which show good model convergence and consistency between parameter estimates of β and α and their true values used in data simulation (Fig. 4.1).

```
library(foreach)  # for parallel computing
library(doParallel) # for parallel computing

numCores <- round(detectCores() / 2) # only use half of the cores for
parallel computing
registerDoParallel(numCores) # setup parallel computing

nmcmc <- 50000 # number of iterations of MCMC computing
chain <- 3    # number of chains

start_time <- Sys.time() # start time of computing
out <- foreach (i=1:chain, .packages='boot') %dopar% {
 cjs_indv_mcmc(c=c, x=x, w=w, nmcmc=nmcmc)
```

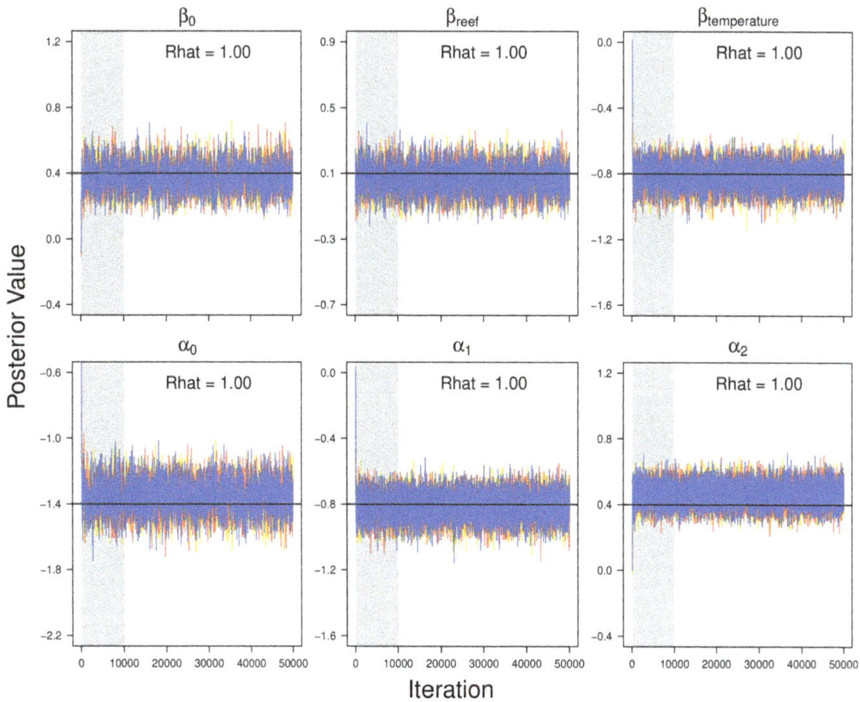

Fig. 4.1 Traceplots showing the posterior chains for β and α in the CJS model with data in individual capture history format. Rhat values are calculated after discarding the first 10,000 iterations. Horizontal lines represent the true values used in data simulation

```
} # i
stopImplicitCluster() # clean up the cluster
end_time <- Sys.time() # end time of computing
time_taken <- end_time - start_time # computing time
```

We can also use the following code to plot the simulated and estimated apparent survival, ϕ, and see good agreement between them (Fig. 4.2). The temporal variation in ϕ is mainly driven by the temporal variation in average temperature.

```
library(vioplot) # for making violin plots

for (i in 1:chain) {
 if (i == 1) {
  phi_post <- out[[i]]$phi[,(nmcmc*0.2+1):nmcmc]
 } else {
  phi_post <- cbind(phi_post, out[[i]]$phi[,(nmcmc*0.2+1):nmcmc])
 }
} # i

par(mfrow=c(1,1))
par(mar=c(5,6,1,1))
```

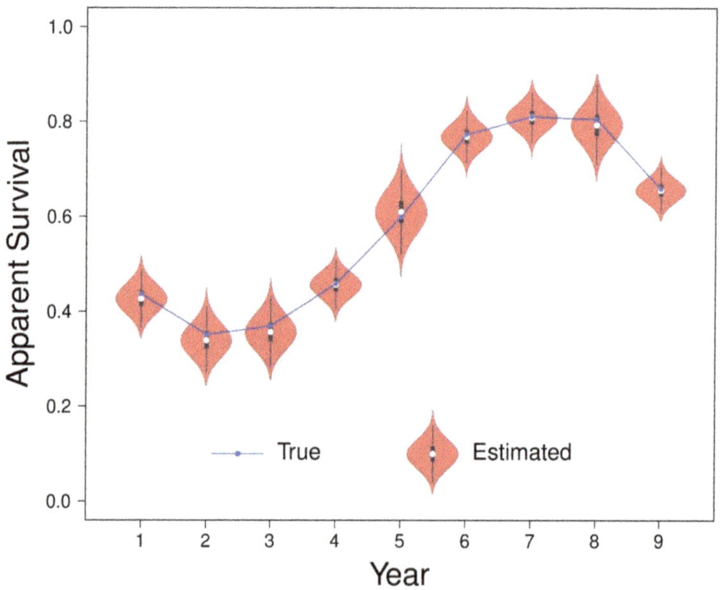

Fig. 4.2 Violin plots showing consistency between the estimated apparent survival in the CJS model with data in individual capture history format and their simulated true values (blue points)

```
plot(1, xlim=c(0.5,nyear-0.5), ylim=c(0,1), type='n', xlab='',
ylab='', axes=F)
vioplot(t(phi_post), add=T, col='lightcoral', rectCol='grey36',
lineCol='grey36', border=NA)
lines(phi ~ c(1:(nyear-1)), type='o', pch=16, cex=0.8, lwd=1.2,
col='royalblue')
axis(1, at=seq(1,nyear,1), cex.axis=1.5)
axis(2, at=seq(0,1,0.2), cex.axis=1.5, las=2)
title(xlab='Year', cex.lab=2.5, line=3.5)
title(ylab='Apparent Survival', cex.lab=2.5, line=3.8)

points(x=2.5, y=0.1, pch=16, cex=1, col='royalblue')
lines(x=c(2.1,2.9), y=c(0.1,0.1), lwd=1.2, col='royalblue')
vioplot(phi_post[1,]-median(phi_post[1,])+0.1, at=5.5, add=T,
col='lightcoral', rectCol='grey36', lineCol='grey36', border=NA)
text(x=3, y=0.1, labels='True', pos=4, cex=1.8)
text(x=6, y=0.1, labels='Estimated', pos=4, cex=1.8)
```

4.2.2 With Data in m-Array

While the CJS model works just fine with individual capture history data, next, we will illustrate how to implement the model with data in an m-array format to greatly accelerate computing. To do so, we first need to use the following code to "extend" the capture histories because the calculation of an m-array is based on the number of individuals that are recaptured in a given year *for the first time* since the previous (re)capture, stratified by the years of (re)capture. Note that code to simulate individual survival history and capture history is the same as what we have seen in Sect. 4.2.1.

```
### Extend individual capture history
for (v in 1:nindv) {
 t1 <- c[v,]
 ycap <- which(t1 == 1)
 ncap <- length(ycap)
 if (ncap == 1) {
  t2 <- t1
 } else {
  t2 <- matrix(, ncap, nyear)
  for (h in 1:(ncap-1)) {
   t2[h, ycap[h]:ycap[h+1]] <- t1[ycap[h]:ycap[h+1]]
  } # h
  t2[ncap, ycap[ncap]] <- 1
```

```
  }
  if (v == 1) {
   c_ext <- t2
  } else {
   c_ext <- rbind(c_ext, t2)
  }
 } # v
c_ext[which(is.na(c_ext))] <- 0

f_ext <- numeric(dim(c_ext)[1])
for (v in 1:dim(c_ext)[1]) {
 f_ext[v] <- min(which(c_ext[v,] == 1))
} # v
```

The extended capture histories of the first three individuals look like the following. The capture history of individual 1 is extended into two rows, as it is captured twice. The 1's highlighted in **bold** are actually the same capture, which now serves as the recapture in the first row and a new capture in the second row. Similarly, the capture history of individual 3 is extended into three rows, as it is captured three times.

	Extended capture history									
	Year 1	Year 2	Year 3	Year 4	Year 5	Year 6	Year 7	Year 8	Year 9	Year 10
Individual 1	1	0	**1**	0	0	0	0	0	0	0
Individual 1	0	0	**1**	0	0	0	0	0	0	0
Individual 2	1	0	0	0	0	0	0	0	0	0
Individual 3	1	0	0	0	1	0	0	0	0	0
Individual 3	0	0	0	0	1	0	0	1	0	0
Individual 3	0	0	0	0	0	0	0	1	0	0

With the extended capture histories, we can now use the following code to convert the data into an m-array format.

```
# Convert individual capture history to m-array
m <- matrix(0, nyear-1, nyear) # m-array of capture-recapture
for (t in 1:(nyear-1)) {
 tt <- c_ext[which(f_ext == t),]
 nband <- dim(tt)[1]
 m[t, t:(nyear-1)] <- colSums(tt, na.rm=T)[(t+1):nyear]
 m[t, nyear] <- nband - sum(m[t, 1:(nyear-1)])
} # t
```

Now, let us imagine a capture-recapture survey that is conducted over 4 years and make up some numbers for illustration. The data will look like the following if it is collated in an m-array format. Cells below the main diagonal of the matrix will always be 0's, as we cannot recapture an individual in the same year, or before, it is captured. The last column is for the number of individuals who are not recaptured after their previous capture.

		Recapture			
		Year 2	Year 3	Year 4	Not recaptured
Capture	Year 1	17	14	0	168
	Year 2	0	40	2	186
	Year 3	0	0	25	225

Each row of the above table follows a multinomial distribution, which has the following cell probabilities. Cells below the main diagonal of the matrix always have a probability of 0, for the reason we just described. The diagonal cells are for next-year recaptures, and their probabilities are relatively simple to calculate. For example, $Pr_{1,\,2}$ is simply the product of the probability of surviving from year 1 to year 2 (ϕ_1) and the probability of being captured in year 2 (p_2). Moving beyond next-year recaptures, the probabilities become more complicated to calculate. For example, $Pr_{1,\,3}$, which is for the individuals that are captured in year 1 and recaptured *for the first time* in year 3, is the product of the probability of surviving from year 1 to year 2 (ϕ_1), the probability of *not* being captured in year 2 ($1 - p_2$), the probability of surviving from year 2 to year 3 (ϕ_2), and the probability of being captured in year 3 (p_3). The probability in the last column is none other than 1 minus the probabilities in the previous cells; this also makes sure that the sum of each row equals 1, a requirement for multinomial distributions.

		Recapture			
		Year 2	Year 3	Year 4	Not recaptured
Capture	Year 1	$Pr_{1,\,2}$	$Pr_{1,\,3}$	$Pr_{1,\,4}$	$1 - Pr_{1,\,2} - Pr_{1,\,3} - Pr_{1,\,4}$
	Year 2	0	$Pr_{2,\,3}$	$Pr_{2,\,4}$	$1 - Pr_{2,\,3} - Pr_{2,\,4}$
	Year 3	0	0	$Pr_{3,\,4}$	$1 - Pr_{3,\,4}$

Now, we first will need to define a function to calculate the cell probability of the m-array, as shown below. Later, we will use this function in the MCMC algorithm.

```
# Define the function to calculate the cell probability of m-array
cell_prob_fun <- function(nyear, phi, p) {
 pmiss <- 1 - p
 cell_prob <- matrix(0, nyear-1, nyear)
 for (t in 1:(nyear-1)) {
  cell_prob[t,t] <- phi[t] * p[t]
 } # t
 for (t1 in 1:(nyear-2)) {
  for (t2 in (t1+1):(nyear-1)) {
   cell_prob[t1,t2] <- prod(phi[t1:t2]) * prod(pmiss[t1:(t2-1)]) * p[t2]
```

```
  } # t2
 } # t1
 for (t in 1:(nyear-1)) {
  cell_prob[t, nyear] <- 1 - sum(cell_prob[t, 1:(nyear-1)])
 } # t

 return(cell_prob)
} # cell_prob_fun
```

We then develop the MCMC algorithm to estimate parameters β, α, and ϕ given data m (the m-array), x and w.

```
cjs_marr_mcmc <- function(m, x, w, nmcmc) {

 # Setup variables
 nyear <- dim(m)[2]
 ncovs <- dim(x)[2]
 npcvs <- dim(w)[2]

 beta_save <- matrix(0, ncovs+1, nmcmc)
 alpha_save <- matrix(0, npcvs+1, nmcmc)
 phi_save <- matrix(0, nyear-1, nmcmc)

 # Prior
 beta_mean <- rep(0, ncovs+1)
 beta_sd <- 2
 alpha_mean <- rep(0, npcvs+1)
 alpha_sd <- 2

 # Starting values
 beta <- rep(0, ncovs+1)
 alpha <- rep(0, npcvs+1)
 phi <- rep(0.5, nyear-1)
 p <- rep(0.5, nyear-1)
 cell_prob <- cell_prob_fun(nyear=nyear, phi=phi, p=p)

 # Tuning factor
 beta_tune <- 0.05
 alpha_tune <- 0.05

 # Begin MCMC loop
 for (k in 1:nmcmc) {
  # Sample beta
```

```
  beta_star <- rnorm(ncovs+1, beta, beta_tune)
  phi_star <- inv.logit(cbind(1,x) %*% beta_star)
  cell_prob_star <- cell_prob_fun(nyear=nyear, phi=phi_star, p=p)
  prob1 <- prob2 <- numeric(nyear-1)
  for (t in 1:(nyear-1)) {
   prob1[t] <- dmultinom(m[t,], size=sum(m[t,]), prob=cell_prob_star
[t,], log=T)
   prob2[t] <- dmultinom(m[t,], size=sum(m[t,]), prob=cell_prob
[t,], log=T)
  } # t
  mh1 <- sum(prob1) +
     sum(dnorm(beta_star, beta_mean, beta_sd, log=T))
  mh2 <- sum(prob2) +
     sum(dnorm(beta     , beta_mean, beta_sd, log=T))
  mh <- exp(mh1 - mh2)
  if (mh > runif(1)) {
   beta <- beta_star
   phi <- phi_star
   cell_prob <- cell_prob_star
  }

  # Sample alpha
  alpha_star <- rnorm(npcvs+1, alpha, alpha_tune)
  p_star <- inv.logit(cbind(1,w) %*% alpha_star)
  cell_prob_star <- cell_prob_fun(nyear=nyear, phi=phi, p=p_star)
  prob1 <- prob2 <- numeric(nyear-1)
  for (t in 1:(nyear-1)) {
   prob1[t] <- dmultinom(m[t,], size=sum(m[t,]), prob=cell_prob_star
[t,], log=T)
   prob2[t] <- dmultinom(m[t,], size=sum(m[t,]), prob=cell_prob
[t,], log=T)
  } # t
  mh1 <- sum(prob1) +
     sum(dnorm(alpha_star, alpha_mean, alpha_sd))
  mh2 <- sum(prob2) +
     sum(dnorm(alpha     , alpha_mean, alpha_sd))
  mh <- exp(mh1 - mh2)
  if (mh > runif(1)) {
   alpha <- alpha_star
   p <- p_star
   cell_prob <- cell_prob_star
  }

  # Save samples
  beta_save[,k] <- beta
```

```
  alpha_save[,k] <- alpha
  phi_save[,k] <- phi
 } # k

 list(beta_save=beta_save, alpha_save=alpha_save,
phi_save=phi_save)
} # cjs_marr_mcmc
```

When using the following code to implement the algorithm, it takes about half a minute to run, meaning that we have improved the computing time of the CJS model by more than 70 times by converting the data into an m-array format. Then using code in Appendix 4.2, we can create traceplots for the monitored parameters, which show good model convergence and consistency between parameter estimates of β and α and their true values used in data simulation (Fig. 4.3).

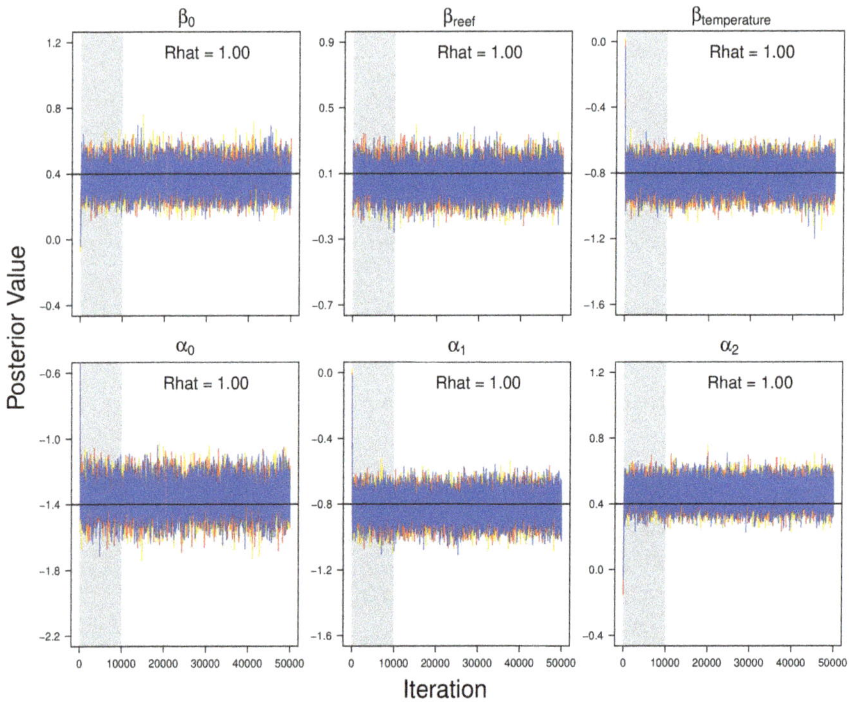

Fig. 4.3 Traceplots showing the posterior chains for β and α in the CJS model with data in m-array format. Rhat values are calculated after discarding the first 10,000 iterations. Horizontal lines represent the true values used in data simulation

```
library(foreach)  # for parallel computing
library(doParallel) # for parallel computing

numCores <- round(detectCores() / 2) # only use half of the cores for
parallel computing
registerDoParallel(numCores) # setup parallel computing

nmcmc <- 50000 # number of iterations of MCMC computing
chain <- 3   # number of chains

start_time <- Sys.time() # start time of computing
out <- foreach (i=1:chain, .packages='boot') %dopar% {
 cjs_marr_mcmc(m=m, x=x, w=w, nmcmc=nmcmc)
} # i
stopImplicitCluster() # clean up the cluster
end_time <- Sys.time() # end time of computing
time_taken <- end_time - start_time # computing time
```

We can also use the following code to plot the estimated apparent survival, ϕ, and compare them to the estimates based on individual capture history data (Fig. 4.4). Estimates based on both approaches are highly consistent with each other as well as the simulated true values.

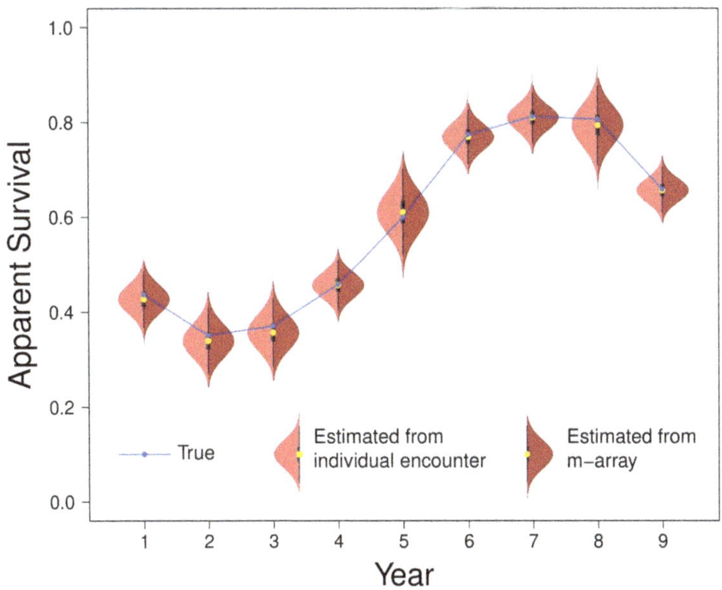

Fig. 4.4 Violin plots showing consistency between the estimated apparent survival based on individual capture history (left half of violin), the estimated apparent survival based on m-array (right half of violin), and their simulated true values (blue points)

```
library(vioplot) # for making violin plots

out_marr <- out
load('results/4.1.1 cjs indv model_output.RData')
out_indv <- output$out

phi_marr_post <- cbind(out_marr[[1]]$phi[,(nmcmc*0.2+1):nmcmc],
out_marr[[2]]$phi[,(nmcmc*0.2+1):nmcmc], out_marr[[3]]$phi[,
(nmcmc*0.2+1):nmcmc])
phi_indv_post <- cbind(out_indv[[1]]$phi[,(nmcmc*0.2+1):nmcmc],
out_indv[[2]]$phi[,(nmcmc*0.2+1):nmcmc], out_indv[[3]]$phi[,
(nmcmc*0.2+1):nmcmc])

par(mfrow=c(1,1))
par(mar=c(5,6,1,1))

plot(1, xlim=c(0.5,nyear-0.5), ylim=c(0,1), type='n', xlab='',
ylab='', axes=F)
vioplot(t(phi_marr_post), col='lightcoral', rectCol='grey36',
lineCol='grey36', colMed='yellow', border=NA, side='left', add=T)
vioplot(t(phi_indv_post), col='firebrick' , rectCol='grey16',
lineCol='grey16', colMed='gold' , border=NA, side='right', add=T)
lines(phi ~ c(1:(nyear-1)), type='o', pch=16, cex=0.8, lwd=1.2,
col='royalblue')
axis(1, at=seq(1,nyear,1), cex.axis=1.5)
axis(2, at=seq(0,1,0.2), cex.axis=1.5, las=2)
title(xlab='Year', cex.lab=2.5, line=3.5)
title(ylab='Apparent Survival', cex.lab=2.5, line=3.8)

points(x=1, y=0.1, pch=16, cex=1, col='royalblue')
lines(x=c(0.6,1.4), y=c(0.1,0.1), lwd=1.2, col='royalblue')
text(x=1.4, y=0.1, labels='True', pos=4, cex=1.6)
vioplot(phi_marr_post[1,]-median(phi_marr_post[1,])+0.1, at=3.4,
col='lightcoral', rectCol='grey36', lineCol='grey36',
colMed='yellow', border=NA, side='left', add=T)
text(x=3.5, y=0.1, labels='Estimated from\nindividual encounter',
pos=4, cex=1.6)
vioplot(phi_indv_post[1,]-median(phi_indv_post[1,])+0.1, at=6.9,
col='firebrick' , rectCol='grey16', lineCol='grey16', colMed='gold'
, border=NA, side='right', add=T)
text(x=7.4, y=0.1, labels='Estimated from\nm-array', pos=4,
cex=1.6)
```

4.3 Multistate Models

The multistate model extends the CJS model to consider different states (rather than just alive or dead) of individuals and transitions between these states (Arnason 1972, 1973; Hestbeck et al. 1991; Nichols et al. 1992; Brownie et al. 1993; Lebreton et al. 2009). States of individuals can be, for example, their age classes, or disease status (healthy or sick). When using the multistate model to address spatial questions, as Arnason (1972, 1973) and many others did, the discrete spatial location of an individual is considered its state, and transition between states basically means movement between sites. To illustrate the multistate model with our fictitious fish, we now assume that its probability of surviving is driven by the size and temperature of the reef it lives in. In addition, it can move between reefs, the chance of which decreases when the distance between two reefs increases. Finally, when a fish is captured by researchers, its spatial location will be recorded. Using a statistical language, we have

$$\left.\begin{array}{c} z_{v,t=f_v} = i_1 \\ z_{v,t>f_v} \begin{cases} \sim Categorical\left(\phi_{i_1,t-1} \times \theta_{i_1,1}, \ldots, \phi_{i_1,t-1} \times \theta_{i_1,n_{site}}, 1 - \phi_{i_1,t-1}\right), & z_{v,t-1} = i_1 \\ = 0, & z_{v,t-1} = 0 \end{cases} \\ \phi_{i_1,t-1} = logit^{-1}\left(\beta_0 + \beta_1 \times x_{i_1,t-1,1} + \ldots + \beta_{n_{covs}} \times x_{i_1,t-1,n_{covs}}\right) \\ \theta_{i_1,i_2} = \dfrac{\eta_{i_1,i_2}}{\sum_{i_2=1}^{n_{site}} \eta_{i_1,i_2}} \\ \eta_{i_1,i_2} = exp(-1 \times \kappa \times d_{i_1,i_2}) \end{array}\right\}, \tag{4.3}$$

in which $z_{v,\,t}$ is the survival-location history of individual v in years t, which equals i_1 when individual v is alive at site i_1 in year t, and 0 if individual v is dead, $\phi_{i_1,t-1}$ is site- and year-specific probability of survival, which is a function of environmental covariates x's with intercept and slope parameters β's, θ_{i_1,i_2} is the standardized movement rate from any site i_1 to any site i_2, including the case when an individual stays at site i_1 (i.e., $i_1 = i_2$), η_{i_1,i_2} is the unstandardized movement rate, and κ is a decay parameter governing the relationship between movement and distance, d_{i_1,i_2}. We have $z_{v,t=f_v} = i_1$ because we always know the location of an individual v in the year of its first capture. In later years, individual v has a probability of $\phi_{i_1,t-1}$ to survive and a probability of θ_{i_1,i_2} to move from site i_1 to site i_2 (i.e., $z_{v,\,t} = i_2$), or a probability of $1 - \phi_{i_1,t-1}$ to die (i.e., $z_{v,\,t} = 0$), given it was alive at site i_1 in the previous year (i.e., $z_{v,\,t\,-\,1} = i_1$); it will always stay dead (i.e., $z_{v,\,t} = 0$) if it was already dead in the previous year (i.e., $z_{v,\,t\,-\,1} = 0$). Note that ϕ is the probability of true survival if all sites of potential habitat are modeled; if some sites are not modeled, which is likely to be the case in reality, ϕ is the probability of apparent survival, which is defined as the situation when an individual survive and stay in one of the modeled sites. Also, we say "modeled" instead of "surveyed" sites in the sense that unsurveyed sites can still be modeled as long as locations and covariates of these sites are available.

Given the survival-location history, the model then specifies the observed capture history as follows:

$$\left.\begin{array}{c} c_{v,t=f_v} = z_{v,t=f_v} = i \\ c_{v,t>f_v} \begin{cases} \sim Bernoulli(p_{i,t}) \times z_{v,t}, & z_{v,t} = i \\ = 0, & z_{v,t} = 0 \end{cases} \\ p_{i,t} = logit^{-1}(\alpha_0 + \alpha_1 \times w_{i,t,1} + \ldots + \alpha_{n_{pcvs}} \times w_{i,t,n_{pcvs}}) \end{array}\right\}, \tag{4.4}$$

in which $c_{v,\ t}$ is the capture history of individual v in year t, which equals i when individual v is captured at site i in year t, and 0 when it is not captured; $p_{i,\ t}$ is site- and year-specific capture probability and is a function of capture covariates w's with intercept and slope parameters α's. We always have $c_{v,t=f_v} = z_{v,t=f_v} = i$ for the initial capture. In later years, $c_{v,t>f_v}$ has a probability of $p_{i,\ t}$ to be i, and a probability of $1 - p_{i,\ t}$ to be 0, given individual v is alive at site i in that year (i.e., $z_{v,\ t} = i$); $c_{v,t>f_v}$ will always be 0 if individual v is dead in that year (i.e., $z_{v,\ t} = 0$).

Next, we can set up some basic values and simulate the individual survival-location status for the multistate model, using the code below. Because it is slow to compute this model when the number of sites is large, we limit the number of sites to 60. We assume that reef size has a positive effect ($\beta_{reef} = 0.6$) and temperature has a negative effect ($\beta_{temperature} = -0.4$) on apparent survival at a logit scale. We also assume a strong tendency of dispersal movement by having $\kappa = 0.5$.

```
library(boot) # for using logit and inv.logit functions
library(LaplacesDemon) # for generating random numbers from
categorical distributions (rcat)

set.seed(6)

# Basic values
nindv <- 10800 # number of captured individuals
nsite <- 60   # number of sites
nyear <- 10   # number of years
ncovs <- 2   # number of covariates
npcvs <- 1   # number of capture covariates

beta <- c( 0.4, 0.6, -0.4) # intercept and slopes for survival
probability
kappa <- 0.5           # distance effect on movement
alpha <- c(-1.4, -0.8)    # intercept and slopes for capture probability

# Read in environmental covariates
load('data/covariates.RData')
x1 <- (log(cov$size) - mean(log(cov$size))) / sd(log(cov$size)) #
standardized log habitat patch size
```

```
x2 <- (cov$temp - mean(cov$temp)) / sd(cov$temp) # standardized
temperature
x <- array(, dim=c(nsite, nyear, ncovs)) # environmental covariates
x[,,1] <- x1[1:nsite, 1:nyear]
x[,,2] <- x2[1:nsite, 1:nyear]

# Simulate data
### Process
phi <- matrix(, nsite, nyear-1) # survival probability
for (t in 1:(nyear-1)) {
 phi[,t] <- inv.logit(cbind(1,x[,t,]) %*% beta)
} # t

lon <- cov$lon[1:nsite] # longitude or easting
lat <- cov$lat[1:nsite] # latitude or northing
d <- matrix(, nsite, nsite) # distance matrix
for (i1 in 1:nsite) {
 for (i2 in 1:nsite) {
  d[i1,i2] <- sqrt((lon[i1] - lon[i2]) ^ 2 + (lat[i1] - lat[i2]) ^ 2)
 } # i2
} # i1
eta <- exp(-1 * kappa * d) # unstandardized colonization rate
theta <- eta / rowSums(eta) #  standardized colonization rate

year_f <- rep(1:(nyear-1), each=nindv/(nyear-1))            # year of
first capture
site_f <- rep(rep(1:nsite, each=nindv/nsite/(nyear-1)), times=nyear-
1) # site of first capture

z <- matrix(0, nindv, nyear) # individual survival-location history
for (v in 1:nindv) {
 z[v,year_f[v]] <- site_f[v]
 for (t in (year_f[v]+1):nyear) {
  if (z[v,t-1] %in% c(1:nsite)) {
   z[v,t] <- rcat(1, p=c(phi[z[v,t-1],t-1] * theta[z[v,t-1],], 1 - phi
[z[v,t-1],t-1]))
  } else {
   z[v,t] <- 0
  }
  if (!(z[v,t] %in% 1:nsite)) {
   z[v,t] <- 0
  }
 } # t
} # v
```

The simulated survival-location histories for the first 5 individuals look like the following. These values mean that individuals 1, 2, and 3 are alive only in year 1; individual 4 survives from year 1 through year 5, during which it moves from site 1 to site 34, then site 12, then site 22, and eventually back to site 12 before it dies; and individual 5 survives the first two years and moves from site 1 to site 14.

	Survival-location history									
	Year 1	Year 2	Year 3	Year 4	Year 5	Year 6	Year 7	Year 8	Year 9	Year 10
Individual 1	1	0	0	0	0	0	0	0	0	0
Individual 2	1	0	0	0	0	0	0	0	0	0
Individual 3	1	0	0	0	0	0	0	0	0	0
Individual 4	1	34	12	22	12	0	0	0	0	0
Individual 5	1	14	0	0	0	0	0	0	0	0

```
### Capture
w <- array(rnorm(nsite * (nyear-1) * npcvs, 0, 1), dim=c(nsite, nyear-1,
npcvs)) # capture covariates

p <- matrix(, nsite, nyear-1) # capture probability
for (t in 1:(nyear-1)) {
 p[,t] <- inv.logit(cbind(1,w[,t,]) %*% alpha)
} # t

c <- matrix(0, nindv, nyear) # individual capture history
for (v in 1:nindv) {
 c[v,year_f[v]] <- z[v,year_f[v]]
 for (t in (year_f[v]+1):nyear) {
  if (z[v,t] %in% c(1:nsite)) {
   c[v,t] <- ifelse(rbinom(1, 1, p[z[v,t],t-1]) == 1, z[v,t], 0)
  } else {
   c[v,t] <- 0
  }
 } # t
} # v
```

We then can use the following code to simulate the individual capture data.

The simulated capture histories for the first 5 individuals look like the following. Individuals 1, 2, 3, and 5 are captured in year 1; and individual 2 is captured in years 1, 2, 3, and 4. We can tell that individual 4 is missed in year 5, and individual 5 is missed in year 2 in this simulation, but we cannot know such a thing in reality.

	Capture history									
	Year 1	Year 2	Year 3	Year 4	Year 5	Year 6	Year 7	Year 8	Year 9	Year 10
Individual 1	1	0	0	0	0	0	0	0	0	0
Individual 2	1	0	0	0	0	0	0	0	0	0
Individual 3	1	0	0	0	0	0	0	0	0	0
Individual 4	1	34	12	22	0	0	0	0	0	0
Individual 5	1	0	0	0	0	0	0	0	0	0

We again will extend the individual capture history data and convert it to an m-array, using the code below:

```
# Extend individual capture history
for (v in 1:nindv) {
 t1 <- c[v,]
 ycap <- which(t1 %in% c(1:nsite))
 ncap <- length(ycap)
 if (ncap == 1) {
  t2 <- t1
 } else {
  t2 <- matrix(, ncap, nyear)
  for (h in 1:(ncap-1)) {
   t2[h, ycap[h]:ycap[h+1]] <- t1[ycap[h]:ycap[h+1]]
  } # h
  t2[ncap, ycap[ncap]] <- t1[ycap[ncap]]
 }
 if (v == 1) {
  c_ext <- t2
 } else {
  c_ext <- rbind(c_ext, t2)
 }
} # v
c_ext[which(is.na(c_ext))] <- 0

year_f_ext <- site_f_ext <- numeric(dim(c_ext)[1])
for (v in 1:dim(c_ext)[1]) {
 year_f_ext[v] <- min(which(c_ext[v,] %in% c(1:nsite)))
 site_f_ext[v] <- c_ext[v, year_f_ext[v]]
} # v

# Converting individual capture history to m-array
for (t in 2:nyear) {
 tt <- paste('site', '_', 1:nsite, ', year', t, sep='')
 if (t == 2) {
  nm <- tt
 } else {
```

```
    nm <- c(nm, tt)
  }
} # t
nm <- c(nm, 'not seen')

m <- array(0, # m-array of capture-recapture
  dim=c(nyear-1, nsite*(nyear-1)+1, nsite),
  dimnames=list(paste('year', 1:(nyear-1), sep='_'), nm, paste
('site', 1:nsite, sep='_')))
for (t1 in 1:(nyear-1)) {
  temp1 <- c_ext[which(year_f_ext == t1),]
  for (i1 in 1:nsite) {
    temp2 <- temp1[which(temp1[,t1] == i1),]
    nband <- length(which(temp1[,t1] == i1))
    for (t2 in (t1+1):nyear) {
      for (i2 in 1:nsite) {
        m[t1,(t2-2)*nsite+i2,i1] <- length(which(temp2[,t2] == i2))
      } # i2
    } # t2
    m[t1,nsite*(nyear-1)+1,i1] <- nband - sum(m[t1,1:(nsite*(nyear-1)),
i1])
  } # i1
} # t1
```

The m-array for the multistate model is relatively complicated. Imagine a survey with two sites and 3 years. The m-array will look like the following. Each row of this matrix follows a multinomial distribution.

		Recapture				
		Site 1, year 2	Site 2, year 2	Site 1, year 3	Site 2, year 3	Not recaptured
Capture	Site 1, year 1	14	16	22	33	167
	Site 2, year 1	11	12	17	18	225
	Site 1, year 2	0	0	40	2	186
	Site 2, year 2	0	0	37	22	154

The cell probabilities of above-mentioned matrix look like the following. Certain cells have 0's because we cannot recapture individuals in the same year or before they are captured. Cell probabilities for next-year recapture, e.g., $Pr_{1,\ 1,\ 2,\ 2}$, the probability of being captured at site 1 in year 1 and recaptured at site 2 in year 2, is the product of the survival probability at site 1 in year 1, $\phi_{1,\ 1}$, the movement rate from site 1 to site 2, $\theta_{1,\ 2}$, and the probability of capture at site 2 in year 2, $p_{2,\ 2}$. Beyond next-year recapture, the cell probability of, for example, $Pr_{1,\ 1,\ 2,\ 3}$, the probability of being captured at site 1 in year 1 and *first-time* recaptured at site 2 in year 3, immediately becomes quite complicated as we have to consider all the possibilities of where these individuals are in the "latent" second year. Thus, $Pr_{1, 1, 2, 3}$ is calculated as $\phi_{1, 1} \times \theta_{1, 1} \times (1 - p_{1, 2}) \times \phi_{1, 2} \times \theta_{1, 2} \times p_{2, 3} + \phi_{1, 1} \times \theta_{1, 2} \times (1 - p_{2, 2}) \times \phi_{2, 2} \times \theta_{2, 2} \times p_{2, 3}$. You can imagine that these calculations become even more complicated with more sites and more years.

		Recapture				
		Site 1, year 2	Site 2, year 2	Site 1, year 3	Site 2, year 3	Not recaptured
Capture	Site 1, year 1	$Pr_{1, 1, 1, 2}$	$Pr_{1, 1, 2, 2}$	$Pr_{1, 1, 1, 3}$	$Pr_{1, 1, 2, 3}$	$1 - Pr_{1, 1, 1, 2} - Pr_{1, 1, 2, 2} - Pr_{1, 1, 1, 3} - Pr_{1, 1, 2, 3}$
	Site 2, year 1	$Pr_{2, 1, 1, 2}$	$Pr_{2, 1, 2, 2}$	$Pr_{2, 1, 1, 3}$	$Pr_{2, 1, 2, 3}$	$1 - Pr_{2, 1, 1, 2} - Pr_{2, 1, 2, 2} - Pr_{2, 1, 1, 3} - Pr_{2, 1, 2, 3}$
	Site 1, year 2	0	0	$Pr_{1, 2, 1, 3}$	$Pr_{1, 2, 2, 3}$	$1 - Pr_{1, 2, 1, 3} - Pr_{1, 2, 2, 3}$
	Site 2, year 2	0	0	$Pr_{2, 2, 1, 3}$	$Pr_{2, 2, 2, 3}$	$1 - Pr_{2, 2, 1, 3} - Pr_{2, 2, 2, 3}$

Despite the apparent complexity, we can use the values in early-occasion cells to calculate the cell probabilities in late-occasion cells of the m-array for multistate models, as shown below. Later we will use this function in the MCMC algorithm.

```
# Define the function to calculate the cell probability of m-array
cell_prob_fun <- function(nsite, nyear, phi, theta, p) {
 q <- 1 - p
 cell_prob_p <- cell_prob_q <- array(0, dim=c(nyear-1, nsite*(nyear-
1)+1, nsite))
 for (i in 1:nsite) {
  for (t in 1:(nyear-1)) {
   cell_prob_p[t,nsite*(t-1)+1:nsite,i] <- phi[i,t] * theta[i,] * p[,
t]
   cell_prob_q[t,nsite*(t-1)+1:nsite,i] <- phi[i,t] * theta[i,] * q[,
t]
  } # t
  for (t1 in 1:(nyear-2)) {
   for (t2 in (t1+1):(nyear-1)) {
    cell_prob_p[t1,nsite*(t2-1)+1:nsite,i] <- ((cell_prob_q[t1,
nsite*(t2-2)+1:nsite,i] * phi[,t2]) %*% theta) * p[,t2]
    cell_prob_q[t1,nsite*(t2-1)+1:nsite,i] <- ((cell_prob_q[t1,
nsite*(t2-2)+1:nsite,i] * phi[,t2]) %*% theta) * q[,t2]
   } # t2
  } # t1
  for (t in 1:(nyear-1)) {
   cell_prob_p[t,nsite*(nyear-1)+1,i] <- 1 - sum(cell_prob_p[t,1:
(nsite*(nyear-1)),i])
  } # t
```

```
  } # i
  cell_prob <- cell_prob_p

  return(cell_prob)
} # cell_prob_fun
```

We then develop the MCMC algorithm to estimate parameters β, κ, and α given data m, x, w, and d.

```
multistate_mcmc <- function(m, x, w, d, nmcmc) {

  # Setup variables
  nsite <- dim(x)[1]
  nyear <- dim(x)[2]
  ncovs <- dim(x)[3]
  npcvs <- dim(w)[3]

  beta_save <- matrix(0, ncovs+1, nmcmc)
  kappa_save <- rep(0, nmcmc)
  alpha_save <- matrix(0, npcvs+1, nmcmc)

  # Prior
  beta_mean <- rep(0, ncovs+1)
  beta_sd <- 2
  log_kappa_mean <- 0
  log_kappa_sd <- 2
  alpha_mean <- rep(0, npcvs+1)
  alpha_sd <- 2

  # Starting values
  beta <- rep(0, ncovs+1)
  kappa <- 1
  alpha <- rep(0, npcvs+1)
  phi <- matrix(0.5, nsite, nyear-1)
  eta <- exp(-1 * kappa * d)
  theta <- eta / rowSums(eta)
  p <- matrix(0.5, nsite, nyear-1) # capture probability
  cell_prob <- cell_prob_fun(nsite=nsite, nyear=nyear, phi=phi,
theta=theta, p=p)

  # Tuning factor
  beta_tune <- 0.035
  kappa_tune <- 0.035
  alpha_tune <- 0.035
```

```
# Begin MCMC loop
for (k in 1:nmcmc) {
  # Sample beta
  beta_star <- rnorm(ncovs+1, beta, beta_tune)
  phi_star <- matrix(, nsite, nyear-1)
  for (t in 1:(nyear-1)) {
   phi_star[,t] <- inv.logit(cbind(1,x[,t,]) %*% beta_star)
  } # t
  cell_prob_star <- cell_prob_fun(nsite=nsite, nyear=nyear,
phi=phi_star, theta=theta, p=p)
  prob1 <- prob2 <- matrix(, nsite, nyear-1)
  for (i in 1:nsite) {
   for (t in 1:(nyear-1)) {
    prob1[i,t] <- dmultinom(m[t,,i], size=sum(m[t,,i]),
prob=cell_prob_star[t,,i], log=T)
    prob2[i,t] <- dmultinom(m[t,,i], size=sum(m[t,,i]),
prob=cell_prob     [t,,i], log=T)
   } # t
  } # i
  mh1 <- sum(prob1) +
      sum(dnorm(beta_star, beta_mean, beta_sd, log=T))
  mh2 <- sum(prob2) +
      sum(dnorm(beta     , beta_mean, beta_sd, log=T))
  mh <- exp(mh1 - mh2)
  if (mh > runif(1)) {
   beta <- beta_star
   phi <- phi_star
   cell_prob <- cell_prob_star
  }

  # Sample kappa
  kappa_star <- exp(rnorm(1, log(kappa), kappa_tune))
  eta_star <- exp(-1 * kappa_star * d)
  theta_star <- eta_star / rowSums(eta_star)
  cell_prob_star <- cell_prob_fun(nsite=nsite, nyear=nyear, phi=phi,
theta=theta_star, p=p)
  prob1 <- prob2 <- matrix(, nsite, nyear-1)
  for (i in 1:nsite) {
   for (t in 1:(nyear-1)) {
    prob1[i,t] <- dmultinom(m[t,,i], size=sum(m[t,,i]),
prob=cell_prob_star[t,,i], log=T)
    prob2[i,t] <- dmultinom(m[t,,i], size=sum(m[t,,i]),
prob=cell_prob     [t,,i], log=T)
   } # t
  } # i
```

```
  mh1 <- sum(prob1) +
      dnorm(log(kappa_star), log_kappa_mean, log_kappa_sd, log=T)
  mh2 <- sum(prob2) +
      dnorm(log(kappa     ), log_kappa_mean, log_kappa_sd, log=T)
  mh <- exp(mh1 - mh2)
  if (mh > runif(1)) {
   kappa <- kappa_star
   eta <- eta_star
   theta <- theta_star
   cell_prob <- cell_prob_star
  }

  # Sample alpha
  alpha_star <- rnorm(npcvs+1, alpha, alpha_tune)
  p_star <- matrix(, nsite, nyear-1)
  for (t in 1:(nyear-1)) {
   p_star[,t] <- inv.logit(cbind(1,w[,t,]) %*% alpha_star)
  } # t
  cell_prob_star <- cell_prob_fun(nsite=nsite, nyear=nyear, phi=phi,
theta=theta, p=p_star)
  prob1 <- prob2 <- matrix(, nsite, nyear-1)
  for (i in 1:nsite) {
   for (t in 1:(nyear-1)) {
    prob1[i,t] <- dmultinom(m[t,,i], size=sum(m[t,,i]),
prob=cell_prob_star[t,,i], log=T)
    prob2[i,t] <- dmultinom(m[t,,i], size=sum(m[t,,i]),
prob=cell_prob     [t,,i], log=T)
   } # t
  } # i
  mh1 <- sum(prob1) +
      sum(dnorm(alpha_star, alpha_mean, alpha_sd, log=T))
  mh2 <- sum(prob2) +
      sum(dnorm(alpha     , alpha_mean, alpha_sd, log=T))
  mh <- exp(mh1 - mh2)
  if (mh > runif(1)) {
   alpha <- alpha_star
   p <- p_star
   cell_prob <- cell_prob_star
  }

  # Save samples
  beta_save[,k] <- beta
  kappa_save[k] <- kappa
  alpha_save[,k] <- alpha
 } # k
```

```
list(beta_save=beta_save, kappa_save=kappa_save,
   alpha_save=alpha_save)
} # multistate_mcmc
```

When using the following code to implement the algorithm, it takes about 7 h to run. Then we can use the code in Appendix 4.3 to create traceplots for the monitored parameters, which show good model convergence and consistency between parameter estimates of β, κ, and α and their true values used in data simulation (Fig. 4.5).

```
library(foreach)  # for parallel computing
library(doParallel) # for parallel computing

numCores <- round(detectCores() / 2) # only use half of the cores for
parallel computing
registerDoParallel(numCores) # setup parallel computing
```

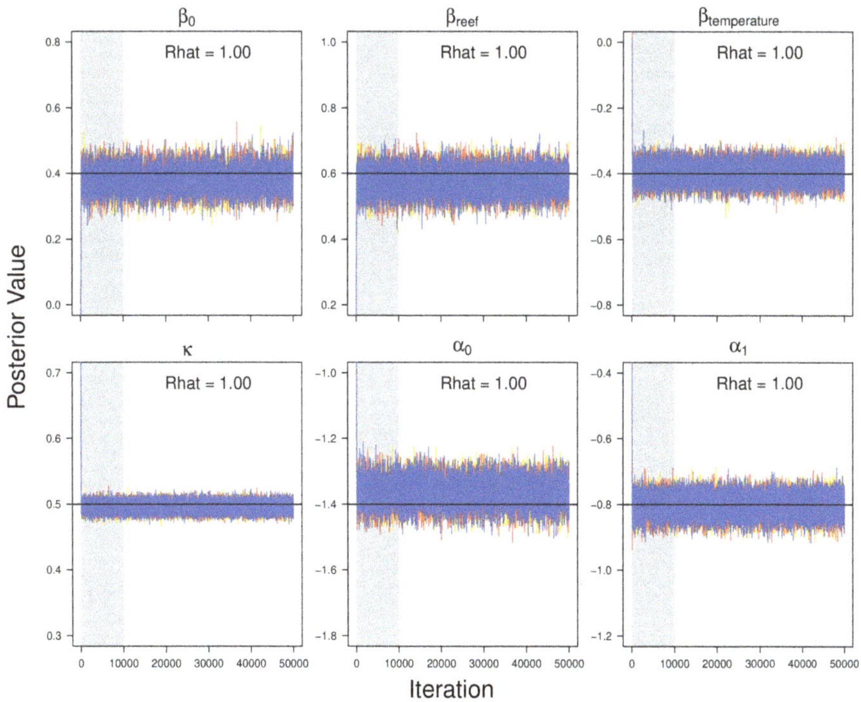

Fig. 4.5 Traceplots showing the posterior chains for β, κ, and α in the multistate model. Rhat values are calculated after discarding the first 10,000 iterations. Horizontal lines represent the true values used in data simulation

```
nmcmc <- 50000 # number of iterations of MCMC computing
chain <- 3   # number of chains

start_time <- Sys.time() # start time of computing
out <- foreach (i=1:chain, .packages='boot') %dopar% {
 multistate_mcmc(m=m, x=x, w=w, d=d, nmcmc=nmcmc)
} # i
stopImplicitCluster() # clean up the cluster
end_time <- Sys.time() # end time of computing
time_taken <- end_time - start_time # computing time
```

Next, let us use the following code to plot the unstandardized movement rate, η, which shows good agreement between the simulated trues values and estimates (Fig. 4.6). Because η is a function of distance, its value is higher (represented by a thicker line) when a pair of sites are closer to each other, and vice versa.

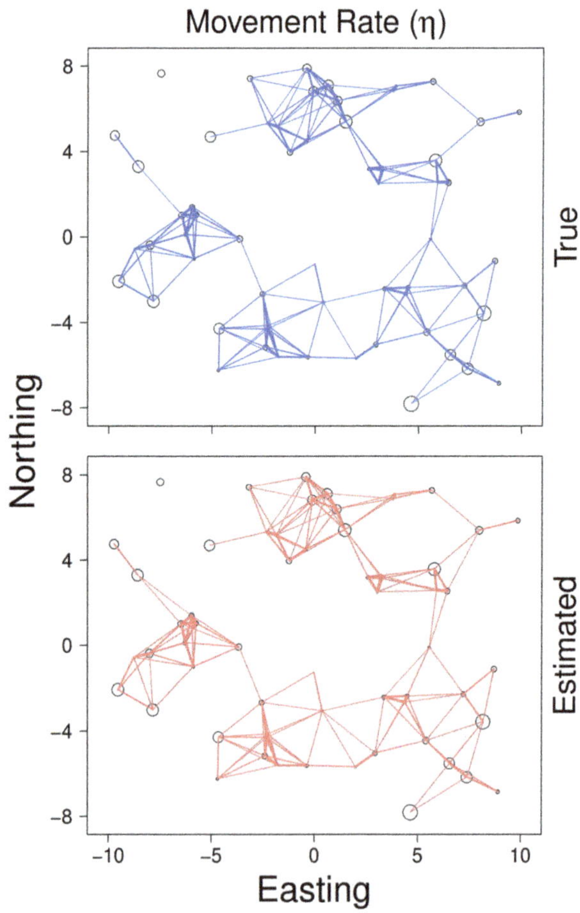

Fig. 4.6 Simulated true (upper row) and estimated (lower row) unstandardized movement rates (η) between sites. A thicker line indicates a higher rate of movement between the pair of sites, and vice versa. Only η's that are greater than 0.2 are plotted

```
kappa_post <- c(out[[1]]$kappa[(nmcmc*0.2+1):nmcmc], out[[2]]$kappa
[(nmcmc*0.2+1):nmcmc], out[[3]]$kappa[(nmcmc*0.2+1):nmcmc])
kappa_med <- median(kappa_post)

eta_true <- exp(-1 * kappa   * d)
eta_pred <- exp(-1 * kappa_med * d)

min_eta <- 0.2
max_lwd <- 5

par(mfrow=c(2,1))
par(mar=c(1,1,1,1))
par(oma=c(4,5,2,2))

plot(lon, lat, xlim=c(-10.2,10.2), ylim=c(-8.2,8.2), axes=F,
xlab='', ylab='',
   pch=21, cex=rowMeans(cov$size)[1:nsite]*4.6, col='grey18',
bg='white')
axis(1, at=seq(-10,10,5), labels=rep('',5))
axis(2, at=seq(-08,08,4), cex.axis=1.6, las=2)
box()
for (i1 in 1:(nsite-1)) {
 for (i2 in (i1+1):nsite) {
  if (eta_true[i1,i2] >= min_eta) {
   lines(x=lon[c(i1,i2)], y=lat[c(i1,i2)], col='royalblue',
lwd=eta_true[i1,i2]*max_lwd)
  }
 } # i2
} # i1
axis(3, at=0, labels=expression(paste('Movement Rate (', eta, ')',
sep='')),
   cex.axis=2.5, tick=F, line=-0.6)
axis(4, at=0, labels='True', cex.axis=2.5, tick=F, line=0.5)

plot(lon, lat, xlim=c(-10.2,10.2), ylim=c(-8.2,8.2), axes=F,
xlab='', ylab='',
   pch=21, cex=rowMeans(cov$size)[1:nsite]*4.6, col='grey18',
bg='white')
axis(1, at=seq(-10,10,5), cex.axis=1.6)
axis(2, at=seq(-08,08,4), cex.axis=1.6, las=2)
box()
for (i1 in 1:(nsite-1)) {
 for (i2 in (i1+1):nsite) {
  if (eta_pred[i1,i2] >= min_eta) {
   lines(x=lon[c(i1,i2)], y=lat[c(i1,i2)], col='lightcoral',
```

```
lwd=eta_pred[i1,i2]*max_lwd)
  }
 } # i2
} # i1
axis(4, at=0, labels='Estimated', cex.axis=2.5, tick=F, line=0.5)

title(xlab='Easting' , line=2.6, cex.lab=2.8, outer=T)
title(ylab='Northing', line=2.4, cex.lab=2.8, outer=T)
```

4.4 Spatial Integrated Population Models

In Chap. 3, we learned that using count data in dynamic N-mixture models can provide some information for demographic processes, but separating the entire BIDE processes with count data only is still challenging. On the other hand, in Sect. 4.3 we learned that using capture-recapture data in multistate models can provide information for survival and movement. Now, we will introduce spatial IPMs that can leverage the strength of count and capture-recapture data. The spatial IPM that we will introduce in Sect. 4.4.1 is none other than a combination of a spatial dynamic N-mixture model (Zhao et al. 2017) as its population sub-model, a reproduction sub-model, and multistate model as its survival-movement sub-model. By combining these models, this spatial IPM eventually allows us to understand the entire BIDE processes and their drivers, making it more advantageous than the spatial dynamic N-mixture model. In later sections of this chapter, we will further extend this spatial IPM to account for age-, body-class-, or gene-class-specific demographic rates.

4.4.1 Without Age-Specific Survival and Movement

The spatial IPM we introduce in this section assumes that adult breeders of our coral reef fish can reproduce juvenile offspring, the number of which depends on the number of adults (i.e., breeding population size), the size, and the temperature of the reef where the fish live. Juveniles cannot reproduce when they are born, but juveniles become adults and thus can reproduce after 1 year, given they survive. At this point, the model assumes that juveniles have the same survival probability and movement rate as adults; survival probability is driven by the number of adults, the size, and the temperature of the reef where the fish lives; the probability of moving between reefs depends on the distance between the reefs. Eventually, the number of adults living in a reef is the sum of the number of adults and juveniles that survive from the last year and stay in the reef, as well as of those that move into the reef. When researchers go to survey a reef, they divide themselves into two groups: while one group does a capture-recapture survey of the fish, the other group independently counts the number of adults and juveniles without worrying about their individual identification. The timing of the surveys has to be perfect such that no adult has died and all juveniles have been born for that breeding season.

Using statistical language, we have the process model that first describes the initial abundance such that:

$$\left.\begin{array}{r}N_{i,1} \sim Poisson(\lambda_{i,1}) \\ \lambda_{i,1} = exp\left(\beta_0^{[0]} + \beta_1^{[0]} \times x_{i,1,1} + \ldots + \beta_{n_{covs}}^{[0]} \times x_{i,1,n_{covs}}\right)\end{array}\right\}, \tag{4.5}$$

in which $N_{i,\ 1}$ is the abundance of adult individuals (i.e., breeding population size) at site i in the first year, and $\lambda_{i,\ 1}$ is the expectation of initial abundance and is a function of covariates x's with intercept and slope parameters $\beta^{[0]}$'s. From the second year (i.e., $t \geq 2$), the process sub-model assumes that the abundance of adults is due to reproduction, survival, and movement (another way of expressing the BIDE processes) such that:

$$\left.\begin{array}{r}N_{i_2,t} \sim Poisson(\lambda_{i_2,t}) \\ \lambda_{i_2,t} = \sum_{i_1=1}^{n_{site}} N_{i_1,t-1} \times \left(1 + \gamma_{i_1,t-1}\right) \times \phi_{i,t-1} \times \theta_{i_1,i_2,t} \\ \gamma_{i_1,t-1} = exp\left(\beta_0^{[\gamma]} + \beta_{density}^{[\gamma]} \times N_{i_1,t-1} + \beta_1^{[\gamma]} \times x_{i_1,t,1} + \ldots + \beta_{n_{covs}}^{[\gamma]} \times x_{i_1,t,n_{covs}}\right) \\ \phi_{i_1,t-1} = logit^{-1}\left(\beta_0^{[\phi]} + \beta_{density}^{[\phi]} \times N_{i_1,t-1} + \beta_1^{[\phi]} \times x_{i_1,t,1} + \ldots + \beta_{n_{covs}}^{[\phi]} \times x_{i_1,t,n_{covs}}\right) \\ \theta_{i_1,i_2} = \frac{\eta_{i_1,i_2}}{\sum_{i_2=1}^{n_{site}} \eta_{i_1,i_2}} \\ \eta_{i_1,i_2} = exp(-1 \times \kappa \times d_{i_1,i_2})\end{array}\right\}, \tag{4.6}$$

in which $\gamma_{i_1,t-1}$ is per capita reproduction rate at site i_1 and year $t - 1$ and is a function of density and environmental covariates x's with intercept and slope parameters $\beta^{[\gamma]}$'s; $\phi_{i_1,t-1}$ is survival probability and is a function of density and environmental covariates with intercept and slope parameters $\beta^{[\phi]}$'s; θ_{i_1,i_2} is the standardized movement rate from sites i_1 to i_2; and η_{i_1,i_2} is the unstandardized movement rate governed by a decay parameter κ and distance between sites i_1 to i_2, denoted d_{i_1,i_2}. Eventually, the abundance of adults at site i_2 in year t is the total number of animals that either stay at site i_2 or move in from any other sites i_1, given they survive to year t. Note that γ and ϕ represent reproduction and true survival if all potential sites are modeled; if some sites are not modeled, these parameters represent recruitment (reproduction plus immigration from unmodeled sites) and apparent survival (true survival minus emigration to unmodeled sites). We can consider modeling all sites, including unsurveyed sites, as long as locations and covariates of these sites are available.

It is worth noting that we will simulate data for this and following spatial IPMs using an individual-based approach to achieve great realism, an idea akin to the observation system simulation experiment (Arnold and Dey 1986; Wikle et al. 2019). This approach allows us to simulate several latent state variables that are useful for simulating capture-recapture and count data. First, we simulate $z_{v,\,t}$, the survival-location status of individual v in year t, and have $\sum_{v=1}^{n_{nind}} I(z_{v,t}=i)=N_{i,t}$, in which n_{nind} is the total number of individuals that are simulated, and I is an indicator function which equals 1 when $z_{v,\,t} = i$, and 0 otherwise. This equation means that abundance at site i in year t is none other than the total number of individuals that are alive at site i in year t, linking the population-level state variable with individual status. Second, we simulate $R_{i,\,t}$, the number of juveniles reproduced at site i in year t, by having $R_{i,\,t}{\sim}Poisson(N_{i,\,t} \times \lambda_{i,\,t})$.

When researchers conduct capture-recapture survey of the animals, they obtain individual capture histories according to the following:

$$\left.\begin{aligned} c_{v,t,j} &\sim Bernoulli\left(p^{[cap]}_{z_{v,t},t,j}\right) \times z_{v,t} \\ p^{[cap]}_{i,t,j} &= logit^{-1}\left(\alpha^{[cap]}_0 + \alpha^{[cap]}_1 \times w_{i,t,j,1} + \ldots + \alpha^{[cap]}_{n_{pcvs}} \times w_{i,t,j,n_{pcvs}}\right) \end{aligned}\right\}, \tag{4.7}$$

in which $c_{v,\,t,\,j}$ is the observed capture history of individual v in year t during the jth visit (we allow n_{reps} visits per year), $p^{[cap]}_{i,t,j}$ is the capture probability and is a function of covariates w's governed by intercept and slope parameters $\alpha^{[cap]}$'s. Note that only individuals that are captured at least once during the study period produce an observed capture history. Individual capture histories are later converted into an m-array format for analysis, as we introduced in Sect. 4.3.

When researchers count the animals, they are assumed to be able to distinguish juveniles from adults and count them separately. The count data follow binomial distributions such that

$$\left.\begin{aligned} y_{i,t,j} &\sim Binomial\left(N_{i,t}, p^{[det]}_{i,t,j}\right) \\ r_{i,t,j} &\sim Binomial\left(R_{i,t}, p^{[det]}_{i,t,j}\right) \\ p^{[det]}_{i,t,j} &= logit^{-1}\left(\alpha^{[det]}_0 + \alpha^{[det]}_1 \times w_{i,t,j,1} + \ldots + \alpha^{[det]}_{n_{pcvs}} \times w_{i,t,j,n_{pcvs}}\right) \end{aligned}\right\}, \tag{4.8}$$

In which $y_{i,\,t,\,j}$ is the counts of adults, $r_{i,\,t,\,j}$ is the counts of juveniles, and $p^{[det]}_{i,t,j}$ is the detection probability and is a function of covariates w's with intercept and slope parameters $\alpha^{[det]}$'s. We let $p^{[det]}_{i,t,j}$ be the same for adults and juveniles for the sake of simplicity, but it can be easily extended to be age specific.

Note that sampling the latent state variable $R_{i,\,t}$ can be challenging. Therefore, instead of having $r_{i,t,j}{\sim}Binomial\left(R_{i,t}, p^{[det]}_{i,t,j}\right)$ in Eq. (4.8), we have $r_{i,t,j}{\sim}Binomial\left(y_{i,t,j}, \frac{\gamma_{i,t}}{\gamma_{i,t}+1}\right)$ in our MCMC algorithm, to sample γ and $\beta^{[\gamma]}$ without sampling $R_{i,\,t}$. This approach is computationally efficient and provides valid parameter estimates.

Next, we will use the code below to simulate data for this spatial IPM without age-specific survival and movement. We assume positive effects of reef size on initial abundance ($\beta^{[0]}_{reef} = 1$), reproduction ($\beta^{[\gamma]}_{reef} = 0.4$) and survival ($\beta^{[\phi]}_{reef} = 0.6$), and negative effects of temperature on initial abundance ($\beta^{[0]}_{temperature} = -0.8$), reproduction ($\beta^{[\gamma]}_{temperature} = -0.2$) and survival ($\beta^{[\phi]}_{temperature} = -0.3$). We also assume negative density dependence on reproduction ($\beta^{[\gamma]}_{density} = -0.3$) and survival ($\beta^{[\phi]}_{density} = -0.5$). We allow long-distance movement by using a relatively weak decay rate ($\kappa = 0.5$).

```
library(boot) # for using logit and inv.logit functions
library(LaplacesDemon) # for generating random numbers from
categorical distributions (rcat)

set.seed(6)

# Basic values
nsite <- 60 # number of sites
nyear <- 10 # number of years
nreps <- 5  # number of within-season replicates
ncovs <- 2  # number of environmental covariates
npcvs <- 2  # number of observational covariates

beta0 <- c(3, 1, -0.8)        # intercept and slopes for initial abundance
(i.e., number of adults)
beta_gamma <- c(-0.8, -0.3, 0.4, -0.2) # intercept and slopes for
reproduction rate
beta_phi  <- c( 0.5, -0.5, 0.6, -0.3) # intercept and slopes for survival
probability
kappa <- 0.5                  # distance effect on movement rate
alpha_det <- c(0.9, -0.4, 0.2)       # intercept and slopes for detection
probability
alpha_cap <- c(-3, -0.8, 0.4)       # intercept and slopes for capture
probability

# Read in environmental covariates
load('data/covariates.RData')
x1 <- (log(cov$size) - mean(log(cov$size))) / sd(log(cov$size)) #
standardized log habitat patch size
x2 <- (cov$temp - mean(cov$temp)) / sd(cov$temp) # standardized
temperature
x <- array(, dim=c(nsite, nyear, ncovs)) # environmental covariates
x[,,1] <- x1[1:nsite, 1:nyear]
x[,,2] <- x2[1:nsite, 1:nyear]
### Standardize covariates for each site
for (i in 1:nsite) {
 for (h in 1:ncovs) {
  x[i,2:nyear,h] <- (x[i,2:nyear,h] - mean(x[i,,h])) / sd(x[i,,h])
 } # h
} # i
```

```
# Simulate data
N <- matrix(, nsite, nyear) # abundance
lambda0 <- exp(cbind(1,x[,1,]) %*% beta0) # expectation of initial
abundance
N[,1] <- rpois(nsite, lambda0) # initial abundance

lon <- cov$lon[1:nsite] # longitude or easting
lat <- cov$lat[1:nsite] # latitude or northing
d <- matrix(, nsite, nsite)  # distance matrix
for (i1 in 1:nsite) {
 for (i2 in 1:nsite) {
  d[i1,i2] <- sqrt((lon[i1] - lon[i2]) ^ 2 + (lat[i1] - lat[i2]) ^ 2)
 } # i2
} # i1
eta <- exp(-1 * kappa * d)  # unstandardized colonization rate
theta <- eta / rowSums(eta) #  standardized colonization rate

nindv <- 15000 # maximum number of individuals for simulation
z <- matrix(0, nindv, nyear) # individual survival-location history
for (i in 1:nsite) {
 if (i == 1) { tt <- rep(i, N[i,1]) } else { tt <- c(tt, rep(i, N[i,1])) }
}
z[1:length(tt),1] <- tt

gamma <- matrix(, nsite, nyear-1) # reproduction rate
phi  <- matrix(, nsite, nyear-1) # survival probability
R   <- matrix(, nsite, nyear-1) # number of offspring

for (t in 2:nyear) {
 gamma[,t-1] <- exp(cbind(1,(N[,t-1]-lambda0)/lambda0,x[,t,]) %*%
beta_gamma)
 R[,t-1] <- rpois(nsite, N[,t-1] * gamma[,t-1])

 for (i in 1:nsite) {
  if (i == 1) { tt <- rep(i, R[i,t-1]) } else { tt <- c(tt, rep(i, R[i,
t-1])) }
 } # i
 z[max(which(rowMeans(z,na.rm=T)>0))+c(1:length(tt)), t-1] <- tt

 phi[,t-1] <- inv.logit(cbind(1,(N[,t-1]-lambda0)/lambda0,x[,t,]) %*
% beta_phi)
 transition_matrix <- matrix(, nsite, nsite+1) # matrix of transition
(survival & movement) probability
 for (i in 1:nsite) {
  transition_matrix[i,1:nsite] <- phi[i,t-1] * theta[i,]
```

```
  transition_matrix[i,nsite+1] <- 1 - phi[i,t-1]
 } # i

 for (v in 1:nindv) {
  if (z[v,t-1] %in% 1:nsite) {
   z[v,t] <- rcat(1, p=transition_matrix[z[v,t-1],])
  }
 } # v
 z[which(!(z %in% 1:nsite))] <- 0

 for (i in 1:nsite) {
  N[i,t] <- length(which(z[,t] == i))
 }
} # t
z[which(!(z %in% 1:nsite))] <- 0
z <- z[which(rowMeans(z, na.rm=T) != 0),] # remove individuals that
never exist

# Simulate count data
w <- array(rnorm(nsite * nyear * nreps * npcvs, 0, 1), dim=c(nsite,
nyear, nreps, npcvs)) # observational covariates

pdet <- array(, dim=c(nsite, nyear, nreps)) # detection probability
for (i in 1:nsite) {
 for (t in 1:nyear) {
  pdet[i,t,] <- inv.logit(cbind(1,w[i,t,,]) %*% alpha_det)
 } # t
} # i

y <- array(, dim=c(nsite, nyear, nreps)) # count data of breeding
population
for (i in 1:nsite) {
 for (t in 1:nyear) {
  y[i,t,] <- rbinom(nreps, N[i,t], pdet[i,t,])
 } # t
} # i

r <- array(, dim=c(nsite, nyear-1, nreps)) # count data of offspring
for (i in 1:nsite) {
 for (t in 1:(nyear-1)) {
  r[i,t,] <- rbinom(nreps, R[i,t], pdet[i,t,])
 } # t
} # i
```

```
# simulate capture-recapture data
pcap_per_visit <- array(, dim=c(nsite, nyear, nreps)) # per visit
capture probability
for (i in 1:nsite) {
 for (t in 1:nyear) {
  pcap_per_visit[i,t,] <- inv.logit(cbind(1,w[i,t,,]) %*% alpha_cap)
 } # t
} # i

c_all <- matrix(, dim(z)[1], nyear) # capture history of all
individuals,
                    # including individuals that have never been captured
for (v in 1:dim(z)[1]) {
 for (t in 1:nyear) {
  if (z[v,t] == 0) {
   c_all[v,t] <- 0
  } else {
   cap_per_visit <- rbinom(nreps, 1, pcap_per_visit[z[v,t],t,])
   c01 <- ifelse(sum(cap_per_visit)==0, 0, 1)
   c_all[v,t] <- c01 * z[v,t]
  }
 } # t
} # v
if_cap <- ifelse(rowSums(c_all) > 0, 1, 0)

c <- c_all[which(if_cap==1),] # capture history of individuals that
have ever been captured
f <- numeric(dim(c)[1]) # year of first capture
for (v in 1:length(f)) {
 f[v] <- min(which(c[v,] %in% c(1:nsite)))
} # v
c <- c[which(f %in% c(1:(nyear-1))),]

# Extend individual capture history
for (v in 1:dim(c)[1]) {
 t1 <- c[v,]
 ycap <- which(t1 %in% c(1:nsite))
 ncap <- length(ycap)
 if (ncap == 1) {
  t2 <- t1
 } else {
  t2 <- matrix(, ncap, nyear)
  for (k in 1:(ncap-1)) {
   t2[k, ycap[k]:ycap[k+1]] <- t1[ycap[k]:ycap[k+1]]
  } # k
```

```
  t2[ncap, ycap[ncap]] <- t1[ycap[ncap]]
 }
 if (v == 1) {
  c_ext <- t2
 } else {
  c_ext <- rbind(c_ext, t2)
 }
} # v
c_ext[which(is.na(c_ext))] <- 0

f_ext <- numeric(dim(c_ext)[1]) # year of first capture for the extended
individual capture history
for (v in 1:length(f_ext)) {
 f_ext[v] <- min(which(c_ext[v,] %in% c(1:nsite)))
} # i

# Converting individual capture history to m-array format
for (t in 2:nyear) {
 tt <- paste('site', '_', 1:nsite, ', year', t, sep='')
 if (t == 2) {
  nm <- tt
 } else {
  nm <- c(nm, tt)
 }
} # t
nm <- c(nm, 'not seen')

m <- array(0, # m-array of capture-recapture
 dim=c(nyear-1, nsite*(nyear-1)+1, nsite),
 dimnames=list(paste('year', 1:(nyear-1), sep='_'), nm, paste
('site', 1:nsite, sep='_')))
for (t1 in 1:(nyear-1)) {
 temp1 <- c_ext[which(f_ext == t1),]
 for (i1 in 1:nsite) {
  temp2 <- temp1[which(temp1[,t1] == i1),]
  nband <- length(which(temp1[,t1] == i1))
  if (nband == 1) {
   temp2 <- matrix(temp2, nrow=1, ncol=length(temp2))
  }
  for (t2 in (t1+1):nyear) {
   for (i2 in 1:nsite) {
    m[t1,(t2-2)*nsite+i2,i1] <- length(which(temp2[,t2] == i2))
   } # j
  } # s
  m[t1,nsite*(nyear-1)+1,i1] <- nband - sum(m[t1,1:(nsite*(nyear-1)),
```

```
i1])
 } # i1
} # t1
```

Next, we will develop a function to calculate the cell probabilities in the m-array and then our own custom MCMC algorithm to estimate parameters $\beta^{[0]}$, $\beta^{[\gamma]}$, $\beta^{[\phi]}$, κ, $\alpha^{[cap]}$, and $\alpha^{[det]}$ using data y, r, m, x, w, and d.

```
# Function to calculate the cell probability of m-array
cell_prob_fun <- function(nsite, nyear, phi, theta, pcap) {
 qcap <- 1 - pcap
 cell_prob_p <- cell_prob_q <- array(0, dim=c(nyear-1, nsite*(nyear-
1)+1, nsite))
 for (i in 1:nsite) {
  for (t in 1:(nyear-1)) {
   cell_prob_p[t,nsite*(t-1)+1:nsite,i] <- phi[i,t] * theta[i,] * pcap
[,t]
   cell_prob_q[t,nsite*(t-1)+1:nsite,i] <- phi[i,t] * theta[i,] * qcap
[,t]
  } # t
  for (t1 in 1:(nyear-2)) {
   for (t2 in (t1+1):(nyear-1)) {
    cell_prob_p[t1,nsite*(t2-1)+1:nsite,i] <- ((cell_prob_q[t1,
nsite*(t2-2)+1:nsite,i] * phi[,t2]) %*% theta) * pcap[,t2]
    cell_prob_q[t1,nsite*(t2-1)+1:nsite,i] <- ((cell_prob_q[t1,
nsite*(t2-2)+1:nsite,i] * phi[,t2]) %*% theta) * qcap[,t2]
   } # t2
  } # t1
  for (t in 1:(nyear-1)) {
   cell_prob_p[t,nsite*(nyear-1)+1,i] <- 1 - sum(cell_prob_p[t,1:
(nsite*(nyear-1)),i])
  } # t
 } # i
 cell_prob <- cell_prob_p

 return(cell_prob)
} # cell_prob_fun

sipm_mcmc <- function(y, r, m, x, w, d, nmcmc) {

 # Setup variables
 nsite <- dim(y)[1]
 nyear <- dim(y)[2]
 nreps <- dim(y)[3]
```

```
ncovs <- dim(x)[3]
npcvs <- dim(w)[4]
ymax <- apply(y, 1:2, max)
site_large <- max(which(rowMeans(x[,,1]) == max(rowMeans(x[,,1]))))
site_small <- min(which(rowMeans(x[,,1]) == min(rowMeans(x[,,1]))))

beta0_save <- matrix(0, ncovs+1, nmcmc)
beta_gamma_save <- matrix(0, ncovs+2, nmcmc)
beta_phi_save <- matrix(0, ncovs+2, nmcmc)
kappa_save <- rep(0, nmcmc)
alpha_det_save <- matrix(0, npcvs+1, nmcmc)
alpha_cap_save <- matrix(0, npcvs+1, nmcmc)
N_mean_save <- matrix(0, nyear, nmcmc)
N_large_save <- matrix(0, nyear, nmcmc)
N_small_save <- matrix(0, nyear, nmcmc)
phi_mean_save <- matrix(0, nyear-1, nmcmc)
phi_large_save <- matrix(0, nyear-1, nmcmc)
phi_small_save <- matrix(0, nyear-1, nmcmc)
gamma_mean_save <- matrix(0, nyear-1, nmcmc)
gamma_large_save <- matrix(0, nyear-1, nmcmc)
gamma_small_save <- matrix(0, nyear-1, nmcmc)

# Prior
beta0_mean <- rep(0, ncovs+1)
beta0_sd <- 2
beta_gamma_mean <- rep(0, ncovs+2)
beta_gamma_sd <- 2
beta_phi_mean <- rep(0, ncovs+2)
beta_phi_sd <- 2
log_kappa_mean <- 0
log_kappa_sd <- 2
alpha_det_mean <- rep(0, npcvs+1)
alpha_det_sd <- 2
alpha_cap_mean <- rep(0, npcvs+1)
alpha_cap_sd <- 2

# Starting values
beta0 <- rep(0, ncovs+1)
beta_gamma <- rep(0, ncovs+2)
beta_phi  <- rep(0, ncovs+2)
kappa <- 1
alpha_det <- rep(0, npcvs+1)
alpha_cap <- rep(0, npcvs+1)
lambda0 <- exp(cbind(1,x[,1,]) %*% beta0)
gamma <- matrix(0.5, nsite, nyear-1)
```

```
  phi  <- matrix(0.5, nsite, nyear-1)
  eta <- exp(-1 * kappa * d)
  theta <- eta / rowSums(eta)
  N <- round((ymax + 1) / .5)
  pdet <- array(.5, dim=c(nsite, nyear, nreps))
  pcap_per_visit <- array(.5, dim=c(nsite, nyear-1, nreps))
  pcap <- matrix(, nsite, nyear-1)
  for (i in 1:nsite) {
   for (t in 1:(nyear-1)) {
    pcap[i,t] <- 1 - prod(1 - pcap_per_visit[i,t,])
   } # t
  } # i
  cell_prob <- cell_prob_fun(nsite=nsite, nyear=nyear, phi=phi,
theta=theta, pcap=pcap)

  # Tuning factor
  beta0_tune <- 0.08
  beta_gamma_tune <- 0.02
  beta_phi_tune <- 0.08
  kappa_tune <- 0.035
  alpha_det_tune <- 0.025
  alpha_cap_tune <- 0.05
  N_tune <- 1

  # Begin MCMC loop
  for (k in 1:nmcmc) {
   ### Sample N
   N_star <- matrix(rpois(nsite*nyear, N + N_tune), nsite, nyear)
   mh1 <- apply(dbinom(y, N_star, pdet, log=T), 1:2, sum) +
     dpois(N_star, cbind(lambda0, t(t(N[,-nyear] * (1 + gamma) * phi) %*%
theta)), log=T) +
       dpois(N, N_star+N_tune, log=T)
   mh2 <- apply(dbinom(y, N    , pdet, log=T), 1:2, sum) +
       dpois(N    , cbind(lambda0, t(t(N[,-nyear] * (1 + gamma) * phi) %*%
theta)), log=T) +
       dpois(N_star, N+N_tune, log=T)
   mh <- exp(mh1 - mh2)
   Nkeep <- ((mh > runif(nsite*nyear)) & (N_star >= ymax))
   N[Nkeep] <- N_star[Nkeep]

   ### Sample beta0
   beta0_star <- rnorm(ncovs+1, beta0, beta0_tune)
   lambda0_star <- exp(cbind(1,x[,1,]) %*% beta0_star)
   mh1 <- sum(dpois(N[,1], lambda0_star, log=T)) +
       sum(dnorm(beta0_star, beta0_mean, beta0_sd, log=T))
```

```
  mh2 <- sum(dpois(N[,1], lambda0    , log=T)) +
      sum(dnorm(beta0    , beta0_mean, beta0_sd, log=T))
  mh <- exp(mh1 - mh2)
  if (mh > runif(1)) {
   beta0 <- beta0_star
   lambda0 <- lambda0_star
  }

  # Sample beta_gamma
  beta_gamma_star <- rnorm(ncovs+2, beta_gamma, beta_gamma_tune)
  gamma <- gamma_star <- matrix(, nsite, nyear-1)
  for (t in 2:nyear) {
  gamma   [,t-1] <- exp(cbind(1,(N[,t-1]-lambda0)/lambda0,x[,t,]) %*%
beta_gamma   )
  gamma_star[,t-1] <- exp(cbind(1,(N[,t-1]-lambda0)/lambda0,x[,t,])
%*% beta_gamma_star)
  } # t
  mh1 <- sum(dbinom(r, r + y[,-nyear,], gamma_star / (gamma_star + 1),
log=T)) +
      sum(dpois(N[,-1], t(t(N[,-nyear] * (1 + gamma_star) * phi) %*%
theta), log=T)) +
      sum(dnorm(beta_gamma_star, beta_gamma_mean, beta_gamma_sd,
log=T))
  mh2 <- sum(dbinom(r, r + y[,-nyear,], gamma   / (gamma   + 1), log=T)) +
     sum(dpois(N[,-1], t(t(N[,-nyear] * (1 + gamma   ) * phi) %*% theta),
log=T)) +
     sum(dnorm(beta_gamma   , beta_gamma_mean, beta_gamma_sd, log=T))
  mh <- exp(mh1 - mh2)
  if (mh > runif(1)) {
   beta_gamma <- beta_gamma_star
  }

  # Sample beta_phi
  beta_phi_star <- rnorm(ncovs+2, beta_phi, beta_phi_tune)
  phi <- phi_star <- matrix(, nsite, nyear-1)
  for (t in 2:nyear) {
   phi   [,t-1] <- inv.logit(cbind(1,(N[,t-1]-lambda0)/lambda0,x[,
t,]) %*% beta_phi)
   phi_star[,t-1] <- inv.logit(cbind(1,(N[,t-1]-lambda0)/lambda0,x[,
t,]) %*% beta_phi_star)
  } # t
  cell_prob    <- cell_prob_fun(nsite=nsite, nyear=nyear, phi=phi    ,
theta=theta, pcap=pcap)
  cell_prob_star <- cell_prob_fun(nsite=nsite, nyear=nyear,
phi=phi_star, theta=theta, pcap=pcap)
```

```
  prob1 <- prob2 <- matrix(, nsite, nyear-1)
  for (i in 1:nsite) {
   for (t in 1:(nyear-1)) {
    prob1[i,t] <- dmultinom(m[t,,i], size=sum(m[t,,i]),
prob=cell_prob_star[t,,i], log=T)
    prob2[i,t] <- dmultinom(m[t,,i], size=sum(m[t,,i]),
prob=cell_prob   [t,,i], log=T)
   } # t
  } # i
  mh1 <- sum(prob1) +
      sum(dpois(N[,-1], t(t(N[,-nyear] * (1 + gamma) * phi_star) %*%
theta), log=T)) +
      sum(dnorm(beta_phi_star, beta_phi_mean, beta_phi_sd, log=T))
  mh2 <- sum(prob2) +
     sum(dpois(N[,-1], t(t(N[,-nyear] * (1 + gamma) * phi    ) %*% theta),
log=T)) +
      sum(dnorm(beta_phi    , beta_phi_mean, beta_phi_sd, log=T))
  mh <- exp(mh1 - mh2)
  if (mh > runif(1)) {
   beta_phi <- beta_phi_star
   phi <- phi_star
   cell_prob <- cell_prob_star
  }

  # Sample kappa
  kappa_star <- exp(rnorm(1, log(kappa), kappa_tune))
  eta_star <- exp(-1 * kappa_star * d)
  theta_star <- eta_star / rowSums(eta_star)
  cell_prob_star <- cell_prob_fun(nsite=nsite, nyear=nyear, phi=phi,
theta=theta_star, pcap=pcap)
  prob1 <- prob2 <- matrix(, nsite, nyear-1)
  for (i in 1:nsite) {
   for (t in 1:(nyear-1)) {
    prob1[i,t] <- dmultinom(m[t,,i], size=sum(m[t,,i]),
prob=cell_prob_star[t,,i], log=T)
    prob2[i,t] <- dmultinom(m[t,,i], size=sum(m[t,,i]),
prob=cell_prob   [t,,i], log=T)
   } # t
  } # i
  mh1 <- sum(prob1) +
      sum(dpois(N[,-1], t(t(N[,-nyear] * (1 + gamma) * phi) %*%
theta_star), log=T)) +
      dnorm(log(kappa_star), log_kappa_mean, log_kappa_sd, log=T)
  mh2 <- sum(prob2) +
     sum(dpois(N[,-1], t(t(N[,-nyear] * (1 + gamma) * phi) %*% theta   ),
```

```
log=T)) +
     dnorm(log(kappa   ), log_kappa_mean, log_kappa_sd, log=T)
  mh <- exp(mh1 - mh2)
  if (mh > runif(1)) {
   kappa <- kappa_star
   eta <- eta_star
   theta <- theta_star
   cell_prob <- cell_prob_star
  }

  ### Sample alpha_det
  alpha_det_star <- rnorm(npcvs+1, alpha_det, alpha_det_tune)
  pdet_star <- array(, dim=c(nsite, nyear, nreps))
  for (i in 1:nsite) {
   for (t in 1:nyear) {
    pdet_star[i,t,] <- inv.logit(cbind(1,w[i,t,,]) %*%
alpha_det_star)
   } # t
  } # i
  mh1 <- sum(dbinom(y, N, pdet_star, log=TRUE), na.rm=T) +
     sum(dnorm(alpha_det_star, alpha_det_mean, alpha_det_sd,
log=TRUE))
  mh2 <- sum(dbinom(y, N, pdet   , log=TRUE), na.rm=T) +
     sum(dnorm(alpha_det   , alpha_det_mean, alpha_det_sd, log=TRUE))
  mh <- exp(mh1 - mh2)
  if (mh > runif(1)) {
   alpha_det <- alpha_det_star
   pdet <- pdet_star
  }

  # Sample alpha_pcap
  alpha_cap_star <- rnorm(npcvs+1, alpha_cap, alpha_cap_tune)
  pcap_per_visit_star <- array(, dim=c(nsite, nyear-1, nreps))
  for (i in 1:nsite) {
   for (t in 1:(nyear-1)) {
    pcap_per_visit_star[i,t,] <- inv.logit(cbind(1,w[i,t+1,,]) %*%
alpha_cap_star)
   } # t
  } # i
  pcap_star <- matrix(, nsite, nyear-1)
  for (i in 1:nsite) {
   for (t in 1:(nyear-1)) {
    pcap_star[i,t] <- 1 - prod(1 - pcap_per_visit_star[i,t,])
   } # t
  } # i
```

```
  cell_prob_star <- cell_prob_fun(nsite=nsite, nyear=nyear, phi=phi,
theta=theta, pcap=pcap_star)
  prob1 <- prob2 <- matrix(, nsite, nyear-1)
  for (i in 1:nsite) {
   for (t in 1:(nyear-1)) {
    prob1[i,t] <- dmultinom(m[t,,i], size=sum(m[t,,i]),
prob=cell_prob_star[t,,i], log=T)
    prob2[i,t] <- dmultinom(m[t,,i], size=sum(m[t,,i]),
prob=cell_prob     [t,,i], log=T)
   } # t
  } # i
  mh1 <- sum(prob1) +
         sum(dnorm(alpha_cap_star, alpha_cap_mean, alpha_cap_sd,
log=T))
  mh2 <- sum(prob2) +
         sum(dnorm(alpha_cap     , alpha_cap_mean, alpha_cap_sd, log=T))
  mh <- exp(mh1 - mh2)
  if (mh > runif(1)) {
   alpha_cap <- alpha_cap_star
   pcap_per_visit <- pcap_per_visit_star
   pcap <- pcap_star
   cell_prob <- cell_prob_star
  }

  # Save samples
  beta0_save[,k] <- beta0
  beta_gamma_save[,k] <- beta_gamma
  beta_phi_save[,k] <- beta_phi
  kappa_save[k] <- kappa
  alpha_det_save[,k] <- alpha_det
  alpha_cap_save[,k] <- alpha_cap
  N_mean_save[,k] <- colMeans(N)
  N_large_save[,k] <- N[site_large,]
  N_small_save[,k] <- N[site_small,]
  phi_mean_save[,k] <- colMeans(phi)
  phi_large_save[,k] <- phi[site_large,]
  phi_small_save[,k] <- phi[site_small,]
  gamma_mean_save[,k] <- colMeans(gamma)
  gamma_large_save[,k] <- gamma[site_large,]
  gamma_small_save[,k] <- gamma[site_small,]
 } # k

 list(beta0_save=beta0_save,
   beta_gamma_save=beta_gamma_save, beta_phi_save=beta_phi_save,
kappa_save=kappa_save,
   alpha_det_save=alpha_det_save, alpha_cap_save=alpha_cap_save,
```

```
   N_mean_save=N_mean_save, N_large_save=N_large_save,
N_small_save=N_small_save,
   phi_mean_save=phi_mean_save, phi_large_save=phi_large_save,
phi_small_save=phi_small_save,
   gamma_mean_save=gamma_mean_save,
gamma_large_save=gamma_large_save, gamma_small_save=gamma_small_save)
} # sipm_mcmc
```

When using the following code to implement the MCMC algorithm, the model takes about 8 h to run. Then we can use the code in Appendix 4.4 to create some traceplots, which show good model convergence and agreement between parameter estimates of $\beta^{[0]}$, $\beta^{[\gamma]}$, $\beta^{[\phi]}$, κ, $\alpha^{[cap]}$, and $\alpha^{[det]}$ and their true values (Fig. 4.7). In

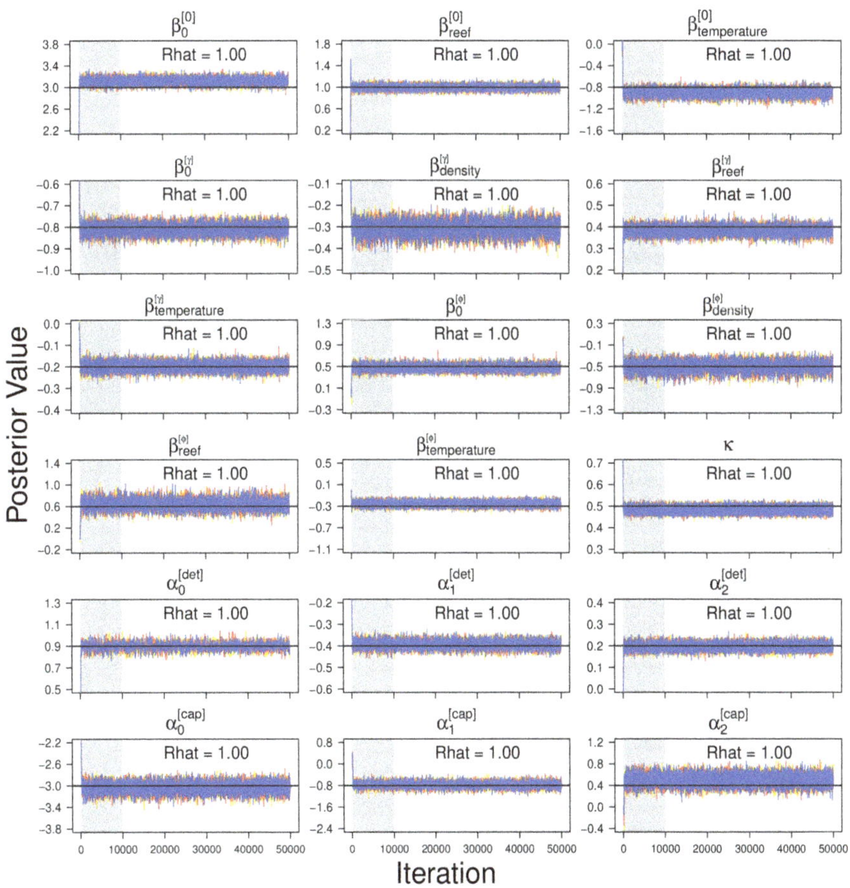

Fig. 4.7 Traceplots showing posterior chains for $\beta^{[0]}$, $\beta^{[\gamma]}$, $\beta^{[\phi]}$, κ, $\alpha^{[det]}$, and $\alpha^{[cap]}$ in the spatial IPM without age-specific survival and movement. Rhat values are calculated after discarding the first 10,000 iterations. Horizontal lines represent the true values used in data simulation

particular, there are no issues in parameter estimates stemming from not sampling the latent individual status $z_{v,\ t}$ and the number of juveniles $R_{i,\ t}$.

```
library(foreach)   # for parallel computing
library(doParallel) # for parallel computing

numCores <- round(detectCores() / 2) # only use half of the cores for
parallel computing
registerDoParallel(numCores) # setup parallel computing

nmcmc <- 50000 # number of iterations of MCMC computing
chain <- 3    # number of chains

start_time <- Sys.time() # start time of computing
out <- foreach (i=1:chain, .packages='boot') %dopar% {
 sipm_mcmc(y=y, r=r, m=m, x=x, w=w, d=d, nmcmc=nmcmc)
} # i
stopImplicitCluster() # clean up the cluster
end_time <- Sys.time() # end time of computing
time_taken <- end_time - start_time # computing time
```

We can also use the following code to plot population-wide and site-level (using the largest and smallest sites as examples) abundance, survival, and reproduction estimates, which are consistent with their simulated true values (Fig. 4.8). For example, the smallest site is empty in the first year, colonized in the second year, goes extinct in the third year, and is colonized again in the fourth year, all of which are well represented by the estimates.

```
library(vioplot) # for making violin plots

for (i in 1:chain) {
 if (i == 1) {
  N_mean <- out[[1]]$N_mean_save
  N_large <- out[[1]]$N_large_save
  N_small <- out[[1]]$N_small_save
  phi_mean <- out[[1]]$phi_mean_save
  phi_large <- out[[1]]$phi_large_save
  phi_small <- out[[1]]$phi_small_save
  gamma_mean <- out[[1]]$gamma_mean_save
  gamma_large <- out[[1]]$gamma_large_save
  gamma_small <- out[[1]]$gamma_small_save
 } else {
  N_mean <- cbind(N_mean, out[[1]]$N_mean_save)
  N_large <- cbind(N_large, out[[1]]$N_large_save)
```

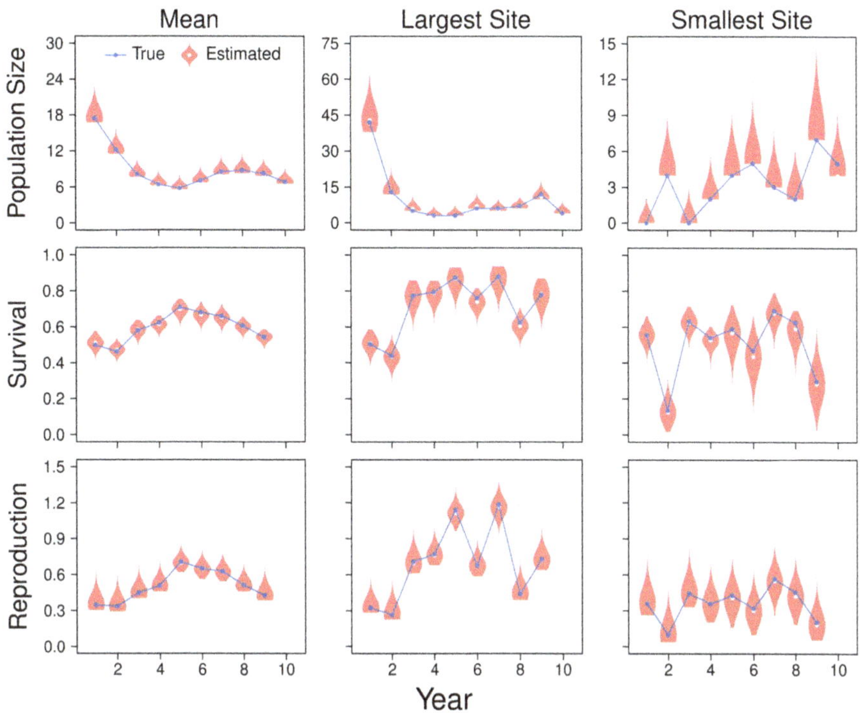

Fig. 4.8 Violin plots showing good agreement between the estimates of the spatial IPM without age-specific survival and movement and simulated true values of the mean population size, survival probability, and reproduction rate at the population level as well as site-level values for the largest and smallest sites

```
   N_small <- cbind(N_small, out[[1]]$N_small_save)
   phi_mean <- cbind(phi_mean, out[[1]]$phi_mean_save)
   phi_large <- cbind(phi_large, out[[1]]$phi_large_save)
   phi_small <- cbind(phi_small, out[[1]]$phi_small_save)
   gamma_mean <- cbind(gamma_mean, out[[1]]$gamma_mean_save)
   gamma_large <- cbind(gamma_large, out[[1]]$gamma_large_save)
   gamma_small <- cbind(gamma_small, out[[1]]$gamma_small_save)
  }
} # i

site_large <- max(which(rowMeans(x[,,1]) == max(rowMeans(x[,,1]))))
site_small <- min(which(rowMeans(x[,,1]) == min(rowMeans(x[,,1]))))

par(mfrow=c(3,3))
par(mar=c(1.5,4,0,0))
par(oma=c(4,3,3,1))
```

```
plot(1, xlim=c(0.5,nyear+0.5), ylim=c(0,30), type='l', axes=F,
xlab='', ylab='')
vioplot(t(N_mean), add=T, col='lightcoral', rectCol=NA, lineCol=NA,
border=NA)
lines(colMeans(N), type='o', pch=16, col='royalblue')
axis(1, at=seq(0,nyear,2), labels=rep('',6))
axis(2, at=seq(0,30,length.out=6), cex.axis=1.6, las=2)
axis(2, at=15, labels='Population Size', cex.axis=2.5, line=3.2,
tick=F)
axis(3, at=5.5, labels='Mean', cex.axis=2.5, line=-.5, tick=F)

points(x=2, y=28, pch=16, cex=1, col='royalblue')
lines(x=c(1.5,2.5), y=c(28,28), lwd=1.2, col='royalblue')
vioplot(rnorm(10000,28,.5), at=5.5, add=T, col='lightcoral',
rectCol=NA, lineCol=NA, border=NA)
text(x=2.5, y=28, labels='True', pos=4, cex=1.8)
text(x=6, y=28, labels='Estimated', pos=4, cex=1.8)

plot(1, xlim=c(0.5,nyear+0.5), ylim=c(0,75), type='l', axes=F,
xlab='', ylab='')
vioplot(t(N_large), add=T, col='lightcoral', rectCol=NA,
lineCol=NA, border=NA)
lines(N[site_large,], type='o', pch=16, col='royalblue')
axis(1, at=seq(0,nyear,2), labels=rep('',6))
axis(2, at=seq(0,75,length.out=6), cex.axis=1.6, las=2)
axis(3, at=5.5, labels='Largest Site', cex.axis=2.5, line=-.5,
tick=F)

plot(1, xlim=c(0.5,nyear+0.5), ylim=c(0,15), type='l', axes=F,
xlab='', ylab='')
vioplot(t(N_small), add=T, col='lightcoral', rectCol=NA,
lineCol=NA, border=NA)
lines(N[site_small,], type='o', pch=16, col='royalblue')
axis(1, at=seq(0,nyear,2), labels=rep('',6))
axis(2, at=seq(0,15,length.out=6), cex.axis=1.6, las=2)
axis(3, at=5.5, labels='Smallest Site', cex.axis=2.5, line=-.5,
tick=F)

plot(1, xlim=c(0.5,nyear+0.5), ylim=c(0,1), type='l', axes=F,
xlab='', ylab='')
vioplot(t(phi_mean), add=T, col='lightcoral', rectCol=NA,
lineCol=NA, border=NA)
lines(colMeans(phi), type='o', pch=16, col='royalblue')
axis(1, at=seq(0,nyear,2), labels=rep('',6))
axis(2, at=seq(0,1,length.out=6), cex.axis=1.6, las=2)
axis(2, at=0.5, labels='Survival', cex.axis=2.5, line=3.2, tick=F)
```

```
plot(1, xlim=c(0.5,nyear+0.5), ylim=c(0,1), type='l', axes=F,
xlab='', ylab='')
vioplot(t(phi_large), add=T, col='lightcoral', rectCol=NA,
lineCol=NA, border=NA)
lines(phi[site_large,], type='o', pch=16, col='royalblue')
axis(1, at=seq(0,nyear,2), labels=rep('',6))
axis(2, at=seq(0,1,length.out=6), labels=rep('',6))

plot(1, xlim=c(0.5,nyear+0.5), ylim=c(0,1), type='l', axes=F,
xlab='', ylab='')
vioplot(t(phi_small), add=T, col='lightcoral', rectCol=NA,
lineCol=NA, border=NA)
lines(phi[site_small,], type='o', pch=16, col='royalblue')
axis(1, at=seq(0,nyear,2), labels=rep('',6))
axis(2, at=seq(0,1,length.out=6), labels=rep('',6))

plot(1, xlim=c(0.5,nyear+0.5), ylim=c(0,1.5), type='l', axes=F,
xlab='', ylab='')
vioplot(t(gamma_mean), add=T, col='lightcoral', rectCol=NA,
lineCol=NA, border=NA)
lines(colMeans(gamma), type='o', pch=16, col='royalblue')
axis(1, at=seq(0,nyear,2), cex.axis=1.6)
axis(2, at=seq(0,1.5,length.out=6), cex.axis=1.6, las=2)
axis(2, at=0.75, labels='Reproduction', cex.axis=2.5, line=3.2,
tick=F)

plot(1, xlim=c(0.5,nyear+0.5), ylim=c(0,1.5), type='l', axes=F,
xlab='', ylab='')
vioplot(t(gamma_large), add=T, col='lightcoral', rectCol=NA,
lineCol=NA, border=NA)
lines(gamma[site_large,], type='o', pch=16, col='royalblue')
axis(1, at=seq(0,nyear,2), cex.axis=1.6)
axis(2, at=seq(0,1.5,length.out=6), labels=rep('',6))

plot(1, xlim=c(0.5,nyear+0.5), ylim=c(0,1.5), type='l', axes=F,
xlab='', ylab='')
vioplot(t(gamma_small), add=T, col='lightcoral', rectCol=NA,
lineCol=NA, border=NA)
lines(gamma[site_small,], type='o', pch=16, col='royalblue')
axis(1, at=seq(0,nyear,2), cex.axis=1.6)
axis(2, at=seq(0,1.5,length.out=6), labels=rep('',6))

title(xlab='Year', cex.lab=3, line=2.4, outer=T)
```

We can also check the correlation between $\beta_0^{[\phi]}$ (mean survival probability at a logit scale) and $\beta_0^{[\gamma]}$ (mean reproduction rate at a log scale), using the code below. We can see there is almost no correlation (0.09) between these parameters (Fig. 4.9), indicating that these parameters are well identifiable. This is a great advantage over dynamic N-mixture models (Sect. 3.6.1; Fig. 3.14), likely due to the incorporation of capture-recapture data.

```
for (i in 1:chain) {
 if (i == 1) {
  beta_phi0_post <- out[[i]]$beta_phi[1, (nmcmc*0.2+1):nmcmc]
  beta_gamma0_post <- out[[i]]$beta_gamma[1, (nmcmc*0.2+1):nmcmc]
 } else {
  beta_phi0_post <- c(beta_phi0_post, out[[i]]$beta_phi[1,
(nmcmc*0.2+1):nmcmc])
  beta_gamma0_post <- c(beta_gamma0_post, out[[i]]$beta_gamma[1,
(nmcmc*0.2+1):nmcmc])
 }
} # i

nplot <- min(c(10000, nmcmc*0.2*chain))
sel <- sort(sample(1:(nmcmc*0.2*chain), nplot, replace=F))
a <- beta_phi0_post[sel]
b <- beta_gamma0_post[sel]
a2 <- (a - min(a)) / (max(a) - min(a))
```

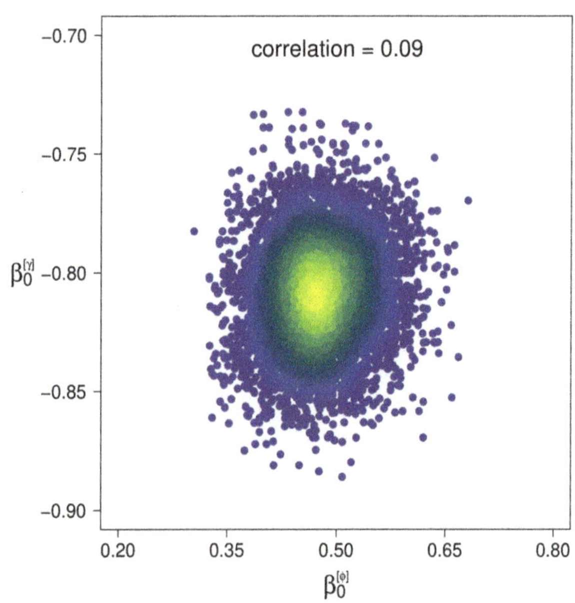

Fig. 4.9 A scatterplot showing the weak correlation between the mean survival probability at a logit scale, $\beta_0^{[\phi]}$, and the mean reproduction rate at a log scale, $\beta_0^{[\gamma]}$, in the spatial IPM without age-specific survival and movement. Each point represents a pair of posterior draws of the two parameters during one iteration of the MCMC algorithm

```
b2 <- (b - min(b)) / (max(b) - min(b))
ab <- matrix(, nplot, nplot)
for (i in 1:nplot) {
 for (j in 1:nplot) {
  ab[i,j] <- sqrt((a2[i] - a2[j]) ^ 2 + (b2[i] - b2[j]) ^ 2)
 } # j
} # i

ab01 <- ifelse(ab < 0.08, 1, 0)
ab_scale <- rowSums(ab01) - min(rowSums(ab01)) + 1
col2 <- c(colorRampPalette(c('navy','seagreen'))(floor(max(ab_scale)
*0.6)),
    colorRampPalette(c('seagreen','yellow'))(ceiling(max(ab_scale)
*0.4)))

par(mfrow=c(1,1))
par(mar=c(5,6.5,1,1))

plot(a, b, col=col2[ab_scale], pch=16, cex=1.4, xlim=c(0.2,0.8),
ylim=c(-0.9,-0.7), axes=F, xlab='', ylab='')
axis(1, at=seq( 0.2, 0.8,length.out=5), cex.axis=1.5)
axis(2, at=seq(-0.9,-0.7,length.out=5), cex.axis=1.5, las=2)
axis(1, at= 0.5, labels=ylab_beta_phi [1], cex.axis=2, line=2.8,
tick=F)
axis(2, at=-0.8, labels=ylab_beta_gamma[1], cex.axis=2, line=3.6,
tick=F, las=2)
box()

text(x=0.5, y=-0.7, labels=paste('correlation = ', round(cor(a, b),
digits=2), sep=''), pos=1, cex=1.8)
```

4.4.2 With Age-Specific Survival and Movement

Next, we will extend the spatial IPM to allow different survival probabilities and movement rates between juvenile and adult individuals, making the model more realistic. Using our coral reef fish as an example, we assume that juveniles have an overall lower survival probability than adults. We also assume that juvenile survival is more sensitive to the number of adult fish, the size, and the temperature of the reef it lives in than adult survival. Eventually, we assume juveniles are more likely to move a long distance than adults. We still assume that all juveniles become adults and thus can reproduce after one year. Statistically speaking, we have the process model that first describes the initial abundance in the same way as in Eq. (4.5):

$$\left.\begin{aligned} N_{i,1} &\sim Poisson(\lambda_{i,1}) \\ \lambda_{i,1} &= exp\left(\beta_0^{[0]} + \beta_1^{[0]} \times x_{i,1,1} + \ldots + \beta_{n_{covs}}^{[0]} \times x_{i,1,n_{covs}}\right) \end{aligned}\right\}, \tag{4.9}$$

in which $N_{i,1}$ is the abundance of adults at site i in the first year; and $\lambda_{i,1}$ is the expectation of initial abundance which is a function of covariates x's with intercept and slope parameters $\beta^{[0]}$'s. From the second year (i.e., $t \geq 2$), the process sub-model specifies the spatiotemporal variation in adult abundance as follows:

$$\left.\begin{aligned} N_{i_2,t} &\sim Poisson(\lambda_{i_2,t}) \\ \lambda_{i_2,t} &= \textstyle\sum_{i_1=1}^{n_{site}} \left(N_{i_1,t-1} \times \gamma_{i_1,t-1} \times \phi_{i,t-1,juv} \times \theta_{i_1,i_2,t,juv} + N_{i_1,t-1} \times \phi_{i,t-1,adu} \times \theta_{i_1,i_2,t,adu}\right) \\ \gamma_{i_1,t-1} &= exp\left(\beta_0^{[\gamma]} + \beta_{density}^{[\gamma]} \times N_{i_1,t-1} + \beta_1^{[\gamma]} \times x_{i_1,t,1} + \ldots + \beta_{n_{covs}}^{[\gamma]} \times x_{i_1,t,n_{covs}}\right) \\ \phi_{i_1,t-1,juv} &= logit^{-1}\left(\beta_{0,juv}^{[\phi]} + \beta_{density,juv}^{[\phi]} \times N_{i_1,t-1} + \beta_{1,juv}^{[\phi]} \times x_{i_1,t,1} + \ldots + \beta_{n_{covs},juv}^{[\phi]} \times x_{i_1,t,n_{covs}}\right) \\ \phi_{i_1,t-1,adu} &= logit^{-1}\left(\beta_{0,adu}^{[\phi]} + \beta_{density,adu}^{[\phi]} \times N_{i_1,t-1} + \beta_{1,adu}^{[\phi]} \times x_{i_1,t,1} + \ldots + \beta_{n_{covs},adu}^{[\phi]} \times x_{i_1,t,n_{covs}}\right) \\ \theta_{i_1,i_2,juv} &= \frac{\eta_{i_1,i_2,juv}}{\sum_{i_2=1}^{n_{site}} \eta_{i_1,i_2},juv} \\ \theta_{i_1,i_2,adu} &= \frac{\eta_{i_1,i_2,adu}}{\sum_{i_2=1}^{n_{site}} \eta_{i_1,i_2,adu}} \\ \eta_{i_1,i_2,juv} &= exp(-1 \times \kappa_{juv} \times d_{i_1,i_2}) \\ \eta_{i_1,i_2,adu} &= exp(-1 \times \kappa_{adu} \times d_{i_1,i_2}) \end{aligned}\right\}, \tag{4.10}$$

in which $\gamma_{i_1,t-1}$ is the per capita reproduction rate at site i_1 in year $t-1$ and is a function of density and environmental covariates x's with intercept and slope parameters $\beta^{[\gamma]}$'s; $\phi_{i_1,t-1,juv}$ and $\phi_{i_1,t-1,adu}$ are juvenile and adult survival probabilities, respectively, and are functions of density and environmental covariates with age-specific intercept and slope parameters $\beta_{\cdot,juv}^{[\phi]}$ and $\beta_{\cdot,adu}^{[\phi]}$'s; $\theta_{i_1,i_2,juv}$ and $\theta_{i_1,i_2,adu}$ are the standardized movement rates from sites i_1 to i_2; and $\eta_{i_1,i_2,juv}$ and $\eta_{i_1,i_2,adu}$ are the unstandardized movement rates governed by decay parameters κ_{juv} and κ_{adu}, respectively. Eventually, the abundance of adults at site i_2 in year t is the total number of adults and juveniles that survive from year $t-1$ and either move from any other sites i_1 or stay at site i_2 (when $i_2 = i_1$). Note that, during data simulation, we have survival-location history $z_{v,t}$ simulated and coded as $z_{v,t} = 1, \ldots, n_{site}$ for juveniles and $z_{v,t} = n_{site} + 1, \ldots, n_{site} \times 2$ for adults that are alive in site $1, \ldots, n_{site}$. We also simulate the number of juvenile offspring $R_{i,t}$.

The capture history data again are obtained in the following way:

$$\left.\begin{aligned} c_{v,t,j} &\sim Bernoulli\left(p_{z_{v,t},t,j}^{[cap]}\right) \times z_{v,t} \\ p_{i,t,j}^{[cap]} &= logit^{-1}\left(\alpha_0^{[cap]} + \alpha_1^{[cap]} \times w_{i,t,j,1} + \ldots + \alpha_{n_{pcvs}}^{[cap]} \times w_{i,t,j,n_{pcvs}}\right) \end{aligned}\right\}, \tag{4.11}$$

in which $c_{v,t,j}$ is the capture history of individual v in year t during the jth visit, $p_{i,t,j}^{[cap]}$ is the capture probability and is a function of covariates w's with intercept and slope parameters $\alpha^{[cap]}$'s. Individual capture history is later converted to an m-array for analysis.

The count data are collected for adults and juveniles separately; they follow binomial distributions such that

$$\left.\begin{aligned} y_{i,t,j} &\sim Binomial\left(N_{i,t}, p_{i,t,j}^{[det]}\right) \\ r_{i,t,j} &\sim Binomial\left(R_{i,t}, p_{i,t,j}^{[det]}\right) \\ p_{i,t,j}^{[det]} &= logit^{-1}\left(\alpha_0^{[det]} + \alpha_1^{[det]} \times w_{i,t,j,1} + \ldots + \alpha_{n_{pcvs}}^{[det]} \times w_{i,t,j,n_{pcvs}}\right) \end{aligned}\right\}, \tag{4.12}$$

In which $y_{i,\,t,\,j}$ is the count of adults, $r_{i,\,t,\,j}$ is the counts of juveniles, and $p_{i,t,j}^{[det]}$ is the detection probability and is a function of covariates w's with intercept and slope parameters $\alpha^{[det]}$'s.

Using the following code, we can simulate data for this spatial IPM with age-specific survival and movement. We assume positive effects of the size of forest fragments on initial abundance ($\beta_{reef}^{[0]} = 1$), reproduction ($\beta_{reef}^{[\gamma]} = 0.4$), adult survival ($\beta_{reef,adu}^{[\phi]} = 0.6$) and juvenile survival ($\beta_{reef,juv}^{[\phi]} = 0.8$), and negative effects of temperature on initial abundance ($\beta_{temperature}^{[0]} = -0.8$), reproduction ($\beta_{temperature}^{[\gamma]} = -0.2$), adult survival ($\beta_{temperature,adu}^{[\phi]} = -0.3$) and juvenile survival ($\beta_{temperature,juv}^{[\phi]} = -0.4$). We also assume negative density dependence on reproduction ($\beta_{density}^{[\gamma]} = -0.3$), adult survival ($\beta_{density,adu}^{[\phi]} = -0.5$) and juvenile survival ($\beta_{density,juv}^{[\phi]} = -0.7$). We assume a stronger rate of decay for adults ($\kappa_{adu} = 2$) than for juveniles ($\kappa_{juv} = 0.5$) to allow a higher probability of long-distance movement for the latter.

```
library(boot) # for using logit and inv.logit functions
library(LaplacesDemon) # for generating random numbers from
categorical distributions (rcat)

set.seed(7)

# Basic values
nsite <- 60 # number of sites
nyear <- 10 # number of years
nreps <- 5  # number of within-season replicates
ncovs <- 2  # number of environmental covariates
npcvs <- 1  # number of observational covariates

beta0 <- c(3.4, 1, -0.8)                # intercept and slopes for initial
abundance (i.e., number of adults)
beta_gamma <- c(-0.7, -0.3, 0.4, -0.2) # intercept and slopes for
reproduction
beta_phi_adu <- c(1 , -0.5, 0.6, -0.3) # intercept and slopes for adult
```

```
survival
beta_phi_juv <- c(0.4, -0.7, 0.8, -0.4) # intercept and slopes for
juvenile survival
kappa_adu <- 2                  # distance effect on adult movement
kappa_juv <- 0.5                # distance effect on juvenile movement
alpha_det <- c( 0.9, -0.4)      # intercept and slopes for detection
probability
alpha_cap <- c(-3 , -0.8)       # intercept and slopes for capture
probability

# Read in environmental covariates
load('data/covariates.RData')
x1 <- (log(cov$size) - mean(log(cov$size))) / sd(log(cov$size)) #
standardized log habitat patch size
x2 <- (cov$temp - mean(cov$temp)) / sd(cov$temp) # standardized
temperature
x <- array(, dim=c(nsite, nyear, ncovs)) # environmental covariates
x[,,1] <- x1[1:nsite, 1:nyear]
x[,,2] <- x2[1:nsite, 1:nyear]
### Standardize covariates for each site
for (i in 1:nsite) {
 for (h in 1:ncovs) {
  x[i,2:nyear,h] <- (x[i,2:nyear,h] - mean(x[i,,h])) / sd(x[i,,h])
 } # h
} # i

# Simulate data
N <- matrix(, nsite, nyear) # abundance
lambda0 <- exp(cbind(1,x[,1,]) %*% beta0) # expectation of initial
abundance
N[,1] <- rpois(nsite, lambda0) # initial abundance

lon <- cov$lon[1:nsite]         # longitude or easting
lat <- cov$lat[1:nsite]         # latitude or northing
d <- matrix(, nsite, nsite)     # distance matrix
for (i1 in 1:nsite) {
 for (i2 in 1:nsite) {
  d[i1,i2] <- sqrt((lon[i1] - lon[i2]) ^ 2 + (lat[i1] - lat[i2]) ^ 2)
 } # i2
} # i1
eta_adu <- exp(-1 * kappa_adu * d)   # unstandardized adult colonization
rate
theta_adu <- eta_adu / rowSums(eta_adu) #  standardized adult
colonization rate
eta_juv <- exp(-1 * kappa_juv * d)   # unstandardized juvenile
```

```
colonization rate
theta_juv <- eta_juv / rowSums(eta_juv) #  standardized juvenile
colonization rate

nindv <- 15000 # maximum number of individuals that ever exist
z <- matrix(0, nindv, nyear) # individual survival-location history
for (i in 1:nsite) {
 if (i == 1) { tt <- rep(i+nsite, N[i,1]) } else { tt <- c(tt, rep(i+nsite,
N[i,1])) }
}
z[1:length(tt),1] <- tt

gamma <- matrix(, nsite, nyear-1)    # reproduction rate
phi_adu <- matrix(, nsite, nyear-1) # adult survival probability
phi_juv <- matrix(, nsite, nyear-1) # juvenile survival probability
R <- matrix(, nsite, nyear-1)        # number of offspring

for (t in 2:nyear) {
 gamma[,t-1] <- exp(cbind(1,(N[,t-1]-lambda0)/lambda0,x[,t,]) %*%
beta_gamma)
 R[,t-1] <- rpois(nsite, N[,t-1] * gamma[,t-1])

 for (i in 1:nsite) {
  if (i == 1) { tt <- rep(i, R[i,t-1]) } else { tt <- c(tt, rep(i, R[i,
t-1])) }
 } # i
 z[max(which(rowMeans(z,na.rm=T)>0))+c(1:length(tt)), t-1] <- tt

 phi_adu[,t-1] <- inv.logit(cbind(1,(N[,t-1]-lambda0)/lambda0,x[,
t,]) %*% beta_phi_adu)
 phi_juv[,t-1] <- inv.logit(cbind(1,(N[,t-1]-lambda0)/lambda0,x[,
t,]) %*% beta_phi_juv)
 transition_matrix <- matrix(, nsite*2, nsite*2+1) # matrix of
transition (survival & movement) probability
 for (i in 1:nsite) {
  transition_matrix[i,1:nsite] <- 0
  transition_matrix[i,(nsite+1):(nsite*2)] <- phi_juv[i,t-1] *
theta_juv[i,]
  transition_matrix[i,nsite*2+1] <- 1 - phi_juv[i,t-1]
  transition_matrix[nsite+i,1:nsite] <- 0
  transition_matrix[nsite+i,(nsite+1):(nsite*2)] <- phi_adu[i,t-1] *
theta_adu[i,]
  transition_matrix[nsite+i,nsite*2+1] <- 1 - phi_adu[i,t-1]
 } # i
```

```
  for (v in 1:nindv) {
   if (z[v,t-1] %in% 1:(nsite*2)) {
    z[v,t] <- rcat(1, p=transition_matrix[z[v,t-1],])
   }
  } # v
  z[which(!(z %in% 1:(nsite*2)))] <- 0

  for (i in 1:nsite) {
   N[i,t] <- length(which(z[,t] %in% c(i,nsite+i)))
  }
} # t
z[which(!(z %in% 1:(nsite*2)))] <- 0
z <- z[which(rowMeans(z, na.rm=T) != 0),]

# Simulate count data
w <- array(rnorm(nsite * nyear * nreps * npcvs, 0, 1), dim=c(nsite,
nyear, nreps, npcvs)) # observational covariates

pdet <- array(, dim=c(nsite, nyear, nreps)) # detection probability
for (i in 1:nsite) {
 for (t in 1:nyear) {
  pdet[i,t,] <- inv.logit(cbind(1,w[i,t,,]) %*% alpha_det)
 } # t
} # i

y <- array(, dim=c(nsite, nyear, nreps)) # count data of breeding
population
for (i in 1:nsite) {
 for (t in 1:nyear) {
  y[i,t,] <- rbinom(nreps, N[i,t], pdet[i,t,])
 } # t
} # i

r <- array(, dim=c(nsite, nyear-1, nreps)) # count data of offspring
for (i in 1:nsite) {
 for (t in 1:(nyear-1)) {
  r[i,t,] <- rbinom(nreps, R[i,t], pdet[i,t,])
 } # t
} # i

# simulate capture-recapture data
pcap_per_visit <- array(, dim=c(nsite*2, nyear, nreps)) # per visit
capture probability
for (i in 1:nsite) {
 for (t in 1:nyear) {
```

```
  pcap_per_visit[i,t,] <- inv.logit(cbind(1,w[i,t,,]) %*% alpha_cap)
 } # t
} # i
pcap_per_visit[(nsite+1):(nsite*2),,] <- pcap_per_visit[1:nsite,,]

c_all <- matrix(, dim(z)[1], nyear) # capture history of all individuals
                    # including individuals that have never been captured
for (v in 1:dim(z)[1]) {
 for (t in 1:nyear) {
  if (z[v,t] == 0) {
   c_all[v,t] <- 0
  } else {
   cap_per_visit <- rbinom(nreps, 1, pcap_per_visit[z[v,t],t,])
   c01 <- ifelse(sum(cap_per_visit)==0, 0, 1)
   c_all[v,t] <- c01 * z[v,t]
  }
 } # t
} # v
if_cap <- ifelse(apply(c_all, 1, sum) > 0, 1, 0)

c <- c_all[which(if_cap==1),] # capture history of individuals that
have ever been captured
f <- numeric(dim(c)[1]) # year of first capture
for (v in 1:length(f)) {
 f[v] <- min(which(c[v,] %in% c(1:(nsite*2))))
} # v
c <- c[which(f %in% c(1:(nyear-1))),]

# Extend individual capture history
for (v in 1:dim(c)[1]) {
 t1 <- c[v,]
 ycap <- which(t1 %in% c(1:(nsite*2)))
 ncap <- length(ycap)
 if (ncap == 1) {
  t2 <- t1
 } else {
  t2 <- matrix(, ncap, nyear)
  for (k in 1:(ncap-1)) {
   t2[k, ycap[k]:ycap[k+1]] <- t1[ycap[k]:ycap[k+1]]
  } # k
  t2[ncap, ycap[ncap]] <- t1[ycap[ncap]]
 }
 if (v == 1) {
  c_ext <- t2
 } else {
```

```
  c_ext <- rbind(c_ext, t2)
 }
} # v
c_ext[which(is.na(c_ext))] <- 0

f_ext <- numeric(dim(c_ext)[1]) # year of first capture for the extended
individual capture history
for (v in 1:length(f_ext)) {
 f_ext[v] <- min(which(c_ext[v,] %in% c(1:(nsite*2))))
} # v

# Converting individual capture history to m-array format
for (t in 2:nyear) {
 t1 <- paste('juv_site', '_', 1:nsite, ', year', t, sep='')
 t2 <- paste('adu_site', '_', 1:nsite, ', year', t, sep='')
 tt <- c(t1, t2)
 if (t == 2) {
  nm1 <- tt
 } else {
  nm1 <- c(nm1, tt)
 }
} # t
nm1 <- c(nm1, 'not seen')
nm2 <- c(paste('juv_site', 1:nsite, sep='_'), paste('adu_site', 1:
nsite, sep='_'))

m <- array(0, # m-array of capture-recapture
 dim=c(nyear-1, nsite*2*(nyear-1)+1, nsite*2),
 dimnames=list(paste('year', 1:(nyear-1), sep='_'), nm1, nm2))
for (t1 in 1:(nyear-1)) {
 temp1 <- c_ext[which(f_ext == t1),]
 for (i1 in 1:(nsite*2)) {
  temp2 <- temp1[which(temp1[,t1] == i1),]
  nband <- length(which(temp1[,t1] == i1))
  if (nband == 1) {
   temp2 <- matrix(temp2, nrow=1, ncol=length(temp2))
  }
  for (t2 in (t1+1):nyear) {
   for (i2 in 1:(nsite*2)) {
    m[t1,(t2-2)*(nsite*2)+i2,i1] <- length(which(temp2[,t2] == i2))
   } # i2
  } # t2
```

```
  m[t1,(nsite*2)*(nyear-1)+1,i1] <- nband - sum(m[t1,1:((nsite*2)*
(nyear-1)),i1])
  } # i1
} # t1
```

Next, we will develop our own function to calculate the cell probabilities of the m-array and a custom MCMC algorithm to estimate parameters $\beta^{[0]}$, $\beta^{[\gamma]}$, $\beta^{[\phi]}$, κ, $\alpha^{[cap]}$, and $\alpha^{[det]}$ using data y, r, m, x, w, and d. We also use this model to illustrate how to conduct posterior predictive checks of model fit (Rubin 1984; Gelman et al. 1996). First, we will need to simulate new data using the complete posterior samples (burn-in discarded). We then will calculate something called discrepancy measures (Gelman et al. 2014; Conn et al. 2018; details see below) for the actual data and the newly simulated data using posterior samples from the same iterations of the MCMC algorithm. Eventually, we will calculate the proportion of discrepancy measures that are higher for the newly simulated data than for the actual data as the posterior predictive p-value; a posterior predictive p-value close to 0.5 indicates a good model fit. For now, we will need to save the posterior samples of parameters that allow for simulating new data and calculating discrepancy measures: N and $p^{[det]}$ for y, γ for r, and the cell probabilities for m (highlighted in **bold**).

```
# Function to calculate the cell probability of m-array
cell_prob_fun <- function(nsite, nyear, phi_adu, phi_juv, theta_adu,
theta_juv, pcap) {
  qcap <- 1 - pcap
  phi_first_year <- rbind(phi_juv, phi_adu)
  phi_from_second_year <- rbind(phi_adu, phi_adu)
  theta_first_year <- rbind(cbind(matrix(0,nsite,nsite),theta_juv),
cbind(matrix(0,nsite,nsite),theta_adu))
  theta_from_second_year <- rbind(cbind(matrix(0,nsite,nsite),
theta_adu), cbind(matrix(0,nsite,nsite),theta_adu))
  cell_prob_p <- cell_prob_q <- array(0, dim=c(nyear-1, (nsite*2)*
(nyear-1)+1, (nsite*2)))
  for (i in 1:(nsite*2)) {
    for (t in 1:(nyear-1)) {
      cell_prob_p[t,(nsite*2)*(t-1)+1:(nsite*2),i] <- phi_first_year[i,
t] * theta_first_year[i,] * pcap[,t]
      cell_prob_q[t,(nsite*2)*(t-1)+1:(nsite*2),i] <- phi_first_year[i,
t] * theta_first_year[i,] * qcap[,t]
    } # t
    for (t1 in 1:(nyear-2)) {
      for (t2 in (t1+1):(nyear-1)) {
        cell_prob_p[t1,(nsite*2)*(t2-1)+1:(nsite*2),i] <-
          ((cell_prob_q[t1,(nsite*2)*(t2-2)+1:(nsite*2),i] *
phi_from_second_year[,t2]) %*% theta_from_second_year) * pcap[,t2]
```

```
    cell_prob_q[t1,(nsite*2)*(t2-1)+1:(nsite*2),i] <-
      ((cell_prob_q[t1,(nsite*2)*(t2-2)+1:(nsite*2),i] *
phi_from_second_year[,t2]) %*% theta_from_second_year) * qcap[,t2]
   } # t2
  } # t1
  for (t in 1:(nyear-1)) {
   cell_prob_p[t,(nsite*2)*(nyear-1)+1,i] <- 1 - sum(cell_prob_p[t,1:
((nsite*2)*(nyear-1)),i])
  } # t
 } # i
 cell_prob <- cell_prob_p

 return(cell_prob)
} # cell_prob_fun

age_sipm_mcmc <- function(y, r, m, x, w, d, nmcmc) {

 # Setup variables
 nsite <- dim(y)[1]
 nyear <- dim(y)[2]
 nreps <- dim(y)[3]
 ncovs <- dim(x)[3]
 npcvs <- dim(w)[4]
 ymax <- apply(y, 1:2, max)
 site_large <- max(which(rowMeans(x[,,1]) == max(rowMeans(x[,,1]))))
 site_small <- min(which(rowMeans(x[,,1]) == min(rowMeans(x[,,1]))))

 beta0_save <- matrix(0, ncovs+1, nmcmc)
 beta_gamma_save <- matrix(0, ncovs+2, nmcmc)
 beta_phi_adu_save <- matrix(0, ncovs+2, nmcmc)
 beta_phi_juv_save <- matrix(0, ncovs+2, nmcmc)
 kappa_adu_save <- rep(0, nmcmc)
 kappa_juv_save <- rep(0, nmcmc)
 alpha_det_save <- matrix(0, npcvs+1, nmcmc)
 alpha_cap_save <- matrix(0, npcvs+1, nmcmc)
 N_mean_save <- matrix(0, nyear, nmcmc)
 N_large_save <- matrix(0, nyear, nmcmc)
 N_small_save <- matrix(0, nyear, nmcmc)
 phi_adu_mean_save <- matrix(0, nyear-1, nmcmc)
 phi_adu_large_save <- matrix(0, nyear-1, nmcmc)
 phi_adu_small_save <- matrix(0, nyear-1, nmcmc)
 phi_juv_mean_save <- matrix(0, nyear-1, nmcmc)
 phi_juv_large_save <- matrix(0, nyear-1, nmcmc)
 phi_juv_small_save <- matrix(0, nyear-1, nmcmc)
 gamma_mean_save <- matrix(0, nyear-1, nmcmc)
```

```
gamma_large_save <- matrix(0, nyear-1, nmcmc)
gamma_small_save <- matrix(0, nyear-1, nmcmc)
N_save <- array(0, dim=c(nsite, nyear, 200)) # for posterior predictive
check of y
pdet_save <- array(0, dim=c(nsite, nyear, nreps, 200)) # for posterior
predictive check of y
gamma_save <- array(0, dim=c(nsite, nyear-1, 200)) # for posterior
predictive check of r
cell_prob_save <- array(0, dim=c(nyear-1, nsite*2*(nyear-1)+1,
nsite*2, 200)) # for posterior predictive check for M-array

# Prior
beta0_mean <- rep(0, ncovs+1)
beta0_sd <- 2
beta_gamma_mean <- rep(0, ncovs+2)
beta_gamma_sd <- 2
beta_phi_adu_mean <- rep(0, ncovs+2)
beta_phi_adu_sd <- 2
beta_phi_juv_mean <- rep(0, ncovs+2)
beta_phi_juv_sd <- 2
log_kappa_adu_mean <- 0
log_kappa_adu_sd <- 2
log_kappa_juv_mean <- 0
log_kappa_juv_sd <- 2
alpha_det_mean <- rep(0, npcvs+1)
alpha_det_sd <- 2
alpha_cap_mean <- rep(0, npcvs+1)
alpha_cap_sd <- 2

# Starting values
beta0 <- rep(0, ncovs+1)
beta_gamma <- rep(0, ncovs+2)
beta_phi_adu <- rep(0, ncovs+2)
beta_phi_juv <- rep(0, ncovs+2)
kappa_adu <- 1
kappa_juv <- 1
alpha_det <- rep(0, npcvs+1)
alpha_cap <- rep(0, npcvs+1)
lambda0 <- exp(cbind(1,x[,1,]) %*% beta0)
gamma <- matrix(0.5, nsite, nyear-1)
phi_adu <- matrix(0.5, nsite, nyear-1)
phi_juv <- matrix(0.5, nsite, nyear-1)
eta_adu <- exp(-1 * kappa_adu * d)
theta_adu <- eta_adu / rowSums(eta_adu)
eta_juv <- exp(-1 * kappa_juv * d)
```

```
 theta_juv <- eta_juv / rowSums(eta_juv)
 N <- round((ymax + 1) / 0.5)
 pdet <- array(0.5, dim=c(nsite, nyear, nreps))
 pcap_per_visit <- array(0.5, dim=c(nsite*2, nyear-1, nreps))
 pcap <- matrix(, nsite*2, nyear-1)
 for (i in 1:(nsite*2)) {
  for (t in 1:(nyear-1)) {
   pcap[i,t] <- 1 - prod(1 - pcap_per_visit[i,t,])
  } # t
 } # i
 cell_prob <- cell_prob_fun(nsite=nsite, nyear=nyear,
phi_adu=phi_adu, phi_juv=phi_juv,
               theta_adu=theta_adu, theta_juv=theta_juv, pcap=pcap)

 # Tuning factor
 beta0_tune <- 0.08
 beta_gamma_tune <- 0.02
 beta_phi_adu_tune <- c(0.10, 0.10, 0.07, 0.07)
 beta_phi_juv_tune <- c(0.10, 0.10, 0.07, 0.07)
 kappa_adu_tune <- 0.035
 kappa_juv_tune <- 0.035
 alpha_det_tune <- c(0.050, 0.035)
 alpha_cap_tune <- 0.050
 N_tune <- 1

 # Begin MCMC loop
 for (k in 1:nmcmc) {
  ### Sample N
  N_star <- matrix(rpois(nsite*nyear, N + N_tune), nsite, nyear)
  mh1 <- apply(dbinom(y, N_star, pdet, log=T), 1:2, sum) +
     dpois(N_star, cbind(lambda0, t(t(N[,-nyear] * gamma * phi_juv) %*%
theta_juv + t(N[,-nyear] * phi_adu) %*% theta_adu)), log=T) +
      dpois(N, N_star+N_tune, log=T)
  mh2 <- apply(dbinom(y, N    , pdet, log=T), 1:2, sum) +
      dpois(N    , cbind(lambda0, t(t(N[,-nyear] * gamma * phi_juv) %*%
theta_juv + t(N[,-nyear] * phi_adu) %*% theta_adu)), log=T) +
      dpois(N_star, N+N_tune, log=T)
  mh <- exp(mh1 - mh2)
  Nkeep <- ((mh > runif(nsite*nyear)) & (N_star >= ymax))
  N[Nkeep] <- N_star[Nkeep]

  ### Sample beta0
  beta0_star <- rnorm(ncovs+1, beta0, beta0_tune)
  lambda0_star <- exp(cbind(1,x[,1,]) %*% beta0_star)
  mh1 <- sum(dpois(N[,1], lambda0_star, log=T)) +
```

```
      sum(dnorm(beta0_star, beta0_mean, beta0_sd, log=T))
  mh2 <- sum(dpois(N[,1], lambda0    , log=T)) +
      sum(dnorm(beta0    , beta0_mean, beta0_sd, log=T))
  mh <- exp(mh1 - mh2)
  if (mh > runif(1)) {
   beta0 <- beta0_star
   lambda0 <- lambda0_star
  }

  # Sample beta_gamma
  beta_gamma_star <- rnorm(ncovs+2, beta_gamma, beta_gamma_tune)
  gamma <- gamma_star <- matrix(, nsite, nyear-1)
  for (t in 2:nyear) {
  gamma   [,t-1] <- exp(cbind(1,(N[,t-1]-lambda0)/lambda0,x[,t,]) %*%
beta_gamma   )
   gamma_star[,t-1] <- exp(cbind(1,(N[,t-1]-lambda0)/lambda0,x[,t,])
%*% beta_gamma_star)
  } # t
  mh1 <- sum(dbinom(r, r + y[,-nyear,], gamma_star / (gamma_star + 1),
log=T)) +
      sum(dpois(N[,-1], t(t(N[,-nyear] * gamma_star * phi_juv) %*%
theta_juv + t(N[,-nyear] * phi_adu) %*% theta_adu), log=T)) +
      sum(dnorm(beta_gamma_star, beta_gamma_mean, beta_gamma_sd,
log=T))
  mh2 <- sum(dbinom(r, r + y[,-nyear,], gamma    / (gamma    + 1), log=T)) +
    sum(dpois(N[,-1], t(t(N[,-nyear] * gamma    * phi_juv) %*% theta_juv
+ t(N[,-nyear] * phi_adu) %*% theta_adu), log=T)) +
     sum(dnorm(beta_gamma    , beta_gamma_mean, beta_gamma_sd, log=T))
  mh <- exp(mh1 - mh2)
  if (mh > runif(1)) {
   beta_gamma <- beta_gamma_star
  }

  # Sample beta_phi_adu & beta_phi_juv
  beta_phi_adu_star <- rnorm(ncovs+2, beta_phi_adu,
beta_phi_adu_tune)
  beta_phi_juv_star <- rnorm(ncovs+2, beta_phi_juv,
beta_phi_juv_tune)
  phi_adu <- phi_adu_star <- phi_juv <- phi_juv_star <- matrix(, nsite,
nyear-1)
  for (t in 2:nyear) {
   phi_adu    [,t-1] <- inv.logit(cbind(1,(N[,t-1]-lambda0)/lambda0,x
[,t,]) %*% beta_phi_adu    )
   phi_adu_star[,t-1] <- inv.logit(cbind(1,(N[,t-1]-lambda0)/
lambda0,x[,t,]) %*% beta_phi_adu_star)
```

```
   phi_juv   [,t-1] <- inv.logit(cbind(1,(N[,t-1]-lambda0)/lambda0,x
[,t,]) %*% beta_phi_juv   )
   phi_juv_star[,t-1] <- inv.logit(cbind(1,(N[,t-1]-lambda0)/
lambda0,x[,t,]) %*% beta_phi_juv_star)
  } # t
  cell_prob   <- cell_prob_fun(nsite=nsite, nyear=nyear,
phi_adu=phi_adu   , phi_juv=phi_juv   , theta_adu=theta_adu,
theta_juv=theta_juv, pcap=pcap)
  cell_prob_star <- cell_prob_fun(nsite=nsite, nyear=nyear,
phi_adu=phi_adu_star, phi_juv=phi_juv_star, theta_adu=theta_adu,
theta_juv=theta_juv, pcap=pcap)
  prob1 <- prob2 <- matrix(, nsite, nyear-1)
  for (i in 1:nsite) {
   for (t in 1:(nyear-1)) {
    prob1[i,t] <- dmultinom(m[t,,i], size=sum(m[t,,i]),
prob=cell_prob_star[t,,i], log=T)
    prob2[i,t] <- dmultinom(m[t,,i], size=sum(m[t,,i]),
prob=cell_prob   [t,,i], log=T)
   } # t
  } # i
  mh1 <- sum(prob1) +
      sum(dpois(N[,-1], t(t(N[,-nyear] * gamma * phi_juv_star) %*%
theta_juv + t(N[,-nyear] * phi_adu_star) %*% theta_adu), log=T)) +
      sum(dnorm(beta_phi_adu_star, beta_phi_adu_mean,
beta_phi_adu_sd, log=T)) +
      sum(dnorm(beta_phi_juv_star, beta_phi_juv_mean,
beta_phi_juv_sd, log=T))
  mh2 <- sum(prob2) +
    sum(dpois(N[,-1], t(t(N[,-nyear] * gamma * phi_juv   ) %*% theta_juv
+ t(N[,-nyear] * phi_adu   ) %*% theta_adu), log=T)) +
      sum(dnorm(beta_phi_adu   , beta_phi_adu_mean, beta_phi_adu_sd,
log=T)) +
      sum(dnorm(beta_phi_juv   , beta_phi_juv_mean, beta_phi_juv_sd,
log=T))
  mh <- exp(mh1 - mh2)
  if (mh > runif(1)) {
   beta_phi_adu <- beta_phi_adu_star
   beta_phi_juv <- beta_phi_juv_star
   phi_adu <- phi_adu_star
   phi_juv <- phi_juv_star
   cell_prob <- cell_prob_star
  }

  # Sample kappa_adu & kappa_juv
  kappa_adu_star <- exp(rnorm(1, log(kappa_adu), kappa_adu_tune))
```

```
  eta_adu_star <- exp(-1 * kappa_adu_star * d)
  theta_adu_star <- eta_adu_star / rowSums(eta_adu_star)
  kappa_juv_star <- exp(rnorm(1, log(kappa_juv), kappa_juv_tune))
  eta_juv_star <- exp(-1 * kappa_juv_star * d)
  theta_juv_star <- eta_juv_star / rowSums(eta_juv_star)
  cell_prob_star <- cell_prob_fun(nsite=nsite, nyear=nyear,
phi_adu=phi_adu, phi_juv=phi_juv, theta_adu=theta_adu_star,
theta_juv=theta_juv_star, pcap=pcap)
  prob1 <- prob2 <- matrix(, nsite, nyear-1)
  for (i in 1:nsite) {
   for (t in 1:(nyear-1)) {
    prob1[i,t] <- dmultinom(m[t,,i], size=sum(m[t,,i]),
prob=cell_prob_star[t,,i], log=T)
    prob2[i,t] <- dmultinom(m[t,,i], size=sum(m[t,,i]),
prob=cell_prob     [t,,i], log=T)
   } # t
  } # i
  mh1 <- sum(prob1) +
      sum(dpois(N[,-1], t(t(N[,-nyear] * gamma * phi_juv) %*%
theta_juv_star + t(N[,-nyear] * phi_adu) %*% theta_adu_star), log=T)) +
      dnorm(log(kappa_adu_star), log_kappa_adu_mean,
log_kappa_adu_sd, log=T) +
      dnorm(log(kappa_juv_star), log_kappa_juv_mean,
log_kappa_juv_sd, log=T)
  mh2 <- sum(prob2) +
      sum(dpois(N[,-1], t(t(N[,-nyear] * gamma * phi_juv) %*% theta_juv
+ t(N[,-nyear] * phi_adu) %*% theta_adu     ), log=T)) +
      dnorm(log(kappa_adu     ), log_kappa_adu_mean, log_kappa_adu_sd,
log=T) +
      dnorm(log(kappa_juv     ), log_kappa_juv_mean, log_kappa_juv_sd,
log=T)
  mh <- exp(mh1 - mh2)
  if (mh > runif(1)) {
   kappa_adu <- kappa_adu_star
   eta_adu <- eta_adu_star
   theta_adu <- theta_adu_star
   kappa_juv <- kappa_juv_star
   eta_juv <- eta_juv_star
   theta_juv <- theta_juv_star
   cell_prob <- cell_prob_star
  }

  ### Sample alpha_det
  alpha_det_star <- rnorm(npcvs+1, alpha_det, alpha_det_tune)
  pdet_star <- array(, dim=c(nsite, nyear, nreps))
```

```
  for (i in 1:nsite) {
   for (t in 1:nyear) {
    pdet_star[i,t,] <- inv.logit(cbind(1,w[i,t,,]) %*%
alpha_det_star)
   } # t
  } # i
  mh1 <- sum(dbinom(y, N, pdet_star, log=TRUE), na.rm=T) +
      sum(dnorm(alpha_det_star, alpha_det_mean, alpha_det_sd,
log=TRUE))
  mh2 <- sum(dbinom(y, N, pdet     , log=TRUE), na.rm=T) +
     sum(dnorm(alpha_det     , alpha_det_mean, alpha_det_sd, log=TRUE))
  mh <- exp(mh1 - mh2)
  if (mh > runif(1)) {
   alpha_det <- alpha_det_star
   pdet <- pdet_star
  }

  # Sample alpha_pcap
  alpha_cap_star <- rnorm(npcvs+1, alpha_cap, alpha_cap_tune)
  pcap_per_visit_star <- array(, dim=c(nsite*2, nyear-1, nreps))
  for (i in 1:nsite) {
   for (t in 1:(nyear-1)) {
    pcap_per_visit_star[i,t,] <- inv.logit(cbind(1,w[i,t+1,,]) %*%
alpha_cap_star)
   } # t
  } # i
  pcap_per_visit_star[(nsite+1):(nsite*2),,] <- pcap_per_visit_star
[1:nsite,,]
  pcap_star <- matrix(, nsite*2, nyear-1)
  for (i in 1:(nsite*2)) {
   for (t in 1:(nyear-1)) {
    pcap_star[i,t] <- 1 - prod(1 - pcap_per_visit_star[i,t,])
   } # t
  } # i
  cell_prob_star <- cell_prob_fun(nsite=nsite, nyear=nyear,
phi_adu=phi_adu, phi_juv=phi_juv, theta_adu=theta_adu,
theta_juv=theta_juv, pcap=pcap_star)
  prob1 <- prob2 <- matrix(, nsite, nyear-1)
  for (i in 1:nsite) {
   for (t in 1:(nyear-1)) {
    prob1[i,t] <- dmultinom(m[t,,i], size=sum(m[t,,i]),
prob=cell_prob_star[t,,i], log=T)
    prob2[i,t] <- dmultinom(m[t,,i], size=sum(m[t,,i]),
prob=cell_prob     [t,,i], log=T)
   } # t
```

```
} # i
mh1 <- sum(prob1) +
   sum(dnorm(alpha_cap_star, alpha_cap_mean, alpha_cap_sd, log=T))
mh2 <- sum(prob2) +
   sum(dnorm(alpha_cap    , alpha_cap_mean, alpha_cap_sd, log=T))
mh <- exp(mh1 - mh2)
if (mh > runif(1)) {
 alpha_cap <- alpha_cap_star
 pcap_per_visit <- pcap_per_visit_star
 pcap <- pcap_star
 cell_prob <- cell_prob_star
}

# Save samples
beta0_save[,k] <- beta0
beta_gamma_save[,k] <- beta_gamma
beta_phi_adu_save[,k] <- beta_phi_adu
beta_phi_juv_save[,k] <- beta_phi_juv
kappa_adu_save[k] <- kappa_adu
kappa_juv_save[k] <- kappa_juv
alpha_det_save[,k] <- alpha_det
alpha_cap_save[,k] <- alpha_cap
N_mean_save[,k] <- colMeans(N)
N_large_save[,k] <- N[site_large,]
N_small_save[,k] <- N[site_small,]
phi_adu_mean_save[,k] <- colMeans(phi_adu)
phi_adu_large_save[,k] <- phi_adu[site_large,]
phi_adu_small_save[,k] <- phi_adu[site_small,]
phi_juv_mean_save[,k] <- colMeans(phi_juv)
phi_juv_large_save[,k] <- phi_juv[site_large,]
phi_juv_small_save[,k] <- phi_juv[site_small,]
gamma_mean_save[,k] <- colMeans(gamma)
gamma_large_save[,k] <- gamma[site_large,]
gamma_small_save[,k] <- gamma[site_small,]
if (nmcmc > 200) {
 if (k > nmcmc-200) {
  N_save[,,k-nmcmc+200] <- N
  pdet_save[,,,k-nmcmc+200] <- pdet
  gamma_save[,,k-nmcmc+200] <- gamma
  cell_prob_save[,,,k-nmcmc+200] <- cell_prob
 }
}
} # k
```

```
 list(beta0_save=beta0_save,
    beta_gamma_save=beta_gamma_save,
    beta_phi_adu_save=beta_phi_adu_save,
beta_phi_juv_save=beta_phi_juv_save,
    kappa_adu_save=kappa_adu_save, kappa_juv_save=kappa_juv_save,
    alpha_det_save=alpha_det_save, alpha_cap_save=alpha_cap_save,
    N_mean_save=N_mean_save, N_large_save=N_large_save,
N_small_save=N_small_save,
    phi_adu_mean_save=phi_adu_mean_save,
phi_adu_large_save=phi_adu_large_save,
phi_adu_small_save=phi_adu_small_save,
    phi_juv_mean_save=phi_juv_mean_save,
phi_juv_large_save=phi_juv_large_save,
phi_juv_small_save=phi_juv_small_save,
    gamma_mean_save=gamma_mean_save,
gamma_large_save=gamma_large_save,
gamma_small_save=gamma_small_save,
    N_save=N_save, pdet_save=pdet_save, gamma_save=gamma_save,
cell_prob_save=cell_prob_save)
} # age_sipm_mcmc
```

Next, let us use the following code to implement the MCMC algorithm, which takes more than 20 h to run. Then we can use the code in Appendix 4.5 to create some traceplots, from which we can see good model convergence and agreement between parameter estimates of $\beta^{[0]}$, $\beta^{[\gamma]}$, $\beta^{[\phi]}$, κ, $\alpha^{[cap]}$, and $\alpha^{[det]}$ and their true values (Fig. 4.10).

```
library(foreach)   # for parallel computing
library(doParallel) # for parallel computing

numCores <- round(detectCores() / 2) # only use half of the cores for
parallel computing
registerDoParallel(numCores) # setup parallel computing

nmcmc <- 50000 # number of iterations of MCMC computing
chain <- 3   # number of chains

start_time <- Sys.time() # start time of computing
out <- foreach (i=1:chain, .packages='boot') %dopar% {
 age_sipm_mcmc(y=y, r=r, m=m, x=x, w=w, d=d, nmcmc=nmcmc)
} # i
stopImplicitCluster() # clean up the cluster
end_time <- Sys.time() # end time of computing
time_taken <- end_time - start_time # computing time
```

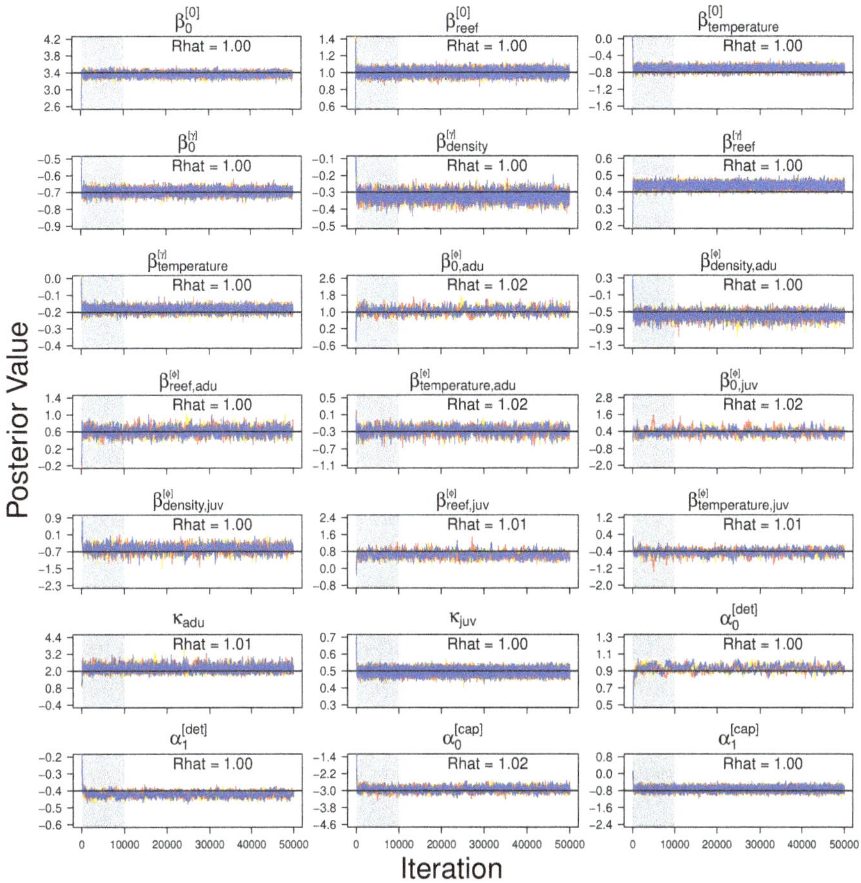

Fig. 4.10 Traceplots showing posterior chains for $\beta^{[0]}$, $\beta^{[\gamma]}$, $\beta^{[\phi]}$, κ, $\alpha^{[det]}$, and $\alpha^{[cap]}$ in the spatial IPM with age-specific survival and movement. Rhat values are calculated after discarding the first 10,000 iterations. Horizontal lines represent the true values used in data simulation

When we plot population-wide and site-level abundance, survival and reproduction values using the following code, we can see considerable agreement between their estimates and the simulated true values (Fig. 4.11):

```
library(vioplot) # for making violin plots

for (i in 1:chain) {
 if (i == 1) {
  N_mean <- out[[1]]$N_mean_save
  N_large <- out[[1]]$N_large_save
  N_small <- out[[1]]$N_small_save
  phi_adu_mean <- out[[1]]$phi_adu_mean_save
```

```
    phi_adu_large <- out[[1]]$phi_adu_large_save
    phi_adu_small <- out[[1]]$phi_adu_small_save
    phi_juv_mean <- out[[1]]$phi_juv_mean_save
    phi_juv_large <- out[[1]]$phi_juv_large_save
    phi_juv_small <- out[[1]]$phi_juv_small_save
    gamma_mean <- out[[1]]$gamma_mean_save
    gamma_large <- out[[1]]$gamma_large_save
    gamma_small <- out[[1]]$gamma_small_save
  } else {
    N_mean <- cbind(N_mean, out[[1]]$N_mean_save)
    N_large <- cbind(N_large, out[[1]]$N_large_save)
    N_small <- cbind(N_small, out[[1]]$N_small_save)
    phi_adu_mean <- cbind(phi_adu_mean, out[[1]]$phi_adu_mean_save)
    phi_adu_large <- cbind(phi_adu_large, out[[1]]$phi_adu_large_save)
    phi_adu_small <- cbind(phi_adu_small, out[[1]]$phi_adu_small_save)
    phi_juv_mean <- cbind(phi_juv_mean, out[[1]]$phi_juv_mean_save)
    phi_juv_large <- cbind(phi_juv_large, out[[1]]$phi_juv_large_save)
    phi_juv_small <- cbind(phi_juv_small, out[[1]]$phi_juv_small_save)
    gamma_mean <- cbind(gamma_mean, out[[1]]$gamma_mean_save)
    gamma_large <- cbind(gamma_large, out[[1]]$gamma_large_save)
    gamma_small <- cbind(gamma_small, out[[1]]$gamma_small_save)
  }
} # i

site_large <- max(which(rowMeans(x[,,1]) == max(rowMeans(x[,,1]))))
site_small <- min(which(rowMeans(x[,,1]) == min(rowMeans(x[,,1]))))

par(mfrow=c(4,3))
par(mar=c(1.5,4,0,0))
par(oma=c(4,3,3,1))

plot(1, xlim=c(0.5,nyear+0.5), ylim=c(0,60), type='l', axes=F,
xlab='', ylab='')
vioplot(t(N_mean), add=T, col='lightcoral', rectCol=NA, lineCol=NA,
border=NA)
lines(colMeans(N), type='o', pch=16, col='royalblue')
axis(1, at=seq(0,nyear,2), labels=rep('',6))
axis(2, at=seq(0,60,length.out=6), cex.axis=1.6, las=2)
axis(2, at=30, labels='Population Size', cex.axis=2.5, line=3.2,
tick=F)
axis(3, at=5.5, labels='Mean', cex.axis=2.5, line=-0.5, tick=F)

points(x=2, y=56, pch=16, cex=1, col='royalblue')
lines(x=c(1.5,2.5), y=c(56,56), lwd=1.2, col='royalblue')
vioplot(rnorm(10000,56,.5), at=5.5, add=T, col='lightcoral',
```

```
rectCol=NA, lineCol=NA, border=NA)
text(x=2.5, y=56, labels='True', pos=4, cex=1.8)
text(x=6, y=56, labels='Estimated', pos=4, cex=1.8)

plot(1, xlim=c(0.5,nyear+0.5), ylim=c(0,100), type='l', axes=F,
xlab='', ylab='')
vioplot(t(N_large), add=T, col='lightcoral', rectCol=NA,
lineCol=NA, border=NA)
lines(N[site_large,], type='o', pch=16, col='royalblue')
axis(1, at=seq(0,nyear,2), labels=rep('',6))
axis(2, at=seq(0,100,length.out=6), cex.axis=1.6, las=2)
axis(3, at=5.5, labels='Largest Site', cex.axis=2.5, line=-0.5,
tick=F)

plot(1, xlim=c(0.5,nyear+0.5), ylim=c(0,30), type='l', axes=F,
xlab='', ylab='')
vioplot(t(N_small), add=T, col='lightcoral', rectCol=NA,
lineCol=NA, border=NA)
lines(N[site_small,], type='o', pch=16, col='royalblue')
axis(1, at=seq(0,nyear,2), labels=rep('',6))
axis(2, at=seq(0,30,length.out=6), cex.axis=1.6, las=2)
axis(3, at=5.5, labels='Smallest Site', cex.axis=2.5, line=-0.5,
tick=F)

plot(1, xlim=c(0.5,nyear+0.5), ylim=c(0,1), type='l', axes=F,
xlab='', ylab='')
vioplot(t(phi_adu_mean), add=T, col='lightcoral', rectCol=NA,
lineCol=NA, border=NA)
lines(colMeans(phi_adu), type='o', pch=16, col='royalblue')
axis(1, at=seq(0,nyear,2), labels=rep('',6))
axis(2, at=seq(0,1,length.out=6), cex.axis=1.6, las=2)
axis(2, at=0.5, labels='Adult Survival', cex.axis=2.5, line=3.2,
tick=F)

plot(1, xlim=c(0.5,nyear+0.5), ylim=c(0,1), type='l', axes=F,
xlab='', ylab='')
vioplot(t(phi_adu_large), add=T, col='lightcoral', rectCol=NA,
lineCol=NA, border=NA)
lines(phi_adu[site_large,], type='o', pch=16, col='royalblue')
axis(1, at=seq(0,nyear,2), labels=rep('',6))
axis(2, at=seq(0,1,length.out=6), labels=rep('',6))

plot(1, xlim=c(0.5,nyear+0.5), ylim=c(0,1), type='l', axes=F,
xlab='', ylab='')
vioplot(t(phi_adu_small), add=T, col='lightcoral', rectCol=NA,
lineCol=NA, border=NA)
```

```
lines(phi_adu[site_small,], type='o', pch=16, col='royalblue')
axis(1, at=seq(0,nyear,2), labels=rep('',6))
axis(2, at=seq(0,1,length.out=6), labels=rep('',6))

plot(1, xlim=c(0.5,nyear+0.5), ylim=c(0,1), type='l', axes=F,
xlab='', ylab='')
vioplot(t(phi_juv_mean), add=T, col='lightcoral', rectCol=NA,
lineCol=NA, border=NA)
lines(colMeans(phi_juv), type='o', pch=16, col='royalblue')
axis(1, at=seq(0,nyear,2), labels=rep('',6))
axis(2, at=seq(0,1,length.out=6), cex.axis=1.6, las=2)
axis(2, at=0.5, labels='Juvenile Survival', cex.axis=2.5, line=3.2,
tick=F)

plot(1, xlim=c(0.5,nyear+0.5), ylim=c(0,1), type='l', axes=F,
xlab='', ylab='')
vioplot(t(phi_juv_large), add=T, col='lightcoral', rectCol=NA,
lineCol=NA, border=NA)
lines(phi_juv[site_large,], type='o', pch=16, col='royalblue')
axis(1, at=seq(0,nyear,2), labels=rep('',6))
axis(2, at=seq(0,1,length.out=6), labels=rep('',6))

plot(1, xlim=c(0.5,nyear+0.5), ylim=c(0,1), type='l', axes=F,
xlab='', ylab='')
vioplot(t(phi_juv_small), add=T, col='lightcoral', rectCol=NA,
lineCol=NA, border=NA)
lines(phi_juv[site_small,], type='o', pch=16, col='royalblue')
axis(1, at=seq(0,nyear,2), labels=rep('',6))
axis(2, at=seq(0,1,length.out=6), labels=rep('',6))

plot(1, xlim=c(0.5,nyear+0.5), ylim=c(0,1.5), type='l', axes=F,
xlab='', ylab='')
vioplot(t(gamma_mean), add=T, col='lightcoral', rectCol=NA,
lineCol=NA, border=NA)
lines(colMeans(gamma), type='o', pch=16, col='royalblue')
axis(1, at=seq(0,nyear,2), cex.axis=1.6)
axis(2, at=seq(0,1.5,length.out=6), cex.axis=1.6, las=2)
axis(2, at=0.75, labels='Reproduction', cex.axis=2.5, line=3.2,
tick=F)

plot(1, xlim=c(0.5,nyear+0.5), ylim=c(0,1.5), type='l', axes=F,
xlab='', ylab='')
vioplot(t(gamma_large), add=T, col='lightcoral', rectCol=NA,
lineCol=NA, border=NA)
lines(gamma[site_large,], type='o', pch=16, col='royalblue')
```

```
axis(1, at=seq(0,nyear,2), cex.axis=1.6)
axis(2, at=seq(0,1.5,length.out=6), labels=rep('',6))

plot(1, xlim=c(0.5,nyear+0.5), ylim=c(0,1.5), type='l', axes=F,
xlab='', ylab='')
vioplot(t(gamma_small), add=T, col='lightcoral', rectCol=NA,
lineCol=NA, border=NA)
lines(gamma[site_small,], type='o', pch=16, col='royalblue')
axis(1, at=seq(0,nyear,2), cex.axis=1.6)
axis(2, at=seq(0,1.5,length.out=6), labels=rep('',6))

title(xlab='Year', cex.lab=3, line=2.4, outer=T)
```

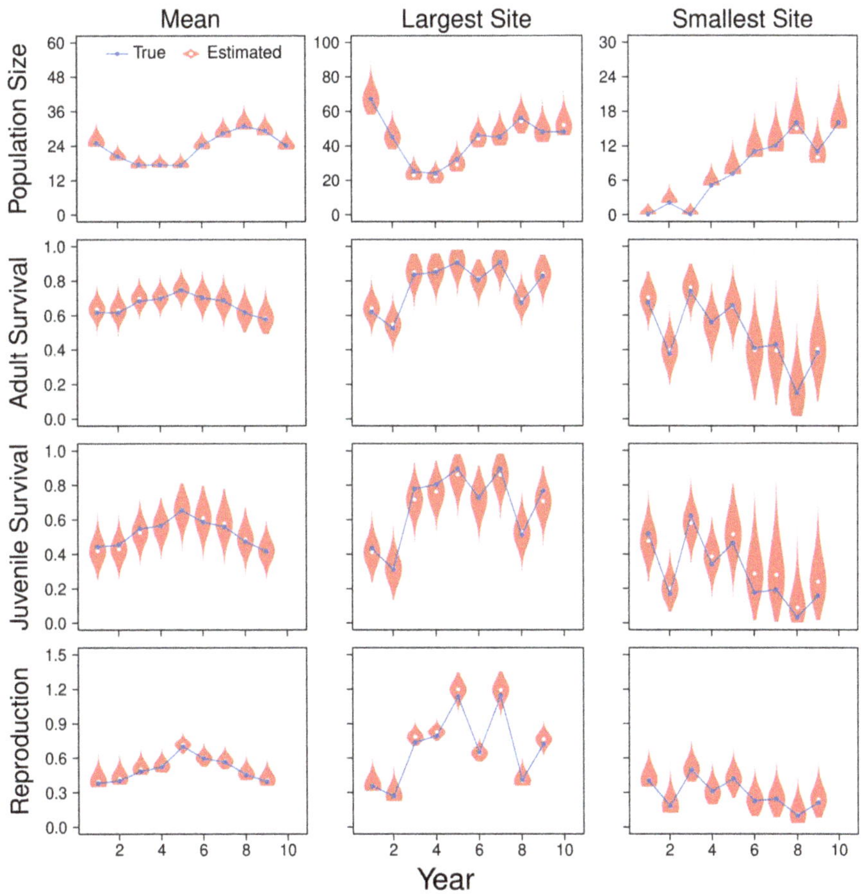

Fig. 4.11 Violin plots showing agreement between the estimates of spatial IPM with age-specific survival and movement and simulated true values of the mean population size, survival, and reproduction at the population level as well as site-level values for the largest and smallest sites

Eventually, we can use the following code to conduct posterior predictive checks. We first need to compute discrepancy measures (Gelman et al. 2014; Conn et al. 2018). We use Chi-squared statistics as the discrepancy measures for the count data (i.e., y and r): for the counts of adults, the discrepancy measure is calculated as $D^{[obs]}_{i,t,j,k} = \frac{\left(y_{i,t,j} - N_{i,t,k} \times p^{[det]}_{i,t,j,k}\right)^2}{Var\left(N_{i,t,k} \times p^{[det]}_{i,t,j,k}\right)}$ for data y, and $D^{[sim]}_{i,t,j,k} = \frac{\left(y^{[sim]}_{i,t,j,k} - N_{i,t,k} \times p^{[det]}_{i,t,j,k}\right)^2}{Var\left(N_{i,t,k} \times p^{[det]}_{i,t,j,k}\right)}$ for simulated data $y^{[sim]}$, in which $N_{i,t,k}$ and $p^{[det]}_{i,t,j,k}$ are the kth posterior sample of the corresponding parameter; for the counts of juveniles, the discrepancy measure is calculated as $D^{[obs]}_{i,t,j,k} = \frac{\left[r_{i,t,j} - \left(r_{i,t,j}+y_{i,t,j}\right) \times \gamma_{i,t,k}/\left(\gamma_{i,t,k}+1\right)\right]^2}{Var\left[\left(r_{i,t,j}+y_{i,t,j}\right) \times \gamma_{i,t,k}/\left(\gamma_{i,t,k}+1\right)\right]}$ for data r, and $D^{[sim]}_{i,t,j,k} = \frac{\left[r^{[sim]}_{i,t,j,k} - \left(r_{i,t,j}+y_{i,t,j}\right) \times \gamma_{i,t,k}/\left(\gamma_{i,t,k}+1\right)\right]^2}{Var\left[\left(r_{i,t,j}+y_{i,t,j}\right) \times \gamma_{i,t,k}/\left(\gamma_{i,t,k}+1\right)\right]}$ for simulated data $r^{[sim]}$, in which $\gamma_{i,t,k}$ is the kth posterior sample. We use the Freeman-Tukey statistic as the discrepancy measure for the capture-recapture data (Brooks et al. 2000): the discrepancy measure is calculated as $D^{[obs]}_{i,t,j,k} = \sum_{f=1}^{F} \left(\sqrt{m_{i,t,f}} - \sqrt{B_{i,t} \times Pr_{i,t,f,k}}\right)^2$ for data m and $D^{[sim]}_{i,t,j,k} = \sum_{f=1}^{F} \left(\sqrt{m^{[sim]}_{i,t,f,k}} - \sqrt{B_{i,t} \times Pr_{i,t,f,k}}\right)^2$ for simulated data $m^{[sim]}$, in which $B_{i,t}$ is the number of captures, and $Pr_{i,t,f,k}$ is the kth posterior sample of the fth cell probability of $m_{i,t,f}$. When these discrepancy measures are computed, a posterior predictive p-value is calculated as the proportion of $D^{[sim]}$ that are larger than $D^{[obs]}$.

```
# Count data
N_post <- array(, dim=c(nsite, nyear, 600))
N_post[,, 1:200] <- out[[1]]$N_save
N_post[,,201:400] <- out[[2]]$N_save
N_post[,,401:600] <- out[[3]]$N_save
pdet_post <- array(, dim=c(nsite, nyear, nreps, 600))
pdet_post[,,, 1:200] <- out[[1]]$pdet_save
pdet_post[,,,201:400] <- out[[2]]$pdet_save
pdet_post[,,,401:600] <- out[[3]]$pdet_save
y_exp <- y_sim <- y_dobs <- y_dsim <- array(, dim=c(nsite, nyear, nreps,
600))
for (i in 1:nsite) {
 for (t in 1:nyear) {
  for (j in 1:nreps) {
   y_exp[i,t,j,] <- N_post[i,t,] * pdet_post[i,t,j,]
   y_sim[i,t,j,] <- rbinom(600, N_post[i,t,], pdet_post[i,t,j,])
   y_dobs[i,t,j,] <- (y  [i,t,j ] - y_exp[i,t,j,]) ^ 2 / var(y_exp[i,t,
j,])
   y_dsim[i,t,j,] <- (y_sim[i,t,j,] - y_exp[i,t,j,]) ^ 2 / var(y_exp[i,
t,j,])
  } # j
```

```
  } # t
} # i
y_ppp <- length(which(y_dsim > y_dobs)) / length(which(y_dsim !=
y_dobs))

# Age-ratio data
gamma_post <- array(, dim=c(nsite, nyear-1, 600))
gamma_post[,, 1:200] <- out[[1]]$gamma_save
gamma_post[,,201:400] <- out[[2]]$gamma_save
gamma_post[,,401:600] <- out[[3]]$gamma_save
r_exp <- r_sim <- r_dobs <- r_dsim <- array(, dim=c(nsite, nyear-1,
nreps, 600))
for (i in 1:nsite) {
  for (t in 1:(nyear-1)) {
    for (j in 1:nreps) {
      r_exp[i,t,j,] <- (r[i,t,j] + y[i,t,j]) * gamma_post[i,t,] /
(gamma_post[i,t,] + 1)
      r_sim[i,t,j,] <- rbinom(600, r[i,t,j] + y[i,t,j], gamma_post[i,t,] /
(gamma_post[i,t,] + 1))
      r_dobs[i,t,j,] <- (r  [i,t,j ] - r_exp[i,t,j,]) ^ 2 / var(r_exp[i,t,
j,])
      r_dsim[i,t,j,] <- (r_sim[i,t,j,] - r_exp[i,t,j,]) ^ 2 / var(r_exp[i,
t,j,])
    } # j
  } # t
} # i
r_ppp <- length(which(r_dsim > r_dobs)) / length(which(r_dsim !=
r_dobs))

# Capture-recapture data
cell_prob_post <- array(, dim=c(nyear-1, nsite*2*(nyear-1)+1,
nsite*2, 600))
cell_prob_post[,,, 1:200] <- out[[1]]$cell_prob_save
cell_prob_post[,,,201:400] <- out[[2]]$cell_prob_save
cell_prob_post[,,,401:600] <- out[[3]]$cell_prob_save
nband <- apply(m, c(1,3), sum)
m_exp <- m_sim <- array(, dim=c(nyear-1, nsite*2*(nyear-1)+1, nsite*2,
600))
for (i in 1:(nsite*2)) {
  for (t in 1:(nyear-1)) {
    for (k in 1:600) {
      m_exp[t,,i,k] <- nband[t,i] * cell_prob_post[t,,i,k]
      m_sim[t,,i,k] <- rmultinom(1, nband[t,i], cell_prob_post[t,,i,k])
    } # k
  } # t
```

```
} # i
m_juv_dobs <- m_juv_dsim <- m_adu_dobs <- m_adu_dsim <- array(, dim=c
(nsite, nyear-1, 600))
for (i in 1:nsite) {
 for (t in 1:(nyear-1)) {
  for (k in 1:600) {
   m_juv_dobs[i,t,k] <- sum((sqrt(m  [t,,i ]) - sqrt(m_exp[t,,i,k])) ^
2)
   m_juv_dsim[i,t,k] <- sum((sqrt(m_sim[t,,i,k]) - sqrt(m_exp[t,,i,
k])) ^ 2)
   m_adu_dobs[i,t,k] <- sum((sqrt(m  [t,,nsite+i ]) - sqrt(m_exp[t,,
nsite+i,k])) ^ 2)
   m_adu_dsim[i,t,k] <- sum((sqrt(m_sim[t,,nsite+i,k]) - sqrt(m_exp
[t,,nsite+i,k])) ^ 2)
  } # k
 } # t
} # i
m_juv_ppp <- length(which(m_juv_dsim > m_juv_dobs)) / length(which
(m_juv_dsim != m_juv_dobs))
m_adu_ppp <- length(which(m_adu_dsim > m_adu_dobs)) / length(which
(m_adu_dsim != m_adu_dobs))
```

Next, we can use the following code to plot the discrepancy measures and posterior predictive p-values. For our model, posterior predictive p-values are 0.51 for adult counts, 0.57 for juvenile counts, 0.50 for adult capture-recapture data, and 0.49 for juvenile capture-recapture data, all are close to 0.5 to indicate a good model fit (Fig. 4.12).

```
col_point <- 'lightcoral'
col_line <- 'royalblue'

par(mfrow=c(2,2))
par(mar=c(2,4,3,1))
par(oma=c(3,3,0,0))

y_sample <- array(c(rep(1,10000),rep(0,nsite*nyear*nreps*600-
10000)), dim=c(nsite, nyear, nreps, 600))
plot(y_dsim[which(y_sample == 1)] ~ y_dobs[which(y_sample == 1)],
  xlim=c(0,600), ylim=c(0,600), axes=F, xlab='', ylab='', pch=16,
cex=1.2, col=col_point)
abline(0, 1, col=col_line, lwd=1.8)
axis(1, at=seq(0,600,length.out=4), cex.axis=1.4)
axis(2, at=seq(0,600,length.out=4), cex.axis=1.4, las=2)
box()
```

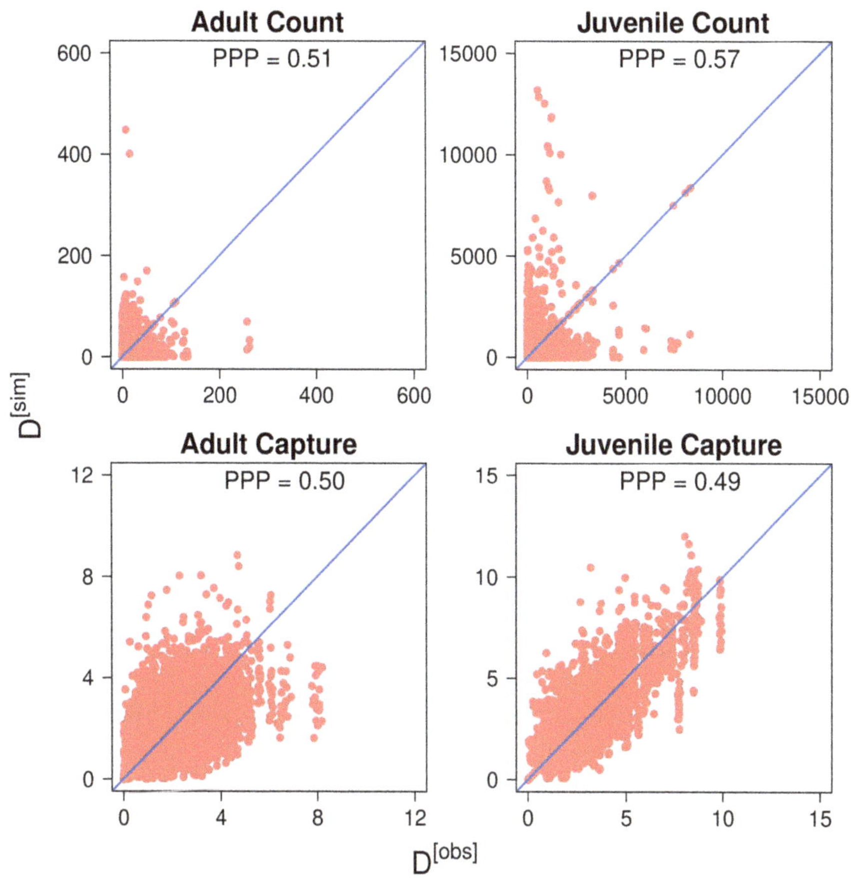

Fig. 4.12 Scatter plots showing that posterior predictive p-values, the proportion of $D^{[sim]}$ that are larger than $D^{[obs]}$, are close to 0.5 for count and capture-recapture data used in the spatial IPM with age-specific survival and movement, indicating a good model fit

```
title(main='Adult Count', cex.main=1.8, line=0.5)
text(x=168, y=584, pos=4, label=paste('PPP = ', format(round(y_ppp,
digits=2), nsmall=2), sep=''), cex=1.6)

r_sample <- array(c(rep(1,10000),rep(0,nsite*(nyear-1)*nreps*600-
10000)), dim=c(nsite, nyear-1, nreps, 600))
plot(r_dsim[which(r_sample == 1)] ~ r_dobs[which(r_sample == 1)],
  xlim=c(0,15000), ylim=c(0,15000), axes=F, xlab='', ylab='',
pch=16, cex=1.2, col=col_point)
abline(0, 1, col=col_line, lwd=1.8)
axis(1, at=seq(0,15000,length.out=4), cex.axis=1.4)
axis(2, at=seq(0,15000,length.out=4), cex.axis=1.4, las=2)
box()
```

```
title(main='Juvenile Count', cex.main=1.8, line=0.5)
text(x=4200, y=14600, pos=4, label=paste('PPP = ', format(round
(r_ppp, digits=2), nsmall=2), sep=''), cex=1.6)

m_adu_sample <- array(c(rep(1,10000),rep(0,nsite*(nyear-1)*600-
10000)), dim=c(nsite, nyear-1, 600))
plot(m_adu_dsim[which(m_adu_sample == 1)] ~ m_adu_dobs[which
(m_adu_sample == 1)],
  xlim=c(0,12), ylim=c(0,12), axes=F, xlab='', ylab='', pch=16,
cex=1.2, col=col_point)
abline(0, 1, col=col_line, lwd=1.8)
axis(1, at=seq(0,12,length.out=4), cex.axis=1.4)
axis(2, at=seq(0,12,length.out=4), cex.axis=1.4, las=2)
box()
title(main='Adult Capture', cex.main=1.8, line=0.5)
text(x=3.78, y=11.68, pos=4, label=paste('PPP = ', format(round
(m_adu_ppp, digits=2), nsmall=2), sep=''), cex=1.6)

m_juv_sample <- array(c(rep(1,10000),rep(0,nsite*(nyear-1)*600-
10000)), dim=c(nsite, nyear-1, 600))
plot(m_juv_dsim[which(m_juv_sample == 1)] ~ m_juv_dobs[which
(m_juv_sample == 1)],
  xlim=c(0,15), ylim=c(0,15), axes=F, xlab='', ylab='', pch=16,
cex=1.2, col=col_point)
abline(0, 1, col=col_line, lwd=1.8)
axis(1, at=seq(0,15,length.out=4), cex.axis=1.4)
axis(2, at=seq(0,15,length.out=4), cex.axis=1.4, las=2)
box()
title(main='Juvenile Capture', cex.main=1.8, line=0.5)
text(x=4.2, y=14.6, pos=4, label=paste('PPP = ', format(round
(m_juv_ppp, digits=2), nsmall=2), sep=''), cex=1.6)

title(xlab=expression(D^"[obs]"), cex.lab=2, outer=T, line=1.6)
title(ylab=expression(D^"[sim]"), cex.lab=2, outer=T, line=0.8)
```

4.4.3 With Body-Class-Specific Reproduction, Survival, and Movement

Next, we will consider a spatial IPM that incorporates information of body-size-class (hereafter body-class). This model is motivated by studies that examine the effect of body class (or more generally individual characteristics) on demography and population structure using non-spatial IPMs (e.g., Gamelon et al. 2021). Using our fictitious fish as an example, we now think about a population in which individuals are either small or large. We assume that newly produced offspring are all small and have to wait until they are one year old to reproduce. As long as a fish is at least 1 year old, it can reproduce regardless of its size, although small fish have a lower reproduction rate than large fish. Small fish also have a lower survival probability, but a higher probability of long-distance movement, than large fish. Eventually, small fish have a chance to grow to large size, but large fish will not become small. For the observation processes, we assume that body class can by correctly identified in both capture-recapture and count surveys. When body class cannot be distinguished in count surveys, some of the parameters in this model may not be identifiable, which warrants further investigation.

Using a statistical language, we can describe the initial abundance as follows:

$$\left.\begin{array}{r} N_{i,1,small} \sim Poisson(\lambda_{i,1}) \\ N_{i,1,large} \sim Poisson(\lambda_{i,1}) \\ \lambda_{i,1} = exp\left(\beta_0^{[0]} + \beta_1^{[0]} \times x_{i,1,1} + \ldots + \beta_{n_{covs}}^{[0]} \times x_{i,1,n_{covs}}\right) \end{array}\right\}, \tag{4.13}$$

in which $N_{i,\ 1,\ small}$ is the abundance of small breeders at site i in the first year; $N_{i,\ 1,\ large}$ is the abundance of large breeders; and $\lambda_{i,\ 1}$ is the expectation of initial abundance for both small and large individuals and is a function of covariates x's with intercept and slope parameters $\beta^{[0]}$'s. Here, we assume even proportion of small and large individuals in the first year, which could be extended to be, for example, small-size skewed. From the second year (i.e., $t \geq 2$), the process sub-model specifies the spatiotemporal variation in abundance as follows:

$$\left.\begin{array}{r} N_{i_2,t,small} \sim Poisson(\lambda_{i_2,t,small}) \\ N_{i_2,t,large} \sim Poisson(\lambda_{i_2,t,large}) \\ \lambda_{i_2,t,small} = \sum_{i_1=1}^{n_{site}} \left\{ \begin{array}{c} \left[N_{i_1,t-1,small} \times (1+\gamma_{i_1,t-1,small}) + N_{i_1,t-1,large} \times \gamma_{i_1,t-1,large}\right] \times \phi_{i,t-1,small} \\ \times (1-\omega) \times \theta_{i_1,i_2,t,small} \end{array} \right\} \\ \lambda_{i_2,t,large} = \sum_{i_1=1}^{n_{site}} \left\{ \begin{array}{c} \left[N_{i_1,t-1,small} \times (1+\gamma_{i_1,t-1,small}) + N_{i_1,t-1,large} \times \gamma_{i_1,t-1,large}\right] \times \phi_{i,t-1,small} \\ \times \omega \times \theta_{i_1,i_2,t,large} \\ + N_{i_1,t-1,large} \times \phi_{i,t-1,large} \times \theta_{i_1,i_2,t,large} \end{array} \right\} \\ \gamma_{i_1,t-1,small} = exp\left(\beta_{0,small}^{[\gamma]} + \beta_{density,small}^{[\gamma]} \times N_{i_1,t-1} + \beta_{1,small}^{[\gamma]} \times x_{i_1,t,1} + \ldots + \beta_{n_{covs},small}^{[\gamma]} \times x_{i_1,t,n_{covs}}\right) \\ \gamma_{i_1,t-1,large} = exp\left(\beta_{0,large}^{[\gamma]} + \beta_{density,large}^{[\gamma]} \times N_{i_1,t-1} + \beta_{1,large}^{[\gamma]} \times x_{i_1,t,1} + \ldots + \beta_{n_{covs},large}^{[\gamma]} \times x_{i_1,t,n_{covs}}\right) \\ \phi_{i_1,t-1,small} = logit^{-1}\left(\beta_{0,small}^{[\phi]} + \beta_{density,small}^{[\phi]} \times N_{i_1,t-1} + \beta_{1,small}^{[\phi]} \times x_{i_1,t,1} + \ldots + \beta_{n_{covs},small}^{[\phi]} \times x_{i_1,t,n_{covs}}\right) \\ \phi_{i_1,t-1,large} = logit^{-1}\left(\beta_{0,large}^{[\phi]} + \beta_{density,large}^{[\phi]} \times N_{i_1,t-1} + \beta_{1,large}^{[\phi]} \times x_{i_1,t,1} + \ldots + \beta_{n_{covs},large}^{[\phi]} \times x_{i_1,t,n_{covs}}\right) \\ \theta_{i_1,i_2,small} = \frac{\eta_{i_1,i_2,small}}{\sum_{i_2=1}^{n_{site}} \eta_{i_1,i_2},small} \\ \theta_{i_1,i_2,large} = \frac{\eta_{i_1,i_2,large}}{\sum_{i_2=1}^{n_{site}} \eta_{i_1,i_2,large}} \\ \eta_{i_1,i_2,small} = exp(-1 \times \kappa_{small} \times d_{i_1,i_2}) \\ \eta_{i_1,i_2,large} = exp(-1 \times \kappa_{large} \times d_{i_1,i_2}) \end{array}\right\}, \tag{4.14}$$

in which $\gamma_{i,\,t\,-\,1,\,small}$ and $\gamma_{i,\,t\,-\,1,\,large}$ are per capita reproduction rates of small and large individuals at site i in year $t-1$, respectively, and are functions of density and environmental covariates x's with intercept and slope parameters $\beta^{[\gamma]}$'s; $\phi_{i,\,t\,-\,1,\,small}$ and $\phi_{i,\,t\,-\,1,\,large}$ are survival probabilities of small and large individuals, respectively, and are functions of density and environmental covariates with intercept and slope parameters $\beta^{[\phi]}$'s; ω is the probability of transitioning from small to large size; $\theta_{i_1,i_2,small}$ and $\theta_{i_1,i_2,large}$ are the standardized movement rates from site i_1 to site i_2 for small and large individuals, respecitvley; $\eta_{i_1,i_2,small}$ and $\eta_{i_1,i_2,large}$ are unstandardized movement rates; and κ_{small} and κ_{large} are the decay parameters governing the effect of distance (d_{i_1,i_2}) on movement. Eventually, the abundance of small/large animals at site i_2 in year t is the total number of animals under the corresponding body class that either stay at site i_2 or move in from any other sites i_1. Individual survival-location history $z_{v,\,t}$ and number of offspring reproduced by small and large breeders, denoted $R_{i,\,t,\,small}$ and $R_{i,\,t,\,large}$, respectively, are also simulated.

The observed capture history data are as before obtained in the following way:

$$\left.\begin{array}{r}c_{v,t,j}\sim Bernoulli\left(p^{[cap]}_{z_{v,t},t,j}\right)\times z_{v,t}\\ p^{[cap]}_{i,t,j}=logit^{-1}\left(\alpha^{[cap]}_0+\alpha^{[cap]}_1\times w_{i,t,j,1}+\ldots+\alpha^{[cap]}_{n_{pcvs}}\times w_{i,t,j,n_{pcvs}}\right)\end{array}\right\}, \quad (4.15)$$

in which $c_{v,\,t,\,j}$ is the capture history of individual v in year t during the jth visit, $p^{[cap]}_{i,t,j}$ is the capture probability and is a function of covariates w's governed by intercept and slope parameters $\alpha^{[cap]}$'s. Note that $z_{v,\,t}$ and $c_{v,\,t,\,j}$ takes values of $1, \ldots, n_{site}$ for small animals and $n_{site}+1, \ldots, n_{site}\times 2$ for large animals to allow body class distinguished in the m-array.

The count data of breeders are collected for small and large individuals separately. When the number of offspring is counted, we assume that we can tell if they are reproduced by small or large breeders. These data follow binomial distributions such that

$$\left.\begin{array}{r}y_{i,t,j,small}\sim Binomial\left(N_{i,t,small},p^{[det]}_{i,t,j}\right)\\ y_{i,t,j,large}\sim Binomial\left(N_{i,t,large},p^{[det]}_{i,t,j}\right)\\ r_{i,t,j,small}\sim Binomial\left(R_{i,t,small},p^{[det]}_{i,t,j}\right)\\ r_{i,t,j,large}\sim Binomial\left(R_{i,t,large},p^{[det]}_{i,t,j}\right)\\ p^{[det]}_{i,t,j}=logit^{-1}\left(\alpha^{[det]}_0+\alpha^{[det]}_1\times w_{i,t,j,1}+\ldots+\alpha^{[det]}_{n_{pcvs}}\times w_{i,t,j,n_{pcvs}}\right)\end{array}\right\}, \quad (4.16)$$

In which $y_{i,\,t,\,j,\,small}$ is the count of small breeders, $y_{i,\,t,\,j,\,large}$ the counts of large breeders, $r_{i,\,t,\,j,\,small}$ the counts of offspring that are reproduced by small breeders, $r_{i,\,t,\,j,\,large}$ the counts of offspring that are reproduced by large breeders, and $p^{[det]}_{i,t,j}$ is the detection probability and is a function of covariates w's with intercept and slope parameters $\alpha^{[det]}$'s.

Using the following code, we can simulate data for this spatial IPM with body-class-specific reproduction, survival, and movement. Because the model is very slow to compute, we reduced the number of sites to 30 in the data simulation. We assume that reef size positively affects initial abundance ($\beta^{[0]}_{reef} = 1$), reproduction of small fish ($\beta^{[\gamma]}_{reef,small} = 0.5$), reproduction of large fish ($\beta^{[\gamma]}_{reef,large} = 0.4$), survival of small fish ($\beta^{[\phi]}_{reef,small} = 0.8$), and survival of large fish ($\beta^{[\phi]}_{reef,large} = 0.6$); temperature negatively affects initial abundance ($\beta^{[0]}_{temperature} = -0.8$), survival of small fish ($\beta^{[\phi]}_{temperature,small} = -0.4$), and survival of large fish ($\beta^{[\phi]}_{temperature,large} = -0.3$) without affecting reproduction; and density negatively affects reproduction of small fish ($\beta^{[\gamma]}_{density,small} = -0.4$), reproduction of large fish ($\beta^{[\gamma]}_{density,large} = -0.3$), survival of small fish ($\beta^{[\phi]}_{density,small} = -0.7$), and survival of large fish ($\beta^{[\phi]}_{density,large} = -0.5$). With these numbers, we assume that small fish are more sensitive to density and environmental variation, in addition to their lower baseline reproduction ($\beta^{[\gamma]}_{0,small} = -1.0$) and survival ($\beta^{[\phi]}_{0,small} = 0.8$), than large fish ($\beta^{[\gamma]}_{0,large} = -0.6$, $\beta^{[\phi]}_{0,large} = 1.0$). We also assume that, every year, small fish have a 70% probability to grow to large size ($\omega = 0.7$). Eventually, we assume a weaker decay for small fish ($\kappa_{small} = 0.5$) to allow a higher probability of long-distance movement of them, in comparison to large fish ($\kappa_{large} = 2$).

```
library(boot) # for using logit and inv.logit functions
library(LaplacesDemon) # for generating random numbers from
categorical distributions (rcat)

set.seed(7)

# Basic values
nsite <- 30 # number of sites
nyear <- 10 # number of years
nreps <- 5  # number of within-season replicates
ncovs <- 2  # number of environmental covariates
npcvs <- 1  # number of observational covariates

beta0 <- c(3.4, 1, -0.8)          # intercept and slopes for initial
abundance
beta_gamma_small <- c(-1.0, -0.4, 0.5)  # intercept and slopes for
reproduction of small individuals
beta_gamma_large <- c(-0.6, -0.3, 0.4)  # intercept and slopes for
reproduction of large individuals
beta_phi_small <- c(0.8, -0.7, 0.8, -0.4) # intercept and slopes for
survival of small individuals
beta_phi_large <- c(1.0, -0.5, 0.6, -0.3) # intercept and slopes for
survival of large individuals
omega <- 0.7                   # growth probability from small to large body
```

```
class
kappa_small <- 0.5              # distance effect on movement of small
individuals
kappa_large <- 2               # distance effect on movement of large
individuals
alpha_det <- c( 0.9, -0.4)          # intercept and slopes for detection
probability
alpha_cap <- c(-3.0, -0.8)          # intercept and slopes for capture
probability

# Read in environmental covariates
load('data/covariates.RData')
x1 <- (log(cov$size) - mean(log(cov$size))) / sd(log(cov$size)) #
standardized log habitat patch size
x2 <- (cov$temp - mean(cov$temp)) / sd(cov$temp) # standardized
temperature
x <- array(, dim=c(nsite, nyear, ncovs)) # environmental covariates
x[,,1] <- x1[1:nsite, 1:nyear]
x[,,2] <- x2[1:nsite, 1:nyear]
### Standardize covariates for each site
for (i in 1:nsite) {
 for (h in 1:ncovs) {
  x[i,2:nyear,h] <- (x[i,2:nyear,h] - mean(x[i,,h])) / sd(x[i,,h])
 } # h
} # i

# Simulate data
lambda0 <- exp(cbind(1,x[,1,]) %*% beta0) # expectation of initial
abundance
N_small <- matrix(, nsite, nyear) # abundance of small individuals
N_large <- matrix(, nsite, nyear) # abundance of large individuals
N_small[,1] <- rpois(nsite, lambda0) # initial abundance of small
individuals
N_large[,1] <- rpois(nsite, lambda0) # initial abundance of large
individuals

lon <- cov$lon[1:nsite] # longitude or easting
lat <- cov$lat[1:nsite] # latitude or northing
d <- matrix(, nsite, nsite) # distance matrix
for (i1 in 1:nsite) {
 for (i2 in 1:nsite) {
  d[i1,i2] <- sqrt((lon[i1] - lon[i2]) ^ 2 + (lat[i1] - lat[i2]) ^ 2)
 } # i2
} # i1
eta_small <- exp(-1 * kappa_small * d)     # unstandardized colonization
```

```
rate of small individuals
theta_small <- eta_small / rowSums(eta_small) #  standardized
colonization rate of small individuals
eta_large <- exp(-1 * kappa_large * d)      # unstandardized colonization
rate of large individuals
theta_large <- eta_large / rowSums(eta_large) #  standardized
colonization rate of large individuals

nindv <- 20000 # maximum number of individuals that ever exist
z <- matrix(0, nindv, nyear) # individual survival-location history
N_acc <- rep(0, nyear) # number of individuals that ever exist
for (i in 1:nsite) {
 if (i == 1) {
  tt <- c(rep(i, N_small[i,1]), rep(i+nsite, N_large[i,1]))
 } else {
  tt <- c(tt, rep(i, N_small[i,1]), rep(i+nsite, N_large[i,1]))
 }
}
z[1:length(tt),1] <- tt
N_acc[1] <- length(tt)

gamma_small <- matrix(, nsite, nyear-1) # reproduction rate or small
individuals
gamma_large <- matrix(, nsite, nyear-1) # reproduction rate or large
individuals
phi_small <- matrix(, nsite, nyear-1) # survival probability of small
individuals
phi_large <- matrix(, nsite, nyear-1) # survival probability of large
individuals
R_small <- matrix(, nsite, nyear-1) # number of individuals reproduced
by small individuals
R_large <- matrix(, nsite, nyear-1) # number of individuals reproduced
by large individuals

for (t in 2:nyear) {
 gamma_small[,t-1] <- exp(cbind(1, (N_small[,t-1]+N_large[,t-1]-
lambda0*2)/(lambda0*2),x[,t,1]) %*% beta_gamma_small)
 gamma_large[,t-1] <- exp(cbind(1, (N_small[,t-1]+N_large[,t-1]-
lambda0*2)/(lambda0*2),x[,t,1]) %*% beta_gamma_large)
 R_small[,t-1] <- rpois(nsite, N_small[,t-1] * gamma_small[,t-1])
 R_large[,t-1] <- rpois(nsite, N_large[,t-1] * gamma_large[,t-1])

 # All offspring are small
 for (i in 1:nsite) {
  if (i == 1) { tt <- rep(i, R_small[i,t-1] + R_large[i,t-1]) } else { tt
```

```
<- c(tt, rep(i, R_small[i,t-1] + R_large[i,t-1])) }
 } # i
 z[N_acc[t-1]+1:length(tt), t-1] <- tt

 phi_small[,t-1] <- inv.logit(cbind(1, (N_small[,t-1]+N_large[,t-1]-
lambda0*2)/(lambda0*2),x[,t,]) %*% beta_phi_small)
 phi_large[,t-1] <- inv.logit(cbind(1, (N_small[,t-1]+N_large[,t-1]-
lambda0*2)/(lambda0*2),x[,t,]) %*% beta_phi_large)
 transition_matrix <- matrix(, nsite*2, nsite*2+1) # matrix of
transition (survival, growth & movement) probability
 for (i in 1:nsite) {
  transition_matrix[i,1:nsite] <- phi_small[i,t-1] * (1 - omega) *
theta_small[i,]
  transition_matrix[i,nsite+1:nsite] <- phi_small[i,t-1] * omega *
theta_large[i,]
  transition_matrix[nsite+i,1:nsite] <- 0
  transition_matrix[nsite+i,nsite+1:nsite] <- phi_large[i,t-1] *
theta_large[i,]
 } # i
 transition_matrix[,nsite*2+1] <- 1 - rowSums(transition_matrix[,1:
(nsite*2)])

 for (v in 1:dim(z)[1]) {
  if (z[v,t-1] == 0) {
   z[v,t] <- 0
  } else {
   z[v,t] <- rcat(1, p=transition_matrix[z[v,t-1],])
  }
 } # v
 z[which(!(z %in% c(1:(nsite*2))))] <- 0

 for (i in 1:nsite) {
  N_small[i,t] <- length(which(z[,t] == i))
  N_large[i,t] <- length(which(z[,t] == nsite+i))
 }
 N_acc[t] <- max(which(rowMeans(z, na.rm=T) > 0))
} # t
z[which(!(z %in% 1:(nsite*2)))] <- 0
z <- z[which(rowMeans(z, na.rm=T) != 0),] # only keep individuals that
ever exist

# Simulate count data
w <- array(rnorm(nsite * nyear * nreps * npcvs, 0, 1), dim=c(nsite,
nyear, nreps, npcvs)) # observational covariates
```

```
pdet <- array(, dim=c(nsite, nyear, nreps)) # detection probability
for (i in 1:nsite) {
 for (t in 1:nyear) {
  pdet[i,t,] <- inv.logit(cbind(1,w[i,t,,]) %*% alpha_det)
 } # t
} # i

y <- array(, dim=c(nsite, nyear, nreps, 2)) # count data of breeders
for (i in 1:nsite) {
 for (t in 1:nyear) {
  y[i,t,,1] <- rbinom(nreps, N_small[i,t], pdet[i,t,]) # count data of
small breeders
  y[i,t,,2] <- rbinom(nreps, N_large[i,t], pdet[i,t,]) # count data of
large breeders
 } # t
} # i

r <- array(, dim=c(nsite, nyear-1, nreps, 2)) # count data of offspring
for (i in 1:nsite) {
 for (t in 1:(nyear-1)) {
  r[i,t,,1] <- rbinom(nreps, R_small[i,t], pdet[i,t,]) # count data of
offspring reproduced by small breeders
  r[i,t,,2] <- rbinom(nreps, R_large[i,t], pdet[i,t,]) # count data of
offspring reproduced by large breeders
 } # t
} # i

# simulate capture-recapture data
pcap_per_visit <- array(, dim=c(nsite*2, nyear, nreps)) # per visit
capture probability
for (i in 1:nsite) {
 for (t in 1:nyear) {
  pcap_per_visit[i,t,] <- inv.logit(cbind(1,w[i,t,,]) %*% alpha_cap)
 } # t
} # i
pcap_per_visit[(nsite+1):(nsite*2),,] <- pcap_per_visit[1:nsite,,]

c_all <- matrix(, dim(z)[1], nyear) # capture history of all individuals
for (v in 1:dim(z)[1]) {
 for (t in 1:nyear) {
  if (z[v,t] == 0) {
   c_all[v,t] <- 0
  } else {
   cap_per_visit <- rbinom(nreps, 1, pcap_per_visit[z[v,t],t,])
   c01 <- ifelse(sum(cap_per_visit)==0, 0, 1)
```

```
    c_all[v,t] <- c01 * z[v,t]
   }
  } # t
} # v
if_cap <- ifelse(apply(c_all, 1, sum) > 0, 1, 0)

c <- c_all[which(if_cap==1),] # capture history of individuals that
have ever been captured
f <- numeric(dim(c)[1]) # year of first capture
for (v in 1:length(f)) {
 f[v] <- min(which(c[v,] %in% c(1:(nsite*2))))
} # v
c <- c[which(f %in% c(1:(nyear-1))),]

# Extend individual capture history
for (v in 1:dim(c)[1]) {
 t1 <- c[v,]
 ycap <- which(t1 %in% c(1:(nsite*2)))
 ncap <- length(ycap)
 if (ncap == 1) {
  t2 <- t1
 } else {
  t2 <- matrix(0, ncap, nyear)
  for (k in 1:(ncap-1)) {
   t2[k, ycap[k]:ycap[k+1]] <- t1[ycap[k]:ycap[k+1]]
  } # k
  t2[ncap, ycap[ncap]] <- t1[ycap[ncap]]
 }
 if (v == 1) {
  c_ext <- t2
 } else {
  c_ext <- rbind(c_ext, t2)
 }
} # v

# Converting individual capture history to m-array format
f_ext <- numeric(dim(c_ext)[1]) # extended year of first capture
for (v in 1:length(f_ext)) {
 f_ext[v] <- min(which(c_ext[v,] %in% c(1:(nsite*2))))
} # v

for (t in 2:nyear) {
 t1 <- paste('small_site', '_', 1:nsite, ', year', t, sep='')
 t2 <- paste('large_site', '_', 1:nsite, ', year', t, sep='')
 tt <- c(t1, t2)
```

```
  if (t == 2) {
   nm1 <- tt
  } else {
   nm1 <- c(nm1, tt)
  }
 } # t
nm1 <- c(nm1, 'not seen')
nm2 <- c(paste('small_site', 1:nsite, sep='_'), paste('large_site',
1:nsite, sep='_'))

m <- array(0, # m-array of capture-recapture
 dim=c(nyear-1, nsite*2*(nyear-1)+1, nsite*2),
 dimnames=list(paste('year', 1:(nyear-1), sep='_'), nm1, nm2))
for (t1 in 1:(nyear-1)) {
 temp1 <- c_ext[which(f_ext == t1),]
 for (i1 in 1:(nsite*2)) {
  temp2 <- temp1[which(temp1[,t1] == i1),]
  nband <- length(which(temp1[,t1] == i1))
  if (nband == 1) {
   temp2 <- matrix(temp2, nrow=1, ncol=length(temp2))
  }
  for (t2 in (t1+1):nyear) {
   for (i2 in 1:(nsite*2)) {
    m[t1,(t2-2)*(nsite*2)+i2,i1] <- length(which(temp2[,t2] == i2))
   } # i2
  } # t2
  m[t1,(nsite*2)*(nyear-1)+1,i1] <- nband - sum(m[t1,1:((nsite*2)*
(nyear-1)),i1])
 } # i1
} # t1
```

Next, we will develop our own function to calculate the cell probabilities of the m-array and custom MCMC algorithm to estimate parameters $\beta^{[0]}$, $\beta^{[\gamma]}$, $\beta^{[\phi]}$, ω, κ, $\alpha^{[cap]}$, $\alpha^{[det]}$, and N using data y, r, m, x, w, d.

```
# Function to calculate the cell probability of m-array
cell_prob_fun <- function(nsite, nyear, phi_small, phi_large, omega,
theta_small, theta_large, pcap) {
 qcap <- 1 - pcap
 phi <- rbind(phi_small, phi_large)
 theta <- rbind(cbind((1-omega)*theta_small,omega*theta_large),
cbind(matrix(0,nsite,nsite),theta_large))
 cell_prob_p <- cell_prob_q <- array(0, dim=c(nyear-1, (nsite*2)*
(nyear-1)+1, (nsite*2)))
```

```
 for (i in 1:(nsite*2)) {
  for (t in 1:(nyear-1)) {
   cell_prob_p[t,(nsite*2)*(t-1)+1:(nsite*2),i] <- phi[i,t] * theta
[i,] * pcap[,t]
   cell_prob_q[t,(nsite*2)*(t-1)+1:(nsite*2),i] <- phi[i,t] * theta
[i,] * qcap[,t]
  } # t
  for (t1 in 1:(nyear-2)) {
   for (t2 in (t1+1):(nyear-1)) {
    cell_prob_p[t1,(nsite*2)*(t2-1)+1:(nsite*2),i] <-
    ((cell_prob_q[t1,(nsite*2)*(t2-2)+1:(nsite*2),i] * phi[,t2]) %*%
theta) * pcap[,t2]
    cell_prob_q[t1,(nsite*2)*(t2-1)+1:(nsite*2),i] <-
    ((cell_prob_q[t1,(nsite*2)*(t2-2)+1:(nsite*2),i] * phi[,t2]) %*%
theta) * qcap[,t2]
   } # t2
  } # t1
  for (t in 1:(nyear-1)) {
   cell_prob_p[t,(nsite*2)*(nyear-1)+1,i] <- 1 - sum(cell_prob_p[t,1:
((nsite*2)*(nyear-1)),i])
  } # t
 } # i
 cell_prob <- cell_prob_p

 return(cell_prob)
} # cell_prob_fun

body_sipm_mcmc <- function(y, r, m, x, w, d, nmcmc) {

 # Setup variables
 nsite <- dim(y)[1]
 nyear <- dim(y)[2]
 nreps <- dim(y)[3]
 ncovs <- dim(x)[3]
 npcvs <- dim(w)[4]
 y_small_max <- apply(y[,,,1], 1:2, max)
 y_large_max <- apply(y[,,,2], 1:2, max)

 beta0_save <- matrix(0, ncovs+1, nmcmc)
 beta_gamma_small_save <- matrix(0, ncovs+1, nmcmc)
 beta_gamma_large_save <- matrix(0, ncovs+1, nmcmc)
 beta_phi_small_save <- matrix(0, ncovs+2, nmcmc)
 beta_phi_large_save <- matrix(0, ncovs+2, nmcmc)
 omega_save <- rep(0, nmcmc)
 kappa_small_save <- rep(0, nmcmc)
```

```
kappa_large_save <- rep(0, nmcmc)
alpha_det_save <- matrix(0, npcvs+1, nmcmc)
alpha_cap_save <- matrix(0, npcvs+1, nmcmc)
N_small_save <- array(0, dim=c(nsite, nyear, 200))
N_large_save <- array(0, dim=c(nsite, nyear, 200))

# Prior
beta0_mean <- rep(0, ncovs+1)
beta0_sd <- 2
beta_gamma_small_mean <- rep(0, ncovs+1)
beta_gamma_small_sd <- 2
beta_gamma_large_mean <- rep(0, ncovs+1)
beta_gamma_large_sd <- 2
beta_phi_small_mean <- rep(0, ncovs+2)
beta_phi_small_sd <- 2
beta_phi_large_mean <- rep(0, ncovs+2)
beta_phi_large_sd <- 2
logit_omega_mean <- 0
logit_omega_sd <- 2
log_kappa_small_mean <- 0
log_kappa_small_sd <- 2
log_kappa_large_mean <- 0
log_kappa_large_sd <- 2
alpha_det_mean <- rep(0, npcvs+1)
alpha_det_sd <- 2
alpha_cap_mean <- rep(0, npcvs+1)
alpha_cap_sd <- 2

# Starting values
beta0 <- rep(0, ncovs+1)
beta_gamma_small <- rep(0, ncovs+1)
beta_gamma_large <- rep(0, ncovs+1)
beta_phi_small <- rep(0, ncovs+2)
beta_phi_large <- rep(0, ncovs+2)
omega <- 0.5
kappa_small <- 1
kappa_large <- 1
alpha_det <- rep(0, npcvs+1)
alpha_cap <- rep(0, npcvs+1)
lambda0 <- exp(cbind(1,x[,1,]) %*% beta0)
gamma_small <- matrix(0.5, nsite, nyear-1)
gamma_large <- matrix(0.5, nsite, nyear-1)
phi_small <- matrix(0.5, nsite, nyear-1)
phi_large <- matrix(0.5, nsite, nyear-1)
eta_small <- exp(-1 * kappa_small * d)
```

```
 theta_small <- eta_small / rowSums(eta_small)
 eta_large <- exp(-1 * kappa_large * d)
 theta_large <- eta_large / rowSums(eta_large)
 N_small <- round((y_small_max + 1) / 0.5)
 N_large <- round((y_large_max + 1) / 0.5)
 pdet <- array(0.5, dim=c(nsite, nyear, nreps))
 pcap_per_visit <- array(0.5, dim=c(nsite*2, nyear-1, nreps))
 pcap <- matrix(, nsite*2, nyear-1)
 for (i in 1:(nsite*2)) {
  for (t in 1:(nyear-1)) {
   pcap[i,t] <- 1 - prod(1 - pcap_per_visit[i,t,])
  } # t
 } # i
 cell_prob <- cell_prob_fun(nsite=nsite, nyear=nyear,
phi_small=phi_small, phi_large=phi_large,
               omega=omega, theta_small=theta_small,
theta_large=theta_large, pcap=pcap)

 # Tuning factor
 beta0_tune <- 0.08
 beta_gamma_small_tune <- c(0.02, 0.02, 0.02)
 beta_gamma_large_tune <- c(0.02, 0.02, 0.02)
 beta_phi_small_tune <- c(0.10, 0.10, 0.07, 0.07)
 beta_phi_large_tune <- c(0.10, 0.10, 0.07, 0.07)
 omega_tune <- 0.05
 kappa_small_tune <- 0.035
 kappa_large_tune <- 0.035
 alpha_det_tune <- c(0.050, 0.035)
 alpha_cap_tune <- 0.050
 N_small_tune <- 1
 N_large_tune <- 1

 # Begin MCMC loop
 for (k in 1:nmcmc) {
  ### Sample N_small
  N_small_star <- matrix(rpois(nsite*nyear, N_small + N_small_tune),
nsite, nyear)
  mh1 <- apply(dbinom(y[,,,1], N_small_star, pdet, log=T), 1:2, sum) +
      dpois(N_small_star, cbind(lambda0, t(t((N_small[,-nyear] * (1 +
gamma_small) + N_large[,-nyear] * gamma_large) * phi_small * (1 -
omega)) %*% theta_small)), log=T) +
      dpois(N_small, N_small_star+N_small_tune, log=T)
  mh2 <- apply(dbinom(y[,,,1], N_small    , pdet, log=T), 1:2, sum) +
      dpois(N_small    , cbind(lambda0, t(t((N_small[,-nyear] * (1 +
gamma_small) + N_large[,-nyear] * gamma_large) * phi_small * (1 -
```

```
omega)) %*% theta_small)), log=T) +
      dpois(N_small_star, N_small+N_small_tune, log=T)
  mh <- exp(mh1 - mh2)
  N_small_keep <- ((mh > runif(nsite*nyear)) & (N_small_star >=
y_small_max))
  N_small[N_small_keep] <- N_small_star[N_small_keep]

  ### Sample N_large
  N_large_star <- matrix(rpois(nsite*nyear, N_large + N_large_tune),
nsite, nyear)
  mh1 <- apply(dbinom(y[,,,2], N_large_star, pdet, log=T), 1:2, sum) +
      dpois(N_large_star, cbind(lambda0, t(t((N_small[,-nyear] * (1 +
gamma_small) + N_large[,-nyear] * gamma_large) * phi_small * omega +
N_large[,-nyear] * phi_large) %*% theta_large)), log=T) +
      dpois(N_large, N_large_star+N_large_tune, log=T)
  mh2 <- apply(dbinom(y[,,,2], N_large     , pdet, log=T), 1:2, sum) +
      dpois(N_large     , cbind(lambda0, t(t((N_small[,-nyear] * (1 +
gamma_small) + N_large[,-nyear] * gamma_large) * phi_small * omega +
N_large[,-nyear] * phi_large) %*% theta_large)), log=T) +
      dpois(N_large_star, N_large+N_large_tune, log=T)
  mh <- exp(mh1 - mh2)
  N_large_keep <- ((mh > runif(nsite*nyear)) & (N_large_star >=
y_large_max))
  N_large[N_large_keep] <- N_large_star[N_large_keep]

  ### Sample beta0
  beta0_star <- rnorm(ncovs+1, beta0, beta0_tune)
  lambda0_star <- exp(cbind(1,x[,1,]) %*% beta0_star)
  mh1 <- sum(dpois(N_small[,1], lambda0_star, log=T)) +
      sum(dpois(N_large[,1], lambda0_star, log=T)) +
      sum(dnorm(beta0_star, beta0_mean, beta0_sd, log=T))
  mh2 <- sum(dpois(N_small[,1], lambda0     , log=T)) +
      sum(dpois(N_large[,1], lambda0     , log=T)) +
      sum(dnorm(beta0     , beta0_mean, beta0_sd, log=T))
  mh <- exp(mh1 - mh2)
  if (mh > runif(1)) {
   beta0 <- beta0_star
   lambda0 <- lambda0_star
  }

  # Sample beta_gamma_small
  beta_gamma_small_star <- rnorm(ncovs+1, beta_gamma_small,
beta_gamma_small_tune)
  gamma_small <- gamma_small_star <- matrix(, nsite, nyear-1)
  for (t in 2:nyear) {
```

```
    gamma_small    [,t-1] <- exp(cbind(1,(N_small[,t-1]+N_large[,t-1]-
lambda0*2)/(lambda0*2),x[,t,1]) %*% beta_gamma_small    )
    gamma_small_star[,t-1] <- exp(cbind(1,(N_small[,t-1]+N_large[,t-
1]-lambda0*2)/(lambda0*2),x[,t,1]) %*% beta_gamma_small_star)
   } # t
  mh1 <- sum(dbinom(r[,,,1], r[,,,1] + y[,-nyear,,1], gamma_small_star
/ (gamma_small_star + 1), log=T)) +
      sum(dpois(N_small[,-1], t(t((N_small[,-nyear] * (1 +
gamma_small_star) + N_large[,-nyear] * gamma_large) * phi_small * (1 -
omega)) %*% theta_small), log=T)) +
      sum(dpois(N_large[,-1], t(t((N_small[,-nyear] * (1 +
gamma_small_star) + N_large[,-nyear] * gamma_large) * phi_small * omega
+ N_large[,-nyear] * phi_large) %*% theta_large), log=T)) +
      sum(dnorm(beta_gamma_small_star, beta_gamma_small_mean,
beta_gamma_small_sd, log=T))
  mh2 <- sum(dbinom(r[,,,1], r[,,,1] + y[,-nyear,,1], gamma_small    /
(gamma_small    + 1), log=T)) +
      sum(dpois(N_small[,-1], t(t((N_small[,-nyear] * (1 + gamma_small
) + N_large[,-nyear] * gamma_large) * phi_small * (1 - omega)) %*%
theta_small), log=T)) +
      sum(dpois(N_large[,-1], t(t((N_small[,-nyear] * (1 + gamma_small
) + N_large[,-nyear] * gamma_large) * phi_small * omega + N_large[,-
nyear] * phi_large) %*% theta_large), log=T)) +
      sum(dnorm(beta_gamma_small    , beta_gamma_small_mean,
beta_gamma_small_sd, log=T))
  mh <- exp(mh1 - mh2)
  if (mh > runif(1)) {
   beta_gamma_small <- beta_gamma_small_star
  }

  # Sample beta_gamma_large
  beta_gamma_large_star <- rnorm(ncovs+1, beta_gamma_large,
beta_gamma_large_tune)
  gamma_large <- gamma_large_star <- matrix(, nsite, nyear-1)
  for (t in 2:nyear) {
   gamma_large    [,t-1] <- exp(cbind(1,(N_small[,t-1]+N_large[,t-1]-
lambda0*2)/(lambda0*2),x[,t,1]) %*% beta_gamma_large    )
   gamma_large_star[,t-1] <- exp(cbind(1,(N_small[,t-1]+N_large[,t-
1]-lambda0*2)/(lambda0*2),x[,t,1]) %*% beta_gamma_large_star)
  } # t
  mh1 <- sum(dbinom(r[,,,2], r[,,,2] + y[,-nyear,,2], gamma_large_star
/ (gamma_large_star + 1), log=T)) +
    sum(dpois(N_small[,-1], t(t((N_small[,-nyear] * (1 + gamma_small)
+ N_large[,-nyear] * gamma_large_star) * phi_small * (1 - omega)) %*%
theta_small), log=T)) +
```

```
    sum(dpois(N_large[,-1], t(t((N_small[,-nyear] * (1 + gamma_small)
+ N_large[,-nyear] * gamma_large_star) * phi_small * omega + N_large[,-
nyear] * phi_large) %*% theta_large), log=T)) +
      sum(dnorm(beta_gamma_large_star, beta_gamma_large_mean,
beta_gamma_large_sd, log=T))
  mh2 <- sum(dbinom(r[,,,2], r[,,,2] + y[,-nyear,,2], gamma_large    /
(gamma_large    + 1), log=T)) +
    sum(dpois(N_small[,-1], t(t((N_small[,-nyear] * (1 + gamma_small)
+ N_large[,-nyear] * gamma_large    ) * phi_small * (1 - omega)) %*%
theta_small), log=T)) +
    sum(dpois(N_large[,-1], t(t((N_small[,-nyear] * (1 + gamma_small)
+ N_large[,-nyear] * gamma_large    ) * phi_small * omega + N_large[,-
nyear] * phi_large) %*% theta_large), log=T)) +
      sum(dnorm(beta_gamma_large    , beta_gamma_large_mean,
beta_gamma_large_sd, log=T))
  mh <- exp(mh1 - mh2)
  if (mh > runif(1)) {
   beta_gamma_large <- beta_gamma_large_star
  }

  # Sample beta_phi_small & beta_phi_large
  beta_phi_small_star <- rnorm(ncovs+2, beta_phi_small,
beta_phi_small_tune)
  beta_phi_large_star <- rnorm(ncovs+2, beta_phi_large,
beta_phi_large_tune)
  phi_small <- phi_small_star <- phi_large <- phi_large_star <- matrix
(, nsite, nyear-1)
  for (t in 2:nyear) {
   phi_small   [,t-1] <- inv.logit(cbind(1, (N_small[,t-1]+N_large[,t-
1]-lambda0*2)/(lambda0*2),x[,t,]) %*% beta_phi_small    )
    phi_small_star[,t-1] <- inv.logit(cbind(1, (N_small[,t-1]+N_large
[,t-1]-lambda0*2)/(lambda0*2),x[,t,]) %*% beta_phi_small_star)
   phi_large   [,t-1] <- inv.logit(cbind(1, (N_small[,t-1]+N_large[,t-
1]-lambda0*2)/(lambda0*2),x[,t,]) %*% beta_phi_large    )
    phi_large_star[,t-1] <- inv.logit(cbind(1, (N_small[,t-1]+N_large
[,t-1]-lambda0*2)/(lambda0*2),x[,t,]) %*% beta_phi_large_star)
  } # t
  cell_prob    <- cell_prob_fun(nsite=nsite, nyear=nyear,
phi_small=phi_small    , phi_large=phi_large    , omega=omega,
theta_small=theta_small, theta_large=theta_large, pcap=pcap)
  cell_prob_star <- cell_prob_fun(nsite=nsite, nyear=nyear,
phi_small=phi_small_star, phi_large=phi_large_star, omega=omega,
theta_small=theta_small, theta_large=theta_large, pcap=pcap)
  prob1 <- prob2 <- matrix(, nsite, nyear-1)
  for (i in 1:nsite) {
```

```
    for (t in 1:(nyear-1)) {
      prob1[i,t] <- dmultinom(m[t,,i], size=sum(m[t,,i]),
prob=cell_prob_star[t,,i], log=T)
      prob2[i,t] <- dmultinom(m[t,,i], size=sum(m[t,,i]),
prob=cell_prob    [t,,i], log=T)
    } # t
  } # i
  mh1 <- sum(prob1) +
      sum(dpois(N_small[,-1], t(t((N_small[,-nyear] * (1 + gamma_small)
+ N_large[,-nyear] * gamma_large) * phi_small_star * (1 - omega)) %*%
theta_small), log=T)) +
      sum(dpois(N_large[,-1], t(t((N_small[,-nyear] * (1 + gamma_small)
+ N_large[,-nyear] * gamma_large) * phi_small_star * omega + N_large[,-
nyear] * phi_large_star) %*% theta_large), log=T)) +
        sum(dnorm(beta_phi_small_star, beta_phi_small_mean,
beta_phi_small_sd, log=T)) +
        sum(dnorm(beta_phi_large_star, beta_phi_large_mean,
beta_phi_large_sd, log=T))
  mh2 <- sum(prob2) +
      sum(dpois(N_small[,-1], t(t((N_small[,-nyear] * (1 + gamma_small)
+ N_large[,-nyear] * gamma_large) * phi_small     * (1 - omega)) %*%
theta_small), log=T)) +
      sum(dpois(N_large[,-1], t(t((N_small[,-nyear] * (1 + gamma_small)
+ N_large[,-nyear] * gamma_large) * phi_small     * omega + N_large[,-
nyear] * phi_large     ) %*% theta_large), log=T)) +
        sum(dnorm(beta_phi_small     , beta_phi_small_mean,
beta_phi_small_sd, log=T)) +
        sum(dnorm(beta_phi_large     , beta_phi_large_mean,
beta_phi_large_sd, log=T))
  mh <- exp(mh1 - mh2)
  if (mh > runif(1)) {
   beta_phi_small <- beta_phi_small_star
   beta_phi_large <- beta_phi_large_star
   phi_small <- phi_small_star
   phi_large <- phi_large_star
  }

  ### Sample omega
  omega_star <- inv.logit(rnorm(1, logit(omega), omega_tune))
  cell_prob     <- cell_prob_fun(nsite=nsite, nyear=nyear,
phi_small=phi_small, phi_large=phi_large, omega=omega     ,
theta_small=theta_small, theta_large=theta_large, pcap=pcap)
  cell_prob_star <- cell_prob_fun(nsite=nsite, nyear=nyear,
phi_small=phi_small, phi_large=phi_large, omega=omega_star,
theta_small=theta_small, theta_large=theta_large, pcap=pcap)
  prob1 <- prob2 <- matrix(, nsite, nyear-1)
```

```
  for (i in 1:nsite) {
   for (t in 1:(nyear-1)) {
    prob1[i,t] <- dmultinom(m[t,,i], size=sum(m[t,,i]),
prob=cell_prob_star[t,,i], log=T)
    prob2[i,t] <- dmultinom(m[t,,i], size=sum(m[t,,i]),
prob=cell_prob   [t,,i], log=T)
   } # t
  } # i
  mh1 <- sum(prob1) +
     sum(dpois(N_small[,-1], t(t((N_small[,-nyear] * (1 + gamma_small)
+ N_large[,-nyear] * gamma_large) * phi_small * (1 - omega_star)) %*%
theta_small), log=T)) +
     sum(dpois(N_large[,-1], t(t((N_small[,-nyear] * (1 + gamma_small)
+ N_large[,-nyear] * gamma_large) * phi_small * omega_star + N_large[,-
nyear] * phi_large) %*% theta_large), log=T)) +
      dnorm(logit(omega_star), logit_omega_mean, logit_omega_sd,
log=T)
  mh2 <- sum(prob2) +
     sum(dpois(N_small[,-1], t(t((N_small[,-nyear] * (1 + gamma_small)
+ N_large[,-nyear] * gamma_large) * phi_small * (1 - omega    )) %*%
theta_small), log=T)) +
     sum(dpois(N_large[,-1], t(t((N_small[,-nyear] * (1 + gamma_small)
+ N_large[,-nyear] * gamma_large) * phi_small * omega    + N_large[,-
nyear] * phi_large) %*% theta_large), log=T)) +
      dnorm(logit(omega   ), logit_omega_mean, logit_omega_sd, log=T)
  mh <- exp(mh1 - mh2)
  if (mh > runif(1)) {
   omega <- omega_star
  }

  # Sample kappa_small & kappa_large
  kappa_small_star <- exp(rnorm(1, log(kappa_small),
kappa_small_tune))
  eta_small_star <- exp(-1 * kappa_small_star * d)
  theta_small_star <- eta_small_star / rowSums(eta_small_star)
  kappa_large_star <- exp(rnorm(1, log(kappa_large),
kappa_large_tune))
  eta_large_star <- exp(-1 * kappa_large_star * d)
  theta_large_star <- eta_large_star / rowSums(eta_large_star)
  cell_prob    <- cell_prob_fun(nsite=nsite, nyear=nyear,
phi_small=phi_small, phi_large=phi_large, omega=omega,
theta_small=theta_small    , theta_large=theta_large    , pcap=pcap)
  cell_prob_star <- cell_prob_fun(nsite=nsite, nyear=nyear,
phi_small=phi_small, phi_large=phi_large, omega=omega,
theta_small=theta_small_star, theta_large=theta_large_star,
pcap=pcap)
```

```
  prob1 <- prob2 <- matrix(, nsite, nyear-1)
  for (i in 1:nsite) {
   for (t in 1:(nyear-1)) {
    prob1[i,t] <- dmultinom(m[t,,i], size=sum(m[t,,i]),
prob=cell_prob_star[t,,i], log=T)
    prob2[i,t] <- dmultinom(m[t,,i], size=sum(m[t,,i]),
prob=cell_prob    [t,,i], log=T)
   } # t
  } # i
  mh1 <- sum(prob1) +
     sum(dpois(N_small[,-1], t(t((N_small[,-nyear] * (1 + gamma_small)
+ N_large[,-nyear] * gamma_large) * phi_small * (1 - omega)) %*%
theta_small_star), log=T)) +
     sum(dpois(N_large[,-1], t(t((N_small[,-nyear] * (1 + gamma_small)
+ N_large[,-nyear] * gamma_large) * phi_small * omega + N_large[,-
nyear] * phi_large) %*% theta_large_star), log=T)) +
      dnorm(log(kappa_small_star), log_kappa_small_mean,
log_kappa_small_sd, log=T) +
      dnorm(log(kappa_large_star), log_kappa_large_mean,
log_kappa_large_sd, log=T)
  mh2 <- sum(prob2) +
     sum(dpois(N_small[,-1], t(t((N_small[,-nyear] * (1 + gamma_small)
+ N_large[,-nyear] * gamma_large) * phi_small * (1 - omega)) %*%
theta_small    ), log=T)) +
     sum(dpois(N_large[,-1], t(t((N_small[,-nyear] * (1 + gamma_small)
+ N_large[,-nyear] * gamma_large) * phi_small * omega + N_large[,-
nyear] * phi_large) %*% theta_large    ), log=T)) +
      dnorm(log(kappa_small    ), log_kappa_small_mean,
log_kappa_small_sd, log=T) +
      dnorm(log(kappa_large    ), log_kappa_large_mean,
log_kappa_large_sd, log=T)
  mh <- exp(mh1 - mh2)
  if (mh > runif(1)) {
   kappa_small <- kappa_small_star
   eta_small <- eta_small_star
   theta_small <- theta_small_star
   kappa_large <- kappa_large_star
   eta_large <- eta_large_star
   theta_large <- theta_large_star
  }

  ### Sample alpha_det
  alpha_det_star <- rnorm(npcvs+1, alpha_det, alpha_det_tune)
  pdet_star <- array(, dim=c(nsite, nyear, nreps))
  for (i in 1:nsite) {
   for (t in 1:nyear) {
```

```
    pdet_star[i,t,] <- inv.logit(cbind(1,w[i,t,,]) %*%
alpha_det_star)
   } # t
  } # i
 mh1 <- sum(dbinom(y[,,,1], N_small, pdet_star, log=TRUE), na.rm=T) +
     sum(dbinom(y[,,,2], N_large, pdet_star, log=TRUE), na.rm=T) +
     sum(dnorm(alpha_det_star, alpha_det_mean, alpha_det_sd,
log=TRUE))
  mh2 <- sum(dbinom(y[,,,1], N_small, pdet   , log=TRUE), na.rm=T) +
     sum(dbinom(y[,,,2], N_large, pdet   , log=TRUE), na.rm=T) +
    sum(dnorm(alpha_det   , alpha_det_mean, alpha_det_sd, log=TRUE))
  mh <- exp(mh1 - mh2)
  if (mh > runif(1)) {
   alpha_det <- alpha_det_star
   pdet <- pdet_star
  }

  # Sample alpha_pcap
  alpha_cap_star <- rnorm(npcvs+1, alpha_cap, alpha_cap_tune)
  pcap_per_visit_star <- array(, dim=c(nsite*2, nyear-1, nreps))
  for (i in 1:nsite) {
   for (t in 1:(nyear-1)) {
    pcap_per_visit_star[i,t,] <- inv.logit(cbind(1,w[i,t+1,,]) %*%
alpha_cap_star)
   } # t
  } # i
  pcap_per_visit_star[(nsite+1):(nsite*2),,] <- pcap_per_visit_star
[1:nsite,,]
  pcap_star <- matrix(, nsite*2, nyear-1)
  for (i in 1:(nsite*2)) {
   for (t in 1:(nyear-1)) {
    pcap_star[i,t] <- 1 - prod(1 - pcap_per_visit_star[i,t,])
   } # t
  } # i
  cell_prob    <- cell_prob_fun(nsite=nsite, nyear=nyear,
phi_small=phi_small, phi_large=phi_large, omega=omega,
theta_small=theta_small, theta_large=theta_large, pcap=pcap    )
  cell_prob_star <- cell_prob_fun(nsite=nsite, nyear=nyear,
phi_small=phi_small, phi_large=phi_large, omega=omega,
theta_small=theta_small, theta_large=theta_large, pcap=pcap_star)
  prob1 <- prob2 <- matrix(, nsite, nyear-1)
  for (i in 1:nsite) {
   for (t in 1:(nyear-1)) {
    prob1[i,t] <- dmultinom(m[t,,i], size=sum(m[t,,i]),
prob=cell_prob_star[t,,i], log=T)
```

```
    prob2[i,t] <- dmultinom(m[t,,i], size=sum(m[t,,i]),
prob=cell_prob   [t,,i], log=T)
   } # t
  } # i
  mh1 <- sum(prob1) +
        sum(dnorm(alpha_cap_star, alpha_cap_mean, alpha_cap_sd,
log=T))
  mh2 <- sum(prob2) +
        sum(dnorm(alpha_cap   , alpha_cap_mean, alpha_cap_sd, log=T))
  mh <- exp(mh1 - mh2)
  if (mh > runif(1)) {
   alpha_cap <- alpha_cap_star
   pcap_per_visit <- pcap_per_visit_star
   pcap <- pcap_star
  }

  # Save samples
  beta0_save[,k] <- beta0
  beta_gamma_small_save[,k] <- beta_gamma_small
  beta_gamma_large_save[,k] <- beta_gamma_large
  beta_phi_small_save[,k] <- beta_phi_small
  beta_phi_large_save[,k] <- beta_phi_large
  omega_save[k] <- omega
  kappa_small_save[k] <- kappa_small
  kappa_large_save[k] <- kappa_large
  alpha_det_save[,k] <- alpha_det
  alpha_cap_save[,k] <- alpha_cap
  if (nmcmc > 200) {
   if (k > nmcmc-200) {
    N_small_save[,,k-nmcmc+200] <- N_small
    N_large_save[,,k-nmcmc+200] <- N_large
   }
  }
 } # k

 list(beta0_save=beta0_save,
    beta_gamma_small_save=beta_gamma_small_save,
beta_gamma_large_save=beta_gamma_large_save,
    beta_phi_small_save=beta_phi_small_save,
beta_phi_large_save=beta_phi_large_save,
    omega_save=omega_save,
    kappa_small_save=kappa_small_save,
kappa_large_save=kappa_large_save,
    alpha_det_save=alpha_det_save, alpha_cap_save=alpha_cap_save,
    N_small_save=N_small_save, N_large_save=N_large_save)
} # body_sipm_mcmc
```

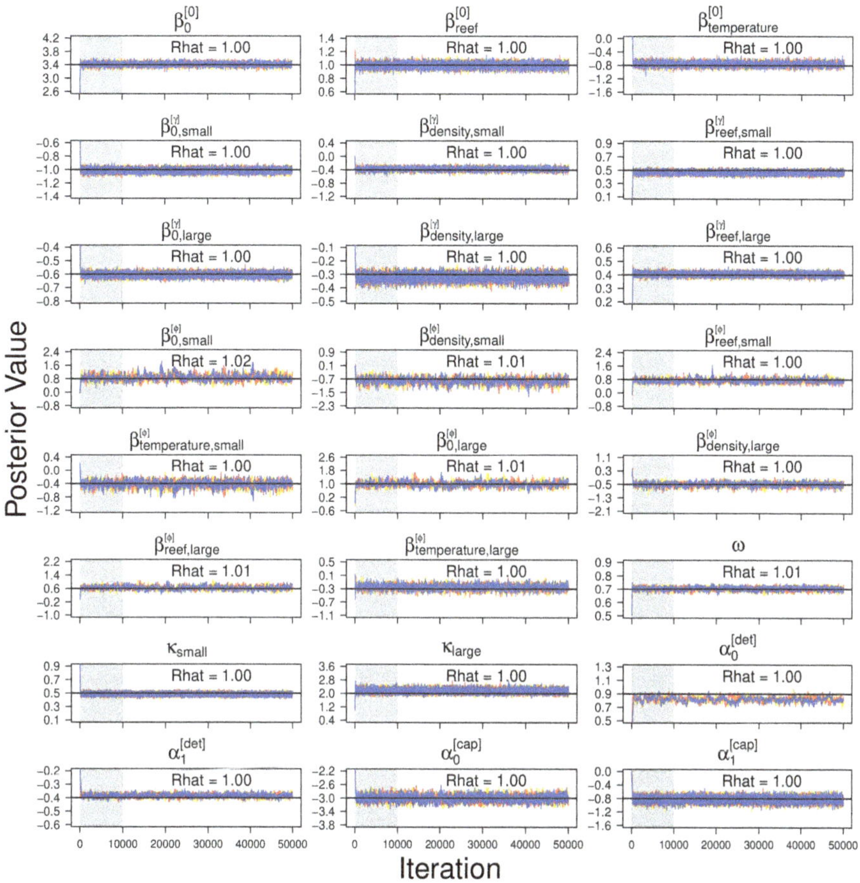

Fig. 4.13 Traceplots showing posterior chains for $\beta^{[0]}$, $\beta^{[\gamma]}$, $\beta^{[\phi]}$, ω, κ, $\alpha^{[cap]}$, and $\alpha^{[det]}$ in the spatial IPM with body-class-specific reproduction, survival, and movement. Rhat values are calculated after discarding the first 10,000 iterations. Horizontal lines represent the true values used in data simulation

Using the following code to implement the MCMC algorithm, it takes about 10 h to run. Then using the code in Appendix 4.6 to create some traceplots, we can see good model convergence and agreement between parameter estimates of $\beta^{[0]}$, $\beta^{[\gamma]}$, $\beta^{[\phi]}$, ω, κ, $\alpha^{[cap]}$, and $\alpha^{[det]}$ and their true values (Fig. 4.13).

```
library(foreach)  # for parallel computing
library(doParallel) # for parallel computing

numCores <- round(detectCores() / 2) # only use half of the cores for
parallel computing
registerDoParallel(numCores) # setup parallel computing
```

```
nmcmc <- 50000 # number of iterations of MCMC computing
chain <- 3    # number of chains

start_time <- Sys.time() # start time of computing
out <- foreach (i=1:chain, .packages='boot') %dopar% {
 body_sipm_mcmc(y=y, r=r, m=m, x=x, w=w, d=d, nmcmc=nmcmc)
} # i
stopImplicitCluster() # clean up the cluster
end_time <- Sys.time() # end time of computing
time_taken <- end_time - start_time # computing time
```

We can also use the following code to create pie plots that represent local population size $N_{i,\ t,\ small} + N_{i,\ t,\ large}$, and the proportion of large animals, $N_{i,\ t,\ large}/(N_{i,\ t,\ small} + N_{i,\ t,\ large})$, in each local population. We can see an overall increase in both local population size and the proportion of large animals; the latter is likely due to higher survival of large animals. These patterns are correctly represented by the estimates (Fig. 4.14).

```
library(plotrix) # for making pie charts at selected positions

N_small_post <- N_large_post <- array(, dim=c(nsite, nyear,
200*chain))
N_small_post[,, 1:200] <- out[[1]]$N_small_save
N_small_post[,,201:400] <- out[[2]]$N_small_save
N_small_post[,,401:600] <- out[[3]]$N_small_save
N_large_post[,, 1:200] <- out[[1]]$N_large_save
N_large_post[,,201:400] <- out[[2]]$N_large_save
N_large_post[,,401:600] <- out[[3]]$N_large_save
N_small_med <- apply(N_small_post, 1:2, median)
N_large_med <- apply(N_large_post, 1:2, median)
years_plot <- c(1, 8)

par(mfrow=c(2,2))
par(mar=c(1,1,1,1))
par(oma=c(4,5,5,2))

for (t in 1:2) {
 plot(1, xlim=c(-11, 11), ylim=c(-8.8, 8.8), type='n', axes=F,
xlab='', ylab='')
 for (i in 1:nsite) {
  if (N_small[i,years_plot[t]] + N_large[i,years_plot[t]] > 0) {
   prop <- c(N_small[i,years_plot[t]], N_large[i,years_plot[t]]) /
(N_small[i,years_plot[t]] + N_large[i,years_plot[t]])
   size <- (N_small[i,years_plot[t]] + N_large[i,years_plot[t]]) ^
```

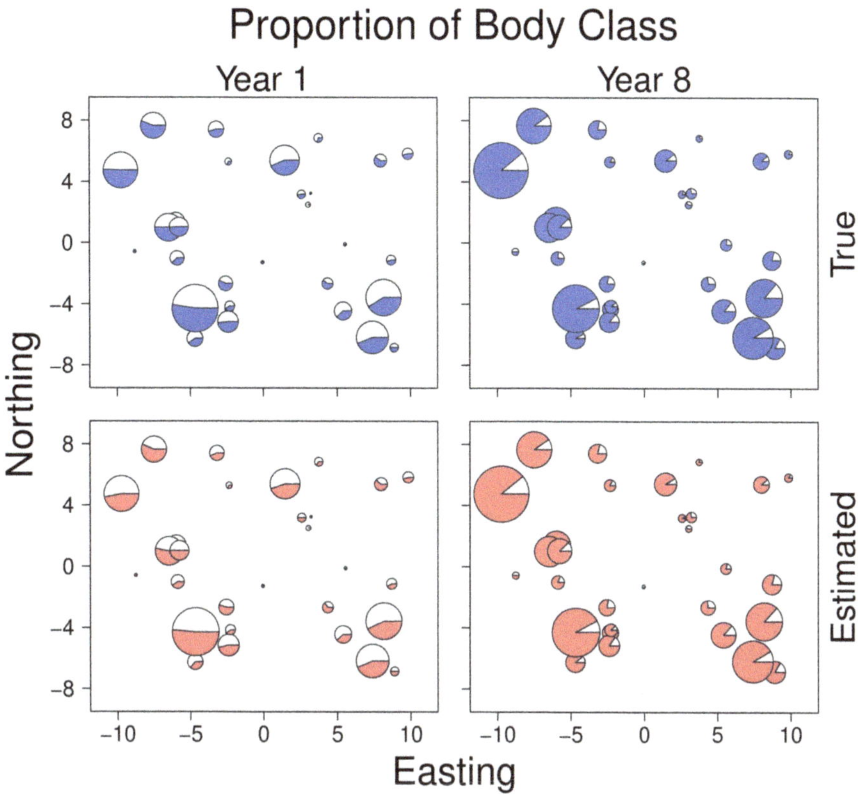

Fig. 4.14 Pie plots showing the consistency between the simulated true (upper row) and estimated (lower row) local population size (proportional to the sizes of the circles) and the proportion of large animals (blue or red) in each local population, both of which vary spatially and temporally (using years 1 and 8 as examples)

```
0.67 / 20
   floating.pie(xpos=cov$lon[i], ypos=cov$lat[i], x=prop,
radius=size, col=c('white','royalblue'), border='grey18')
  }
 } # i
 axis(1, at=seq(-10,10,5), labels=rep('',5))
 if (t == 1) {
  axis(2, at=seq(-8,8,4), las=2, cex.axis=1.6)
 } else {
  axis(2, at=seq(-8,8,4), labels=rep('',5))
 }
 axis(3, at=0, labels=paste('Year', years_plot[t], sep=' '), cex.
axis=2.5, line=-0.8, tick=F)
 if (t == 2) {
  axis(4, at=0, labels='True', cex.axis=2.5, line=0.5, tick=F)
```

```
  }
  box()
} # t

for (t in 1:2) {
  plot(1, xlim=c(-11, 11), ylim=c(-8.8, 8.8), type='n', axes=F,
xlab='', ylab='')
  for (i in 1:nsite) {
    if (N_small_med[i,years_plot[t]] + N_large_med[i,years_plot[t]] >
0) {
      prop <- c(N_small_med[i,years_plot[t]], N_large_med[i,years_plot
[t]]) / (N_small_med[i,years_plot[t]] + N_large_med[i,years_plot
[t]])
      size <- (N_small_med[i,years_plot[t]] + N_large_med[i,years_plot
[t]]) ^ 0.67 / 20
      floating.pie(xpos=cov$lon[i], ypos=cov$lat[i], x=prop,
radius=size, col=c('white','lightcoral'), border='grey18')
    }
  } # i
  axis(1, at=seq(-10,10,5), cex.axis=1.6)
  if (t == 1) {
    axis(2, at=seq(-8,8,4), las=2, cex.axis=1.6)
  } else {
    axis(2, at=seq(-8,8,4), labels=rep('',5))
  }
  if (t == 2) {
    axis(4, at=0, labels='Estimated', cex.axis=2.5, line=0.5, tick=F)
  }
  box()
} # t

title(xlab='Easting' , cex.lab=2.8, line=2.6, outer=T)
title(ylab='Northing', cex.lab=2.8, line=2.4, outer=T)
title(main=expression('Proportion of Body Class'), line=3.2, cex.
main=3, outer=T)
```

4.4.4 With Gene-Class-Specific Reproduction, Survival, and Movement

Studies have shown that genetic variation can influence demography and eventually population structures (Gade et al. 2022). Therefore, next, we will consider a spatial IPM that incorporates information on gene-class. For our coral reef fish, we now have a population with six gene classes, ordered from 1 to 6. While gene class 1 represents extreme scramblers that are inferior in competition but are good at searching for suitable habitats, gene class 6 represents extreme contesters that are superior in competition but are bad at searching for suitable habitats. We assume that contesters have a higher reproduction rate and survival probability than scramblers, but the advantages of contesters over scramblers are smaller in larger reefs, as larger reefs are more difficult to defend than smaller ones. While both contesters and scramblers are reluctant to move away from larger reefs, scramblers are more likely to move long distances than contesters. Other gene classes between these two extremes have the characteristics of both scramblers and contesters, the extent of which depends on how close they are from class 1 or 6 (e.g., class 2 is a lot more similar to class 1 than to class 6, and class 4 is a bit more similar to class 6 than to class 1). While this model considers more than two classes, it does not allow transition between classes because the gene of an individual normally does not change over time. Eventually, we assume that gene classes can be distinguished in both capture-recapture and count surveys. The model performance needs to be further investigated when gene classes cannot be distinguished in count surveys.

Using a statistical language, we can describe the initial abundance as follows:

$$\left.\begin{array}{r}N_{i,1,g} \sim Poisson(\lambda_{i,1}) \\ \lambda_{i,1} = exp\left(\beta_0^{[0]} + \beta_1^{[0]} \times x_{i,1,1} + \ldots + \beta_{n_{covs}}^{[0]} \times x_{i,1,n_{covs}}\right)\end{array}\right\}, \tag{4.17}$$

in which $N_{i,\,1,\,g}$ is the abundance of individuals (breeders) at site i in the first year that are in gene class g, and $\lambda_{i,\,1}$ is the expectation of initial abundance for any gene class and is a function of covariates x's with intercept and slope parameters $\beta^{[0]}$'s. Here, we assume even proportions of gene classes in the first year, which could be extended to be uneven. From the second year (i.e., $t \geq 2$), the process sub-model specifies the spatiotemporal variation in abundance as follows:

$$\left.\begin{array}{r}N_{i_2,t,g} \sim Poisson\left(\lambda_{i_2,t,g}\right) \\ \lambda_{i_2,t,g} = \sum_{i_1=1}^{n_{site}} N_{i_1,t-1,g} \times \left(1 + \gamma_{i_1,t-1,g}\right) \times \phi_{i,t-1,g} \times \theta_{i_1,i_2,t-1,g} \\ \gamma_{i_1,t-1,g} = exp\left(\begin{array}{c}\beta_0^{[\gamma]} + \beta_{density}^{[\gamma]} \times \sum_{g=1}^{n_{gene}} N_{i_1,t-1,g} + \beta_1^{[\gamma]} \times x_{i_1,t,1} + \ldots \\ +\beta_{gene}^{[\gamma]} \times g + \beta_{1\times gene}^{[\gamma]} \times x_{i_1,t,1} \times g + \ldots\end{array}\right) \\ \phi_{i_1,t-1,g} = logit^{-1}\left(\begin{array}{c}\beta_0^{[\phi]} + \beta_{density}^{[\phi]} \times \sum_{g=1}^{n_{gene}} N_{i_1,t-1,g} + \beta_1^{[\phi]} \times x_{i_1,t,1} + \ldots \\ +\beta_{gene}^{[\phi]} \times g + \beta_{1\times gene}^{[\phi]} \times x_{i_1,t,1} \times g + \ldots\end{array}\right) \\ \theta_{i_1,i_2,t-1,g} = \frac{\eta_{i_1,i_2,t-1,g}}{\sum_{i_2=1}^{n_{site}} \eta_{i_1,i_2,t-1,g}} \\ \eta_{i_1,i_2,t-1,g} = exp\left(-1 \times \kappa_{i_1,t-1,g} \times d_{i_1,i_2}\right) \\ \kappa_{i_1,t-1,g} = exp\left(\begin{array}{c}\beta_0^{[\kappa]} + \beta_{density}^{[\kappa]} \times \sum_{g=1}^{n_{gene}} N_{i_1,t-1,g} + \beta_1^{[\kappa]} \times x_{i_1,t,1} + \ldots \\ +\beta_{gene}^{[\kappa]} \times g + \beta_{1\times gene}^{[\kappa]} \times x_{i_1,t,1} \times g + \ldots\end{array}\right)\end{array}\right\}, \tag{4.18}$$

in which $\gamma_{i_1,t-1,g}$ is per capita reproduction rates of individuals at site i_1 in year $t-1$ that are in gene class g; $\gamma_{i_1,t-1,g}$ is a function of the summed density of all gene classes, environmental covariates x's, gene class, and interactions between environmental covariates and gene class with intercept and slope parameters $\beta^{[\gamma]}$'s; $\phi_{i_1,t-1,g}$ is survival probability and is a function of density, environment, and gene class with intercept and slope parameters $\beta^{[\phi]}$'s; $\theta_{i_1,i_2,t-1,g}$ is the standardized movement rate from site i_1 to site i_2 in year $t-1$ for individuals that are in gene class g, $\eta_{i_1,i_2,t-1,g}$ is the unstandardized movement rates, and $\kappa_{i,t-1,g}$ is the decay parameter which is a function of density, environment, and gene class with intercept and slope parameters $\beta^{[\kappa]}$'s. Eventually, the abundance of animals in gene class g at site i_2 in year t, $N_{i_2,t,g}$, is the total number of animals in the corresponding gene class that either survive and stay at site i_2 or move in from any other sites i_1. We also simulate individual survival-location history $z_{v,t}$ and number of offspring reproduced by each gene class $R_{i,t,g}$.

The observed capture history data are as before obtained in the following way:

$$\left.\begin{aligned} c_{v,t,j} &\sim Bernoulli\left(p^{[cap]}_{z_{v,t},t,j}\right) \times z_{v,t} \\ p^{[cap]}_{i,t,j} &= logit^{-1}\left(\alpha^{[cap]}_0 + \alpha^{[cap]}_1 \times w_{i,t,j,1} + \ldots + \alpha^{[cap]}_{n_{pcvs}} \times w_{i,t,j,n_{pcvs}}\right) \end{aligned}\right\}, \tag{4.19}$$

in which $c_{v,t,j}$ is the capture history of individual v in year t during the jth visit, $p^{[cap]}_{i,t,j}$ is the capture probability and is a function of covariates w's governed by intercept and slope parameters $\alpha^{[cap]}$'s. Gene class is assumed to be known when an individual is captured. Because the model is very slow to compute, we convert the individual capture history into a truncated m-array that only contains next-year recaptures. Using a truncated m-array does not seem to lead to bias, but may increase uncertainty in parameter estimates, which warrants further investigation.

We assume that gene class can be distinguished in counts of breeders and offspring such that

$$\left.\begin{aligned} y_{i,t,j,g} &\sim Binomial\left(N_{i,t,g}, p^{[det]}_{i,t,j}\right) \\ r_{i,t,j,g} &\sim Binomial\left(R_{i,t,g}, p^{[det]}_{i,t,j}\right) \\ p^{[det]}_{i,t,j} &= logit^{-1}\left(\alpha^{[det]}_0 + \alpha^{[det]}_1 \times w_{i,t,j,1} + \ldots + \alpha^{[det]}_{n_{pcvs}} \times w_{i,t,j,n_{pcvs}}\right) \end{aligned}\right\}, \tag{4.20}$$

in which $y_{i,t,j,g}$ is the counts of breeders at site i in year t during the jth visit that are in gene class g, $r_{i,t,j,g}$ is the count of offspring reproduced by breeders in gene class g, and $p^{[det]}_{i,t,j}$ is the detection probability and is a function of covariates w's with intercept and slope parameters $\alpha^{[det]}$'s.

Next, let us use the following code to simulate data for this spatial IPM with gene-class-specific reproduction, survival, and movement. We assume that reef size positively affects initial abundance ($\beta^{[0]}_{reef} = 1$), reproduction ($\beta^{[\gamma]}_{reef} = 0.4$), survival ($\beta^{[\phi]}_{reef} = 0.6$), and the decay parameter of movement ($\beta^{[\kappa]}_{reef} = 0.6$); temperature negatively affects initial abundance ($\beta^{[0]}_{temperature} = -0.8$), reproduction ($\beta^{[\gamma]}_{temperature} = -0.2$), survival ($\beta^{[\phi]}_{temperature} = -0.3$); density negatively affects reproduction (

$\beta_{density}^{[\gamma]} = -0.3$), survival ($\beta_{density}^{[\phi]} = -0.5$), and the decay parameter of movement ($\beta_{density}^{[\kappa]} = -0.6$); gene class positively affects reproduction ($\beta_{gene}^{[\gamma]} = 0.4$), survival ($\beta_{gene}^{[\phi]} = 0.6$), and the decay parameter of movement ($\beta_{gene}^{[\kappa]} = 0.8$); and the interaction between reef size and gene class negatively affects reproduction ($\beta_{reef \times gene}^{[\gamma]} = -0.3$), survival ($\beta_{reef \times gene}^{[\phi]} = -0.4$), and the decay parameter of movement ($\beta_{reef \times gene}^{[\kappa]} = -0.4$). Because the model is very slow to compute, we reduce the number of sites to 30.

```
library(boot) # for using logit and inv.logit functions
library(LaplacesDemon) # for generating random numbers from
categorical distributions (rcat)

set.seed(8)

# Basic values
nsite <- 30 # number of sites
nyear <- 10 # number of years
nreps <- 5  # number of within-season replicates
ngene <- 6  # number of classes of genetic variation
ncovs <- 2  # number of environmental covariates
npcvs <- 1  # number of observational covariates

beta0 <- c(3.4, 1, -0.8)              # intercept and slopes for initial
abundance
beta_gamma <- c(-1.2, -0.3, 0.4, -0.2, 0.4, -0.3) # intercept and slopes
for reproduction
beta_phi  <- c( 1.0, -0.5, 0.6, -0.3, 0.6, -0.4) # intercept and slopes
for apparent survival
beta_kappa <- c(-0.8, -0.6, 0.6, 0.8, -0.4)    # intercept and slopes for
movement
alpha_det <- c( 1 , -0.4)              # intercept and slopes for detection
probability
alpha_cap <- c(-1.4, -0.8)              # intercept and slopes for capture
probability

# Read in environmental covariates
load('data/covariates.RData')
x1 <- (log(cov$size) - mean(log(cov$size))) / sd(log(cov$size)) #
standardized log habitat patch size
x2 <- (cov$temp - mean(cov$temp)) / sd(cov$temp) # standardized
temperature
x <- array(, dim=c(nsite, nyear, ncovs)) # environmental covariates
x[,,1] <- x1[1:nsite, 1:nyear]
```

```
x[,,2] <- x2[1:nsite, 1:nyear]
### Standardize covariates for each site
for (i in 1:nsite) {
 for (h in 1:ncovs) {
  x[i,2:nyear,h] <- (x[i,2:nyear,h] - mean(x[i,,h])) / sd(x[i,,h])
 } # h
} # i

# Simulate data
N <- array(, dim=c(nsite, nyear, ngene)) # abundance for each site, year
and gene class
lambda0 <- exp(cbind(1,x[,1,]) %*% beta0) # expectation of initial
abundance
for (i in 1:nsite) {
 N[i,1,] <- rpois(ngene, lambda0[i]/ngene) # initial abundance
} # i

lon <- cov$lon[1:nsite] # longitude or easting
lat <- cov$lat[1:nsite] # latitude or northing
d <- matrix(, nsite, nsite) # distance matrix
for (i1 in 1:nsite) {
 for (i2 in 1:nsite) {
  d[i1,i2] <- sqrt((lon[i1] - lon[i2]) ^ 2 + (lat[i1] - lat[i2]) ^ 2)
 } # i2
} # i1

nindv <- 15000 # maximum number of individuals that ever exist
z <- matrix(0, nindv, nyear) # individual survival-location history
gene <- numeric(nindv) # gene class for each individual
for (i in 1:nsite) {
 for (g in 1:ngene) {
  if (i == 1 & g == 1) {
   tz <- rep(i, N[i,1,g])
   tg <- rep(g, N[i,1,g])
  } else {
   tz <- c(tz, rep(i, N[i,1,g]))
   tg <- c(tg, rep(g, N[i,1,g]))
  }
 } # g
} # i
z[1:length(tz),1] <- tz
gene[1:length(tg)] <- tg

gamma <- array(, dim=c(nsite, nyear-1, ngene)) # reproduction rate
phi  <- array(, dim=c(nsite, nyear-1, ngene)) # survival probability
```

```
kappa <- array(, dim=c(nsite, nyear-1, ngene)) # distance effect on
movement
eta  <- array(, dim=c(nsite, nsite, nyear-1, ngene)) # unstandardized
movement rate
theta <- array(, dim=c(nsite, nsite, nyear-1, ngene)) #  standardized
movement rate
R   <- array(, dim=c(nsite, nyear-1, ngene)) # number of reproduced
individuals

for (t in 2:nyear) {
 for (i in 1:nsite) {
  for (g in 1:ngene) {
   x_temp1 <- (sum(N[i,t-1,])-lambda0[i])/lambda0[i]
   x_temp2 <- x[i,t,1]
   x_temp3 <- x[i,t,2]
   x_temp4 <- (g - 3.5) / 1.7 # standardize gene class with hypothesized
mean (3.5) and SD (1.7)
   x_temp5 <- x_temp2 * x_temp4
   x_temp <- c(1, x_temp1, x_temp2, x_temp3, x_temp4, x_temp5)
   gamma[i,t-1,g] <- exp(x_temp %*% beta_gamma)
   phi[i,t-1,g] <- inv.logit(x_temp %*% beta_phi)
   kappa[i,t-1,g] <- exp(x_temp[-4] %*% beta_kappa)
   eta[i,,t-1,g] <- exp(-1 * kappa[i,t-1,g] * d[i,])
   theta[i,,t-1,g] <- eta[i,,t-1,g] / sum(eta[i,,t-1,g])
   R[i,t-1,g] <- rpois(1, N[i,t-1,g] * gamma[i,t-1,g])
   if (i == 1 & g == 1) {
    tz <- rep(i, R[i,t-1,g])
    tg <- rep(g, R[i,t-1,g])
   } else {
    tz <- c(tz, rep(i, R[i,t-1,g]))
    tg <- c(tg, rep(g, R[i,t-1,g]))
   }
  } # g
 } # i
 z[max(which(rowMeans(z,na.rm=T)>0))+c(1:length(tz)), t-1] <- tz
 gene[max(which(gene>0))+c(1:length(tg))] <- tg

 for (v in 1:nindv) {
  if (z[v,t-1] %in% 1:nsite) {
   z[v,t] <- rcat(1, c(phi[z[v,t-1],t-1,gene[v]] * theta[z[v,t-1],,t-
1,gene[v]], 1 - phi[z[v,t-1],t-1,gene[v]]))
  }
 } # v
 z[which(!(z %in% 1:nsite))] <- 0
```

```
 for (i in 1:nsite) {
  for (g in 1:ngene) {
   N[i,t,g] <- length(which(z[,t] == i & gene == g))
  } # g
 } # i
} # t

exit <- which(rowMeans(z, na.rm=T) > 0) # indicating if an individual
ever exists
z <- z[exit,]    # only keep individuals that ever exist
gene <- gene[exit] # only keep individuals that ever exist

# Simulate count data
w <- array(rnorm(nsite * nyear * nreps * npcvs, 0, 1), dim=c(nsite,
nyear, nreps, npcvs)) # observational covariates

pdet <- array(, dim=c(nsite, nyear, nreps)) # detection probability
for (i in 1:nsite) {
 for (t in 1:nyear) {
  pdet[i,t,] <- inv.logit(cbind(1,w[i,t,,]) %*% alpha_det)
 } # t
} # i

y <- array(, dim=c(nsite, nyear, nreps, ngene)) # count data of breeders
for (i in 1:nsite) {
 for (t in 1:nyear) {
  for (g in 1:ngene) {
   y[i,t,,g] <- rbinom(nreps, N[i,t,g], pdet[i,t,])
  } # g
 } # t
} # i

r <- array(, dim=c(nsite, nyear-1, nreps, ngene)) # count data of
offspring
for (i in 1:nsite) {
 for (t in 1:(nyear-1)) {
  for (g in 1:ngene) {
   r[i,t,,g] <- rbinom(nreps, R[i,t,g], pdet[i,t,])
  } # g
 } # t
} # i

# simulate capture-recapture data
pcap_per_visit <- array(, dim=c(nsite, nyear, nreps)) # per visit
capture probability
```

```
for (i in 1:nsite) {
 for (t in 1:nyear) {
  pcap_per_visit[i,t,] <- inv.logit(cbind(1,w[i,t,,]) %*% alpha_cap)
 } # t
} # i

c_all <- matrix(, dim(z)[1], nyear) # capture history of all individuals
for (v in 1:dim(z)[1]) {
 for (t in 1:nyear) {
  if (z[v,t] == 0) {
   c_all[v,t] <- 0
  } else {
   cap_per_visit <- rbinom(nreps, 1, pcap_per_visit[z[v,t],t,])
   c01 <- ifelse(sum(cap_per_visit)==0, 0, 1)
   c_all[v,t] <- c01 * z[v,t]
  }
 } # t
} # v
if_cap <- ifelse(apply(c_all, 1, sum) > 0, 1, 0)

c <- c_all[which(if_cap==1),] # capture history of individuals that
have ever been captured
f <- numeric(dim(c)[1]) # year of first capture
for (v in 1:length(f)) {
 f[v] <- min(which(c[v,] > 0))
} # v
c <- c[which(f %in% c(1:(nyear-1))),]

gene_cap <- gene[which(if_cap==1)]
gene_cap <- gene_cap[which(f %in% c(1:(nyear-1)))]

# Extend individual capture history
for (v in 1:dim(c)[1]) {
 t1 <- c[v,]
 t1_gene <- gene_cap[v]
 ycap <- which(t1 %in% c(1:nsite))
 ncap <- length(ycap)
 if (ncap == 1) {
  t2 <- t1
  t2_gene <- t1_gene
 } else {
  t2 <- matrix(, ncap, nyear)
  for (k in 1:(ncap-1)) {
   t2[k, ycap[k]:ycap[k+1]] <- t1[ycap[k]:ycap[k+1]]
  } # k
```

```
  t2[ncap, ycap[ncap]] <- t1[ycap[ncap]]
  t2_gene <- rep(t1_gene, ncap)
 }
 if (v == 1) {
  c_ext <- t2
  gene_ext <- t2_gene
 } else {
  c_ext <- rbind(c_ext, t2)
  gene_ext <- c(gene_ext, t2_gene)
 }
} # v
c_ext[which(!(c_ext %in% 1:nsite))] <- 0

# Converting individual capture history to m-array format
f_ext <- numeric(dim(c_ext)[1])
for (v in 1:length(f_ext)) {
 f_ext[v] <- min(which(c_ext[v,] %in% c(1:nsite)))
} # v

nm <- matrix(, ngene, nsite)
for (k in 1:ngene) {
 for (i in 1:nsite) {
  nm[k,i] <- paste('site ', i, ', gene ', k, sep='')
 } # i
} # k
nm <- as.vector(nm)

m <- array(0, # truncated m-array of capture-recapture (i.e., next-year
recapture only)
      dim=c(nsite*ngene, nsite+1, nyear-1),
      dimnames=list(nm, c(paste('site',1:nsite,sep=' '),'not seen'),
paste('year',1:(nyear-1),sep=' ')))
for (t in 1:(nyear-1)) {
 t1c <- c_ext[which(f_ext == t),]
 t1g <- gene_ext[which(f_ext == t)]
 for (i in 1:nsite) {
  for (g in 1:ngene) {
   nband <- length(which(t1c[,t] == i & t1g == g))
   if (nband > 0) {
    t2 <- matrix(t1c[which(t1c[,t] == i & t1g == g),], nrow=nband,
ncol=nyear)
    for (j in 1:nsite) {
     m[ngene*(i-1)+g,j,t] <- length(which(t2[,t+1] == j))
    } # j
    m[ngene*(i-1)+g,nsite+1,t] <- nband - sum(m[ngene*(i-1)+g,1:
```

```
nsite,t])
    }
   } # g
  } # i
} # t
```

Next, we will develop our own function to calculate the cell probabilities of the truncated m-array and custom MCMC algorithm to estimate parameters $\beta^{[0]}$, $\beta^{[\gamma]}$, $\beta^{[\phi]}$, $\beta^{[\kappa]}$, $\alpha^{[cap]}$, $\alpha^{[det]}$, and N using data y, r, m, x, w, d.

```
# Function to calculate the cell probability of truncated m-array
cell_prob_fun <- function(nsite, ngene, nyear, phi, theta, pcap) {
 cell_prob <- array(0, dim=c(nsite*ngene, nsite+1, nyear-1))
 for (t in 1:(nyear-1)) {
  for (i in 1:nsite) {
   for (g in 1:ngene) {
    cell_prob[ngene*(i-1)+g,1:nsite,t] <- phi[i,t,g] * theta[i,,t,g] *
pcap[,t]
     cell_prob[ngene*(i-1)+g,nsite+1,t] <- 1 - sum(cell_prob[ngene*
(i-1)+g,1:nsite,t])
   } # g
  } # i
 } # t

 return(cell_prob)
} # cell_prob_fun

gene_sipm_mcmc <- function(y, r, m, x, w, d, nmcmc) {

 # Setup variables
 nsite <- dim(y)[1]
 nyear <- dim(y)[2]
 nreps <- dim(y)[3]
 ngene <- dim(y)[4]
 ncovs <- dim(x)[3]
 npcvs <- dim(w)[4]
 ymax <- apply(y, c(1,2,4), max)
 xgene <- (1:ngene - 3.5) / 1.7

 beta0_save <- matrix(0, ncovs+1, nmcmc)
 beta_gamma_save <- matrix(0, ncovs+4, nmcmc)
 beta_phi_save <- matrix(0, ncovs+4, nmcmc)
 beta_kappa_save <- matrix(0, ncovs+3, nmcmc)
 alpha_det_save <- matrix(0, npcvs+1, nmcmc)
```

```
alpha_cap_save <- matrix(0, npcvs+1, nmcmc)
N_save <- array(0, dim=c(nsite, nyear, ngene, 500))

# Prior
beta0_mean <- rep(0, ncovs+1)
beta0_sd <- 2
beta_gamma_mean <- rep(0, ncovs+4)
beta_gamma_sd <- 2
beta_phi_mean <- rep(0, ncovs+4)
beta_phi_sd <- 2
beta_kappa_mean <- rep(0, ncovs+3)
beta_kappa_sd <- 2
alpha_det_mean <- rep(0, npcvs+1)
alpha_det_sd <- 2
alpha_cap_mean <- rep(0, npcvs+1)
alpha_cap_sd <- 2

# Starting values
beta0 <- rep(0, ncovs+1)
beta_gamma <- rep(0, ncovs+4)
beta_phi  <- rep(0, ncovs+4)
beta_kappa <- rep(0, ncovs+3)
alpha_det <- rep(0, npcvs+1)
alpha_cap <- rep(0, npcvs+1)
lambda0 <- rep(1, nsite)
gamma <- array(1, dim=c(nsite, nyear-1, ngene))
phi <- array(0.5, dim=c(nsite, nyear-1, ngene))
kappa <- array(1, dim=c(nsite, nyear-1, ngene))
eta <- theta <- array(, dim=c(nsite, nsite, nyear-1, ngene))
for (i in 1:nsite) {
 for (t in 1:(nyear-1)) {
  for (g in 1:ngene) {
   eta[i,,t,g] <- exp(-1 * kappa[i,t,g] * d[i,])
   theta[i,,t,g] <- eta[i,,t,g] / sum(eta[i,,t,g])
  } # g
 } # t
} # i
N <- ymax + 1
pdet <- array(0.5, dim=c(nsite, nyear, nreps))
pcap_per_visit <- array(0.5, dim=c(nsite, nyear-1, nreps))
pcap <- matrix(, nsite, nyear-1)
for (i in 1:nsite) {
 for (t in 1:(nyear-1)) {
  pcap[i,t] <- 1 - prod(1 - pcap_per_visit[i,t,])
 } # t
```

```
 } # i
 cell_prob <- cell_prob_fun(nsite=nsite, ngene=ngene, nyear=nyear,
phi=phi, theta=theta, pcap=pcap)

 # Tuning factor
 beta0_tune <- 0.08
 beta_gamma_tune <- 0.02
 beta_phi_tune <- 0.08
 beta_kappa_tune <- 0.04
 alpha_det_tune <- 0.025
 alpha_cap_tune <- 0.05
 N_tune <- 1

 # Begin MCMC loop
 for (k in 1:nmcmc) {
  ### Sample N
  N_star <- array(rpois(nsite*nyear*ngene, N + N_tune), dim=c(nsite,
nyear, ngene))
  prob1 <- prob2 <- array(, dim=c(nsite, nyear, nreps, ngene))
  for (j in 1:nreps) {
   for (g in 1:ngene) {
    prob1[,,j,g] <- dbinom(y[,,j,g], N_star[,,g], pdet[,,j], log=T)
    prob2[,,j,g] <- dbinom(y[,,j,g], N     [,,g], pdet[,,j], log=T)
   } # g
  } # j
  Nexp <- array(, dim=c(nsite, nyear, ngene))
  Nexp[,1,] <- lambda0 / ngene
  for (t in 2:nyear) {
   for (g in 1:ngene) {
    Nexp[,t,g] <- (N[,t-1,g] * (1 + gamma[,t-1,g]) * phi[,t-1,g]) %*%
theta[,,t-1,g] + 1e-6
   } # g
  } # t
  mh1 <- apply(prob1, c(1,2,4), sum) +
      dpois(N_star, Nexp, log=T) +
      dpois(N, N_star+N_tune, log=T)
  mh2 <- apply(prob2, c(1,2,4), sum) +
      dpois(N     , Nexp, log=T) +
      dpois(N_star, N+N_tune, log=T)
  mhN <- exp(mh1 - mh2)
  Nkeep <- ((mhN > runif(nsite*nyear*ngene)) & (N_star >= ymax))
  N[Nkeep] <- N_star[Nkeep]

  ### Sample beta0
  beta0_star <- rnorm(ncovs+1, beta0, beta0_tune)
```

```
  lambda0_star <- exp(cbind(1,x[,1,]) %*% beta0_star)
  mh1 <- sum(dpois(N[,1,], lambda0_star / ngene, log=T)) +
     sum(dnorm(beta0_star, beta0_mean, beta0_sd, log=T))
  mh2 <- sum(dpois(N[,1,], lambda0     / ngene, log=T)) +
     sum(dnorm(beta0     , beta0_mean, beta0_sd, log=T))
  mhB0 <- exp(mh1 - mh2)
  if (mhB0 > runif(1)) {
   beta0 <- beta0_star
   lambda0 <- lambda0_star
  }

  # Sample beta_gamma
  beta_gamma_star <- rnorm(ncovs+4, beta_gamma, beta_gamma_tune)
  gamma <- gamma_star <- array(, dim=c(nsite, nyear-1, ngene))
  for (t in 2:nyear) {
   for (g in 1:ngene) {
    x_temp1 <- (rowSums(N[,t-1,]) - lambda0) / lambda0
    x_temp2 <- x[,t,1]
    x_temp3 <- x[,t,2]
    x_temp4 <- (g - 3.5) / 1.7
    x_temp5 <- x_temp2 * x_temp4
    x_temp <- cbind(1, x_temp1, x_temp2, x_temp3, x_temp4, x_temp5)
    gamma     [,t-1,g] <- exp(x_temp %*% beta_gamma      )
    gamma_star[,t-1,g] <- exp(x_temp %*% beta_gamma_star)
   } # g
  } # t
  Nexp <- Nexp_star <- array(, dim=c(nsite, nyear-1, ngene))
  for (t in 1:(nyear-1)) {
   for (g in 1:ngene) {
   Nexp     [,t,g] <- (N[,t,g] * (1 + gamma     [,t,g]) * phi[,t,g]) %*% theta
[,,t,g] + 1e-6
   Nexp_star[,t,g] <- (N[,t,g] * (1 + gamma_star[,t,g]) * phi[,t,g]) %*%
theta[,,t,g] + 1e-6
   } # g
  } # t
  prob1 <- prob2 <- numeric(nreps)
  for (j in 1:nreps) {
   prob1[j] <- sum(dbinom(r[,,j,], r[,,j,] + y[,-nyear,j,], gamma_star
/ (gamma_star + 1), log=T))
   prob2[j] <- sum(dbinom(r[,,j,], r[,,j,] + y[,-nyear,j,], gamma      /
(gamma      + 1), log=T))
  } # j
  mh1 <- sum(prob1) +
         sum(dpois(N[,-1,], Nexp_star, log=T)) +
         sum(dnorm(beta_gamma_star, beta_gamma_mean, beta_gamma_sd,
```

```
log=T))
  mh2 <- sum(prob2) +
        sum(dpois(N[,-1,], Nexp   , log=T)) +
        sum(dnorm(beta_gamma   , beta_gamma_mean, beta_gamma_sd,
log=T))
  mhBG <- exp(mh1 - mh2)
  if (mhBG > runif(1)) {
   beta_gamma <- beta_gamma_star
   gamma <- gamma_star
  }

  # Sample beta_phi
  beta_phi_star <- rnorm(ncovs+4, beta_phi, beta_phi_tune)
  phi <- phi_star <- array(, dim=c(nsite, nyear-1, ngene))
  for (t in 2:nyear) {
   for (g in 1:ngene) {
    x_temp1 <- (rowSums(N[,t-1,]) - lambda0) / lambda0
    x_temp2 <- x[,t,1]
    x_temp3 <- x[,t,2]
    x_temp4 <- (g - 3.5) / 1.7
    x_temp5 <- x_temp2 * x_temp4
    x_temp <- cbind(1, x_temp1, x_temp2, x_temp3, x_temp4, x_temp5)
    phi   [,t-1,g] <- inv.logit(x_temp %*% beta_phi   )
    phi_star[,t-1,g] <- inv.logit(x_temp %*% beta_phi_star)
   } # g
  } # t
  cell_prob   <- cell_prob_fun(nsite=nsite, ngene=ngene,
nyear=nyear, phi=phi   , theta=theta, pcap=pcap)
  cell_prob_star <- cell_prob_fun(nsite=nsite, ngene=ngene,
nyear=nyear, phi=phi_star, theta=theta, pcap=pcap)
  prob1 <- prob2 <- matrix(, nsite*ngene, nyear-1)
  for (i in 1:nsite) {
   for (t in 1:(nyear-1)) {
    for (g in 1:ngene) {
     prob1[ngene*(i-1)+g,t] <- dmultinom(m[ngene*(i-1)+g,,t],
size=sum(m[ngene*(i-1)+g,,t]), prob=cell_prob_star[ngene*(i-1)+g,,
t], log=T)
     prob2[ngene*(i-1)+g,t] <- dmultinom(m[ngene*(i-1)+g,,t],
size=sum(m[ngene*(i-1)+g,,t]), prob=cell_prob   [ngene*(i-1)+g,,t],
log=T)
    } # g
   } # t
  } # i
  Nexp <- Nexp_star <- array(, dim=c(nsite, nyear-1, ngene))
  for (t in 1:(nyear-1)) {
   for (g in 1:ngene) {
```

```
    Nexp   [,t,g] <- (N[,t,g] * (1 + gamma[,t,g]) * phi   [,t,g]) %*% theta
[,,t,g] + 1e-6
    Nexp_star[,t,g] <- (N[,t,g] * (1 + gamma[,t,g]) * phi_star[,t,g]) %*%
theta[,,t,g] + 1e-6
    } # g
  } # t
  mh1 <- sum(prob1) +
         sum(dpois(N[,-1,], Nexp_star, log=T)) +
         sum(dnorm(beta_phi_star, beta_phi_mean, beta_phi_sd, log=T))
  mh2 <- sum(prob2) +
         sum(dpois(N[,-1,], Nexp     , log=T)) +
         sum(dnorm(beta_phi     , beta_phi_mean, beta_phi_sd, log=T))
  mhBO <- exp(mh1 - mh2)
  if (mhBO > runif(1)) {
   beta_phi <- beta_phi_star
   phi <- phi_star
  }

  # Sample beta_kappa
  beta_kappa_star <- rnorm(ncovs+3, beta_kappa, beta_kappa_tune)
  kappa <- kappa_star <- array(, dim=c(nsite, nyear-1, ngene))
  eta <- theta <- eta_star <- theta_star <- array(, dim=c(nsite, nsite,
nyear-1, ngene))
  for (i in 1:nsite) {
   for (t in 1:(nyear-1)) {
    for (g in 1:ngene) {
     x_temp1 <- (sum(N[i,t,]) - lambda0[i]) / lambda0[i]
     x_temp2 <- x[i,t+1,1]
     x_temp3 <- (g - 3.5) / 1.7
     x_temp4 <- x_temp2 * x_temp3
     x_temp <- c(1, x_temp1, x_temp2, x_temp3, x_temp4)
     kappa    [i,t,g] <- exp(x_temp %*% beta_kappa     )
     kappa_star[i,t,g] <- exp(x_temp %*% beta_kappa_star)
     eta    [i,,t,g] <- exp(-1 * kappa    [i,t,g] * d[i,])
     eta_star[i,,t,g] <- exp(-1 * kappa_star[i,t,g] * d[i,])
     theta    [i,,t,g] <- eta    [i,,t,g] / sum(eta    [i,,t,g])
     theta_star[i,,t,g] <- eta_star[i,,t,g] / sum(eta_star[i,,t,g])
    } # g
   } # t
  } # i
  cell_prob    <- cell_prob_fun(nsite=nsite, ngene=ngene,
nyear=nyear, phi=phi, theta=theta    , pcap=pcap)
  cell_prob_star <- cell_prob_fun(nsite=nsite, ngene=ngene,
nyear=nyear, phi=phi, theta=theta_star, pcap=pcap)
  prob1 <- prob2 <- matrix(, nsite*ngene, nyear-1)
```

```
  for (i in 1:nsite) {
   for (t in 1:(nyear-1)) {
    for (g in 1:ngene) {
     prob1[ngene*(i-1)+g,t] <- dmultinom(m[ngene*(i-1)+g,,t],
size=sum(m[ngene*(i-1)+g,,t]), prob=cell_prob_star[ngene*(i-1)+g,,
t], log=T)
     prob2[ngene*(i-1)+g,t] <- dmultinom(m[ngene*(i-1)+g,,t],
size=sum(m[ngene*(i-1)+g,,t]), prob=cell_prob   [ngene*(i-1)+g,,t],
log=T)
    } # g
   } # t
  } # i
  Nexp <- Nexp_star <- array(, dim=c(nsite, nyear-1, ngene))
  for (t in 1:(nyear-1)) {
   for (g in 1:ngene) {
    Nexp   [,t,g] <- (N[,t,g] * (1 + gamma[,t,g]) * phi[,t,g]) %*% theta
[,,t,g] + 1e-6
    Nexp_star[,t,g] <- (N[,t,g] * (1 + gamma[,t,g]) * phi[,t,g]) %*%
theta_star[,,t,g] + 1e-6
   } # g
  } # t
  mh1 <- sum(prob1) +
         sum(dpois(N[,-1,], Nexp_star, log=T)) +
         sum(dnorm(beta_kappa_star, beta_kappa_mean, beta_kappa_sd,
log=T))
  mh2 <- sum(prob2) +
         sum(dpois(N[,-1,], Nexp    , log=T)) +
         sum(dnorm(beta_kappa     , beta_kappa_mean, beta_kappa_sd,
log=T))
  mhBK <- exp(mh1 - mh2)
  if (mhBK > runif(1)) {
   beta_kappa <- beta_kappa_star
   kappa <- kappa_star
   eta <- eta_star
   theta <- theta_star
  }

  ### Sample alpha_det
  alpha_det_star <- rnorm(npcvs+1, alpha_det, alpha_det_tune)
  pdet_star <- array(, dim=c(nsite, nyear, nreps))
  for (i in 1:nsite) {
   for (t in 1:nyear) {
    pdet_star[i,t,] <- inv.logit(cbind(1,w[i,t,,]) %*%
alpha_det_star)
   } # t
```

```
  } # i
  prob1 <- prob2 <- array(, dim=c(nsite, nyear, nreps, ngene))
  for (j in 1:nreps) {
   for (g in 1:ngene) {
    prob1[,,j,g] <- dbinom(y[,,j,g], N[,,g], pdet_star[,,j], log=T)
    prob2[,,j,g] <- dbinom(y[,,j,g], N[,,g], pdet     [,,j], log=T)
   } # g
  } # j
  mh1 <- sum(prob1) +
         sum(dnorm(alpha_det_star, alpha_det_mean, alpha_det_sd,
log=TRUE))
  mh2 <- sum(prob2) +
         sum(dnorm(alpha_det     , alpha_det_mean, alpha_det_sd,
log=TRUE))
  mhAD <- exp(mh1 - mh2)
  if (mhAD > runif(1)) {
   alpha_det <- alpha_det_star
   pdet <- pdet_star
  }

  # Sample alpha_pcap
  alpha_cap_star <- rnorm(npcvs+1, alpha_cap, alpha_cap_tune)
  pcap_per_visit_star <- array(, dim=c(nsite, nyear-1, nreps))
  for (i in 1:nsite) {
   for (t in 1:(nyear-1)) {
    pcap_per_visit_star[i,t,] <- inv.logit(cbind(1,w[i,t+1,,]) %*%
alpha_cap_star)
   } # t
  } # i
  pcap_star <- matrix(, nsite, nyear-1)
  for (i in 1:nsite) {
   for (t in 1:(nyear-1)) {
    pcap_star[i,t] <- 1 - prod(1 - pcap_per_visit_star[i,t,])
   } # t
  } # i
  cell_prob      <- cell_prob_fun(nsite=nsite, ngene=ngene,
nyear=nyear, phi=phi, theta=theta, pcap=pcap     )
  cell_prob_star <- cell_prob_fun(nsite=nsite, ngene=ngene,
nyear=nyear, phi=phi, theta=theta, pcap=pcap_star)
  prob1 <- prob2 <- matrix(, nsite*ngene, nyear-1)
  for (i in 1:nsite) {
   for (t in 1:(nyear-1)) {
    for (g in 1:ngene) {
     prob1[ngene*(i-1)+g,t] <- dmultinom(m[ngene*(i-1)+g,,t],
size=sum(m[ngene*(i-1)+g,,t]), prob=cell_prob_star[ngene*(i-1)+g,,
t], log=T)
```

```
      prob2[ngene*(i-1)+g,t] <- dmultinom(m[ngene*(i-1)+g,,t],
size=sum(m[ngene*(i-1)+g,,t]), prob=cell_prob  [ngene*(i-1)+g,,t],
log=T)
     }#g
    }#t
  }#i
  mh1 <- sum(prob1) +
        sum(dnorm(alpha_cap_star, alpha_cap_mean, alpha_cap_sd,
log=T))
  mh2 <- sum(prob2) +
        sum(dnorm(alpha_cap   , alpha_cap_mean, alpha_cap_sd, log=T))
  mh <- exp(mh1 - mh2)
  if (mh > runif(1)) {
   alpha_cap <- alpha_cap_star
   pcap_per_visit <- pcap_per_visit_star
   pcap <- pcap_star
  }

  # Save samples
  beta0_save[,k] <- beta0
  beta_gamma_save[,k] <- beta_gamma
  beta_phi_save[,k] <- beta_phi
  beta_kappa_save[,k] <- beta_kappa
  alpha_det_save[,k] <- alpha_det
  alpha_cap_save[,k] <- alpha_cap
  if (nmcmc > 500) {
   if (k > nmcmc-500) {
    N_save[,,,k-nmcmc+500] <- N
   }
  }
 }#k

 list(beta0_save=beta0_save,
   beta_gamma_save=beta_gamma_save, beta_phi_save=beta_phi_save,
   beta_kappa_save=beta_kappa_save,
   alpha_det_save=alpha_det_save, alpha_cap_save=alpha_cap_save,
   N_save=N_save)
} # gene_sipm_mcmc
```

Using the following code to implement the MCMC algorithm, it takes about 5 h to run. Then using the code in Appendix 4.7 to create some traceplots, we can see good model convergence and agreement between parameter estimates of $\beta^{[0]}$, $\beta^{[\gamma]}$, $\beta^{[\phi]}$, $\beta^{[\kappa]}$, $\alpha^{[cap]}$, and $\alpha^{[det]}$ and their true values (Fig. 4.15).

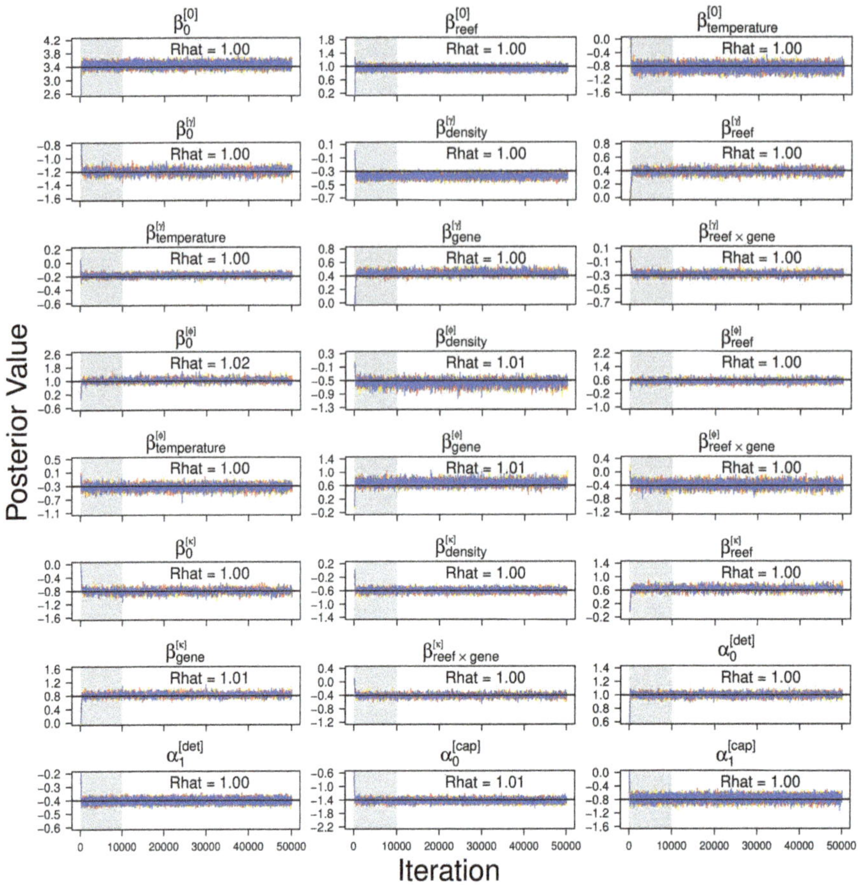

Fig. 4.15 Traceplots showing posterior chains for $\beta^{[0]}$, $\beta^{[\gamma]}$, $\beta^{[\phi]}$, $\beta^{[\kappa]}$, $\alpha^{[cap]}$, and $\alpha^{[det]}$ in the spatial IPM with gene-class-specific reproduction, survival, and movement. Rhat values are calculated after discarding the first 10,000 iterations. Horizontal lines represent the true values used in data simulation

```
library(foreach)  # for parallel computing
library(doParallel) # for parallel computing

numCores <- round(detectCores() / 2) # only use half of the cores for
parallel computing
registerDoParallel(numCores) # setup parallel computing

nmcmc <- 50000 # number of iterations of MCMC computing
chain <- 3    # number of chains

start_time <- Sys.time() # start time of computing
out <- foreach (i=1:chain, .packages='boot') %dopar% {
```

```
  gene_sipm_mcmc(y=y, r=r, m=m, x=x, w=w, d=d, nmcmc=nmcmc)
} # i
stopImplicitCluster() # clean up the cluster
end_time <- Sys.time() # end time of computing
time_taken <- end_time - start_time # computing time
```

We can also use the following code to create pie plots that represent local population size, $\sum_{g=1}^{n_{gene}} N_{i,g,t}$, and the proportion of animals under each gene class, $N_{i,g,t}/\sum_{g=1}^{n_{gene}} N_{i,g,t}$, for each population. We can see an overall increase in both local population size and the proportion of contesters; the latter is likely due to higher reproduction and survival of contesters. These patterns are correctly represented by the estimates (Fig. 4.16).

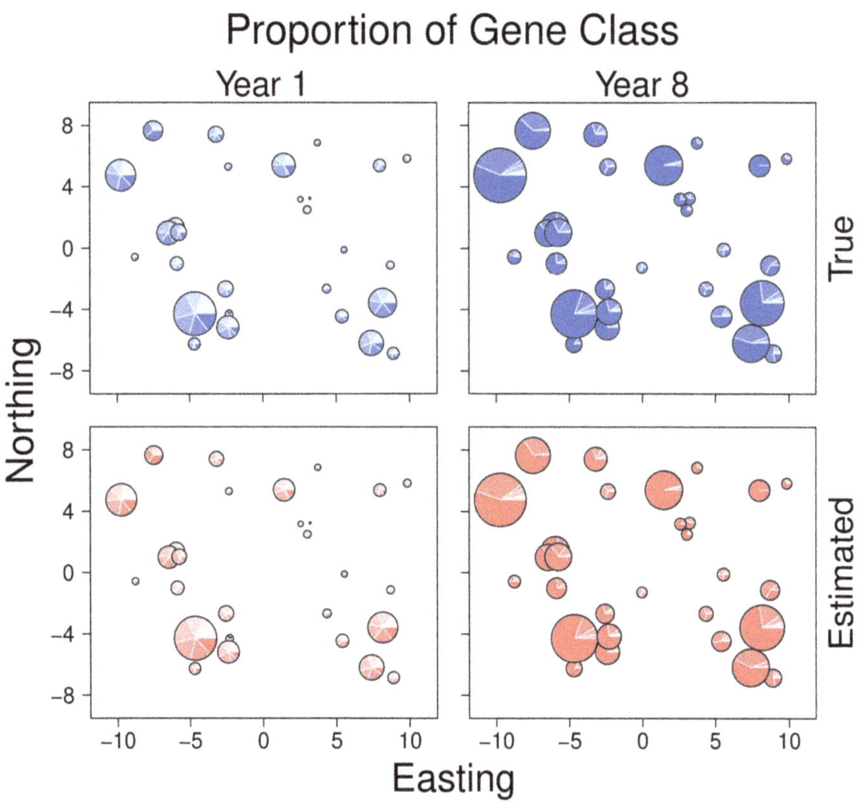

Fig. 4.16 Pie plots showing the consistency between the simulated true (upper row) and estimated (lower row) local population size (proportional to the sizes of circles) and the proportion of contesters (deeper blue or red) in each local population, both of which vary spatially and temporally (using years 1 and 8 as examples)

```
library(plotrix) # for making pie charts at selected positions

N_post <- array(, dim=c(nsite, nyear, ngene, 1500))
N_post[,,,   1: 500] <- out[[1]]$N
N_post[,,, 501:1000] <- out[[2]]$N
N_post[,,,1001:1500] <- out[[3]]$N
N_med <- apply(N_post, 1:3, median)
years_plot <- c(1, 8)

par(mfrow=c(2,2))
par(mar=c(1,1,1,1))
par(oma=c(4,5,5,2))

for (t in 1:2) {
 plot(1, xlim=c(-11, 11), ylim=c(-8.8, 8.8), type='n', axes=F,
xlab='', ylab='')
 for (i in 1:nsite) {
  tt <- N[i,years_plot[t],]
  if (sum(tt) > 0) {
   prop <- tt / sum(tt)
   size <- sum(N[i,years_plot[t],]) ^ 0.67 / 14
   cols <- colorRampPalette(c('white','royalblue'))(ngene+1)[-1]
   floating.pie(xpos=cov$lon[i], ypos=cov$lat[i], x=prop,
radius=size, col=cols, border=cols[1], lwd=1)
   draw.circle(x=cov$lon[i], y=cov$lat[i], radius=size, nv=100,
border='grey18', col=NA, lwd=1.2)
  }
 } # i
 axis(1, at=seq(-10,10,5), labels=rep('',5))
 if (t == 1) {
  axis(2, at=seq(-8,8,4), las=2, cex.axis=1.6)
 } else {
  axis(2, at=seq(-8,8,4), labels=rep('',5))
 }
 axis(3, at=0, labels=paste('Year', years_plot[t], sep=' '), cex.
axis=2.5, line=-0.8, tick=F)
 if (t == 2) {
  axis(4, at=0, labels='True', cex.axis=2.5, line=0.5, tick=F)
 }
 box()
} # t

for (t in 1:2) {
 plot(1, xlim=c(-11, 11), ylim=c(-8.8, 8.8), type='n', axes=F,
xlab='', ylab='')
```

```
 for (i in 1:nsite) {
  tt <- N_med[i,years_plot[t],]
  if (sum(tt) > 0) {
   prop <- tt / sum(tt)
   size <- sum(N_med[i,years_plot[t],]) ^ 0.67 / 14
   cols <- colorRampPalette(c('white','lightcoral'))(ngene+1)[-1]
   floating.pie(xpos=cov$lon[i], ypos=cov$lat[i], x=prop,
radius=size, col=cols, border=cols[1], lwd=1)
   draw.circle(x=cov$lon[i], y=cov$lat[i], radius=size, nv=100,
border='grey18', col=NA, lwd=1.2)
  }
 } # i
 axis(1, at=seq(-10,10,5), cex.axis=1.6)
 if (t == 1) {
  axis(2, at=seq(-8,8,4), las=2, cex.axis=1.6)
 } else {
  axis(2, at=seq(-8,8,4), labels=rep('',5))
 }
 if (t == 2) {
  axis(4, at=0, labels='Estimated', cex.axis=2.5, line=0.5, tick=F)
 }
 box()
} # t

title(xlab='Easting' , cex.lab=2.8, line=2.6, outer=T)
title(ylab='Northing', cex.lab=2.8, line=2.4, outer=T)
title(main=expression('Proportion of Gene Class'), line=3.2, cex.
main=3, outer=T)
```

4.5 Summary and Additional Reading

In this chapter, we talked about spatial IPMs that are capable of estimating the entire BIDE processes and their drivers by jointly analyzing count and capture-recapture data. While we have focused on models that utilize live-encounter data (i.e., recapture or resighting of live animals), survival and movement could also be estimated with dead-encounter (i.e., recovery of dead animals; Brownie et al. 1978, 1985; Schwarz et al. 1993; Lindberg et al. 2001) or a combination of live- and dead-encounter data (Burnham 1993). Therefore, there is a great potential to use dead-encounter data in spatial IPMs. There has been a boost in the applications of non-spatial (e.g., Schaub et al. 2013, 2015; Robinson et al. 2018; Saunders et al. 2018; Gamelon et al. 2021; Lewis et al. 2023) and to a lesser extent spatial IPMs (Gade et al. 2022) during the last decade or so, all of which are worth reading. Or, readers can refer to Schaub and Kéry (2022) for a comprehensive introduction to the methodology and applications of IPMs. While spatial IPMs are quite powerful tools for understanding the BIDE processes, spatial capture-recapture (SCR) models push it even further to link individual habitat use with BIDE, which we will detail in the next chapter.

References

Arnason AN (1972) Parameter estimates from mark-recapture experiments on two populations subject to migration and death. Res Popul Ecol (Kyoto) 13:97–113
Arnason AN (1973) The estimation of population size, migration rates and survival in a stratified population. Res Popul Ecol (Kyoto) 15:1
Arnold CP, Dey CH (1986) Observing-systems simulation experiments: past, present and future. Bull Am Meteorol Soc 67:687–695
Besbeas P, Freeman SN, Morgan BJT, Catchpole EA (2002) Integrating mark-recapture-recovery and census data to estimate animal abundance and demographic parameters. Biometrics 58:540–547
Borysiewicz RS, Morgan BJT, Hénaux V, Bregnballe T, Lebreton JD, Gimenez O (2009) An integrated analysis of multisite recruitment, mark–recapture–recovery and multisite census data. In: Thomson DL et al (eds) Modeling demographic processes in marked populations. Environmental and ecological statistics, vol 3. Springer, pp 579–591
Brooks SP, Catchpole EA, Morgan BJT, Barry SC (2000) On the Bayesian analysis of ring-recovery data. Biometrics 56:951–956
Brooks SP, King R, Morgan BJT (2004) A Bayesian approach to combining animal abundance and demographic data. Anim Biodivers Conserv 27:515–529
Brownie C, Anderson DR, Burnham KP, Robson DR (1978) Statistical inference from band recovery data: a handbook. US Fish & Wildlife Service
Brownie C, Anderson DR, Burnham KP, Robson DS (1985) Statistical inference from band recovery data: a handbook, 2nd edn. US Fish & Wildlife Service
Brownie C, Hines JE, Nichols JD, Pollock KH, Hestbeck JB (1993) Capture-recapture studies for multiple strata including non-Markovian transition probabilities. Biometrics 49:1173–1187
Burnham KP (1993) A theory for combined analysis of ring recovery and recapture data. In: Lebreton JD, North PM (eds) Marked individuals in study of bird populations, pp 199–213
Conn PB, Johnson DS, Williams PJ, Melin SR, Hooten MB (2018) A guide to Bayesian model checking for ecologists. Ecol Monogr 88:526–542
Cormack RM (1964) Estimates of survival from the sighting of marked animals. Biometrika 51: 429–438
Fournier D, Archibald CP (1982) A general theory for analyzing catch at age data. Can J Fish Aquatic Sci 39:1195–1207
Frost F, McCrea R, King R, Gimenez O, Zipkin E (2023) Integrated population models: achieving their potential. J Stat Theory Pract 17:6
Gade MR, Zhao Q, Peterman WE (2022) Spatial variation in demographic processes and the potential role of hybridization for the future. Landsc Ecol 37:2671–2687
Gamelon M, Nater CR, Baubet É, Besnard A, Touzot L, Gaillard JM, Lebreton JD, Gimenez O (2021) Efficient use of harvest data: a size-class-structured integrated population model for exploited populations. Ecography 44:1296–1310
Gelman A, Meng XL, Stern H (1996) Posterior predictive assessment of model fitness via realized discrepancies. Stat Sin 6:733–760
Gelman A, Carlin JB, Stern HS, Dunson DB, Vehtari A, Rubin DB (2014) Bayesian data analysis. Chapman & Hall/CRC

Hestbeck JB, Nichols JD, Malecki RA (1991) Estimates of movement and site fidelity using mark-resight data of wintering Canada geese. Ecology 72:523–533

Jolly GM (1965) Explicit estimates from capture-recapture data with both death and immigration-stochastic model. Biometrika 52:225–248

Lebreton JD, Burnham KP, Clobert J, Anderson DR (1992) Modeling survival and testing biological hypotheses using marked animals: a unified approach with case studies. Ecol Monogr 62:67–118

Lebreton JD, Nichols JD, Barker RJ, Pradel R, Spendelow JA (2009) Modeling individual animal histories with multistate capture-recapture models. Adv Ecol Res 41:87–173

Lewis WB, Cooper RJ, Chandler RB, Chitwood RW, Cline MH, Hallworth MT, Hatt JL, Hepinstall-Cymerman J, Kaiser SA, Rodenhouse NL, Sillett TS, Stodola KW, Webster MS, Holmes RT (2023) Climate-mediated population dynamics of a migratory songbird differ between the trailing edge and range core. Ecol Monogr 93:e1559

Lindberg MS, Kendall WL, Hines JE, Anderson MG (2001) Combining band recovery data and Pollock's robust design to model temporary and permanent emigration. Biometrics 57:273–281

McCrea RS, Morgan BJT, Gimenez O, Besbeas P, Lebreton JD, Bregnballe T (2010) Multi-site integrated population modelling. J Agric Biol Environ Stat 15:539–561

Nichols JD, Sauer JR, Pollock KH, Hestbeck JB (1992) Estimating transition probabilities for stage-based population projection matrices using capture-recapture data. Ecology 73:306–312

Péron G, Crochét PA, Doherty PF, Lebreton JD (2010) Studying dispersal at the landscape scale: efficient combination of population surveys and capture-recapture data. Ecology 91:3365–3375

Rakhimberdiev E, Karagicheva J, Saveliev A, Loonstra AHJ, Verhoeven MA, Hooijmeijer JCEW, Schaub M, Piersma T (2022) Misidentification errors in reencounters result in biased estimates of survival probability from CJS models: Evidence and a solution using the robust design. Methods Ecol Evol 13:1106–1118

Robinson OJ, Ruiz-Gutierrez V, Fink D, Meese RJ, Holyoak M, Cooch EG (2018) Using citizen science data in integrated population models to inform conservation. Biol Conserv 227:361–368

Rubin DB (1984) Bayesianly justifiable and relevant frequency calculations for the applied statistician. Ann Statistics 12:1151–1172

Saunders SP, Cuthbert FJ, Zipkin EF (2018) Evaluating population viability and efficacy of conservation management using integrated population models (J Rhodes, Ed.). J Appl Ecol 55:1380–1392

Schaub M, Abadi F (2011) Integrated population models: a novel analysis framework for deeper insights into population dynamics. J Ornithol 152:S227–S237

Schaub M, Kéry M (2022) Integrated population models: theory and ecological applications with R and JAGS. Academic Press

Schaub M, Jakober H, Stauber W (2013) Strong contribution of immigration to local population regulation: evidence from a migratory passerine. Ecology 94:1828–1838

Schaub M, von Hirschheydt J, Grüebler MU (2015) Differential contribution of demographic rate synchrony to population synchrony in barn swallows. J Anim Ecol 84:1530–1541

Schwarz CJ, Stobo WT (1999) Estimation and effects of tag-misread rates in capture-recapture studies. Can J Fish Aquatic Sci 56:551–559

Schwarz CJ, Schweigert JF, Arnason AN (1993) Estimating migration rates using tag-recovery data. Biometrics 49:177–193

Seber GAF (1965) A note on the multiple-recapture census. Biometrika 52:249–260

Tavecchia G, Besbeas P, Coulson T, Morgan BJT, Clutton-Brock TH (2009) Estimating population size and hidden demographic parameters with state-space modeling. Am Naturalist 173:722–733

Wikle CK, Zammit-Mangion A, Cressie N (2019) Spatio-temporal Statistics with R. CRC Press

Zhao Q (2020) On the sampling design of spatially explicit integrated population models. Methods Ecol Evol 11:1207–1220

Zhao Q, Royle JA, Boomer GS (2017) Spatially explicit dynamic N-mixture models. Popul Ecol 59:293–300

Zhao Q, Boomer GS, Royle JA (2019) Integrated modeling predicts shifts in waterbird population dynamics under climate change. Ecography 42:1470–1481

Spatial Capture-Recapture Models 5

5.1 Introduction

In this chapter, we will introduce static spatial capture-recapture (SCR) models (Efford 2004; Borchers and Efford 2008; Royle and Young 2008) and their dynamic extensions (Gardner et al. 2010; Ergon and Gardner 2014; Chandler and Clark 2014). These are truly individual-based models in the sense that they consider the uniqueness of each individual, such as their spatial locations (and thus environment) and characteristics. Furthermore, dynamic SCR models allow interactions between individuals through local density-dependent processes, alongside individual habitat use, to influence demography, and eventually population size.

While both the multistate models (Arnason 1972, 1973) introduced in the last chapter and SCR models in this chapter consider the spatial locations of the captured animals, they work on different spatial resolutions and thus changes in spatial locations have different ecological meanings in these models. In multistate models, sites are considered to be larger than the territory of the study species; therefore, changes in spatial locations (sites) represent dispersal movement (i.e., abandoning the old territory and establishing a new one in a different area). In SCR models, sites are considered to be smaller than the territory; therefore, an individual animal can be at multiple sites in the same year, which simply represents habitat use within its territory. Dispersal movement is considered in dynamic SCR models, but is represented by changes in the latent centroid of the territory across years (Ergon and Gardner 2014, Chandler and Clark 2014; Sutherland et al. 2015; Royle et al. 2016; details see below) rather than locational changes between sites. In addition to movement, dynamic SCR models also allow the inference of survival and recruitment (Gardner et al. 2010; Chandler and Clark 2014). Thus, dynamic SCR models are capable of linking individual habitat use with the BIDE processes (Royle et al. 2013a, 2018).

Supplementary Information The online version contains supplementary material available at https://doi.org/10.1007/978-3-031-64518-1_5.

Q. Zhao, *Bayesian Analysis of Spatially Structured Population Dynamics*, Analysis and Synthesis 253, https://doi.org/10.1007/978-3-031-64518-1_5

Next, we will introduce a series of SCR models with increasing ecological realism. We will start this chapter by first introducing the static SCR model (also called the single-season SCR model; Efford 2004; Royle and Young 2008) which allows the inference of habitat use without considering the BIDE processes. We then will introduce several dynamic SCR models (also called multi-season SCR models or open population SCR models) by allowing survival, recruitment, and movement (i.e., BIDE) across years (Gardner et al. 2010; Ergon and Gardner 2014; Chandler and Clark 2014). Eventually, we will introduce an advanced dynamic SCR model that considers unique genetic information of each individual. We will end this chapter with a brief summary of other extensions of SCR models.

5.2 Static Spatial Capture-Recapture Models

The static or single-season SCR model first assumes that the amount of time that an animal uses a given site is driven by the distance between this site and a latent centroid of the animal's territory as well as environmental conditions at the site (Efford 2004; Royle and Young 2008; Royle et al. 2013a). It is considered static because we only consider a time period (e.g., a breeding season) that is short enough so that the number of animals and their territories do not change. Also note that the model we introduce here is different from classic static SCR models in two ways. First, while classic SCR models assume habitat uses are only related to the distance from the centroid (Efford 2004; Royle and Young 2008), we also consider the effect of environmental conditions on habitat uses (Royle et al. 2013a; Hooten et al. 2024). Second, we adopt a hierarchical form of the model to allow separation of the true ecological processes (i.e., the amount of time of habitat uses) and observations (i.e., the frequency of captures). This is made possible by assuming some kind of replicates in observations. In later sections of dynamic SCR model, we can see the importance of estimating the true amount of time of habitat use because it allows inference of, for example, the effect of local density dependence on demographic rates.

Now, let us imagine a fictitious red panda species that lives in a forest with discrete bamboo patches (and a reminder that red panda, like giant panda, almost exclusively eats bamboo). The animal is more likely to use a bamboo patch if the patch is closer to the centroid of its territory, the patch is larger, or the average temperature at the patch is lower, and vice versa. Statistically speaking, we have

$$\upsilon_{v,i} = exp\left(-1 \times \kappa \times dist_{v,i} + \beta_0 + \beta_1 \times x_{i,1} + \ldots + \beta_{n_{covs}} \times x_{i,n_{covs}}\right), \tag{5.1}$$

in which $\upsilon_{v,\,i}$ represents the true amount of time that individual v uses site i, κ is a decay parameter representing the effect of the distance between the centroid of the territory of individual v and site i, denoted $dist_{v,\,i}$, on the logarithm of $\upsilon_{v,\,i}$, and β's are intercept and slope parameters representing the effects of environmental covariates x's on the logarithm of $\upsilon_{v,\,i}$.

Note that we consider $v = 1, \ldots, n_{indv}$ individuals, some of which do not necessarily exists. Thus, we are using data augmentation here (Royle et al. 2007, 2013b; Lewis et al. 2023) because we cannot know if an individual exists or not if we fail to capture it, and we need to rely on our model to figure that out. The number of individuals that actually exist (i.e., population size) is calculated as $N = \sum_{v=1}^{n_{indv}} z_v$, in which $z_v = 1$ if individual v exists and $z_v = 0$ otherwise; z_v follows a Bernoulli distribution such that

$$z_v \sim Bernoulli(\psi), \tag{5.2}$$

in which ψ is the probability of existence. With data augmentation, parameter $\upsilon_{v,\,i}$ in fact represents some kind of hypothesized habitat use, as it can be none-zero even for individuals who do not exist. However, later we will see that we have a chance to observe an individual only if it exists (i.e., $z_v = 1$), and will never observe an individual if it does not exist (i.e., $z_v = 0$).

Now, we can deploy some camera traps to "capture" (i.e., take photos of) the red panda with the capability of identifying each individual, either from their natural color patterns or tags we put on them earlier. The cameras are deployed in some bamboo patches in the study area during the study season (e.g., May–June), and we assume that multiple cameras are deployed in each bamboo patch simultaneously to allow replicated observations of the same ecological process. Eventually, we count the number of times for which each red panda is captured in each bamboo patch by each camera as our data. A red panda is likely to be captured more times in a bamboo patch if it spends more time in that patch, and vice versa. But of course, a red panda has a chance to be captured only if it exists, and it will never be captured if it does not exist. Even though these cameras are recording the habitat use of the same animal, they can have different chances to capture the animal due to different observation conditions (e.g., angle of the camera). Lastly, as camera traps are deployed only in some, but not all, bamboo patches, we can capture the animals in a bamboo patch only if a camera is deployed there. Using statistical language, we have

$$\left.\begin{array}{r} y_{v,i,j} \sim Poisson\left(\upsilon_{v,i} \times z_v \times p_{i,j} \times o_i\right) \\ p_{i,j} = exp\left(\alpha_1 \times w_{i,j,1} + \ldots + \alpha_{n_{pcvs}} \times w_{i,j,n_{pcvs}}\right) \end{array}\right\}, \tag{5.3}$$

in which $y_{v,\,i,\,j}$ represents the number of times for which individual v is captured at site i by camera j, z_v is the indicator of individual existence, as described above, $p_{i,\,j}$ is the detection rate of camera j at site i, and is a function of observation covariates w's with slope parameters α's, and o_i is another indicator that equals 1 if camera is deployed at site i and 0 otherwise. A Poisson distribution is a reasonable choice here because $\upsilon_{v,\,i}$ is a continuous variable and will not impose an upper limited of the discrete variable $y_{v,\,i,\,j}$.

Next, let us use the following code to simulate data for the static SCR model. We assume a relatively strong decay parameter ($\kappa = 2.5$) to limit the size of territories. We also assume that bamboo patch size and temperature have positive ($\beta_{bamboo} = 0.6$) and negative ($\beta_{temperature} = -0.3$) effects on the true amount of time of habitat use on the log scale, respectively.

```
library(boot) # for using logit and inv.logit functions

set.seed(8)
```

```
# Basic values
nindv <- 100 # number of individuals (with data augmentation)
nsite <- 360 # number of sites
nreps <- 5  # number of within-season replicates
ncovs <- 2  # number of environmental covariates
npcvs <- 4  # number of observational covariates

psi <- 0.8              # probability of existence
kappa <- 2.5             # distance effect on habitat use
beta <- c(2, 0.6, -0.3)       # intercept and slopes for habitat use
alpha <- c(0.8, -0.4, 0.2, -0.1) # slopes for detection rate
lon_min <- -10            # lower bound of longitude or easting
lon_max <- +10            # upper bound of longitude or easting
lat_min <- - 8           # lower bound of latitude or northing
lat_max <- + 8           # upper bound of latitude or northing

# Define functions
crossdist <- function(ss, ll) { # function for calculating distance
 c1 <- complex(real=ss[,1], imaginary=ss[,2])
 c2 <- complex(real=ll[,1], imaginary=ll[,2])
 dist <- outer(c1, c2, function(z1, z2) Mod(z1 - z2))
} # crossdist

# Read in environmental covariates
load('data/covariates.RData')
ll <- cbind(cov$lon, cov$lat) # coordinates of sites

x1 <- (log(cov$size) - mean(log(cov$size))) / sd(log(cov$size)) #
standardized log habitat patch size
x2 <- (cov$temp - mean(cov$temp)) / sd(cov$temp) # standardized
temperature
x <- cbind(x1[,1], x2[,1]) # environmental covariates

# Simulate data
### Process
ss <- matrix(, nindv, 2) # cooridnates of animal territory centroids
ss[,1] <- runif(nindv, lon_min, lon_max)
ss[,2] <- runif(nindv, lat_min, lat_max)

d <- crossdist(ss, ll) # distance between territory centroids and sites
upsilon <- matrix(0, nindv, nsite) # habitat use (i.e., time spent at a
site)
for (v in 1:nindv) {
 upsilon[v,] <- exp(-1 * kappa * d[v,] + cbind(1, x) %*% beta)
} # v
```

```
z <- rbinom(nindv, 1, psi) # indicator of individual existence
N <- sum(z) # population size, i.e., number of individuals that actually
exist

### Spatial capture-recapture data
w <- array(rnorm(nsite * nreps * npcvs, 0, 1), dim=c(nsite, nreps,
npcvs)) # observational covariates

p <- matrix(0, nsite, nreps) # detection rate
for (i in 1:nsite) {
 p[i,] <- exp(w[i,,] %*% alpha)
} # i

o <- ifelse(ll[,1] <= 8 & ll[,1] >= -8 & ll[,2] <= 6 & ll[,2] >= -6, 1, 0) #
indicator of sites with camera traps

y <- array(0, dim=c(nindv, nsite, nreps)) # frequency of capture data
for (v in 1:nindv) {
 for (i in 1:nsite) {
  y[v,i,] <- rpois(nreps, upsilon[v,i] * z[v] * p[i,] * o[i])
 } # i
} # v
```

The above code first allows us to simulate the coordinates of the territory centroids of the animals, which is an $n_{indv} \times 2$ matrix. The first three rows of this matrix look like the following:

	Easting	Northing
Individual 1	−0.6740952	5.758689
Individual 2	−5.8435337	−5.795853
Individual 3	5.9931592	−2.279983

The code then simulates the true amount of time of habitat use for each animal at each site, which is in an $n_{indv} \times n_{site}$ matrix. The first three rows and five columns of this matrix give the amount of time of habitat use of individuals 1–3 at sites 1–5, which look like the following (rounded to three decimal places). We can see a lot of 0.000's in the following table, which is not surprising since each individual's territory only covers a handful of sites.

	Site 1	Site 2	Site 3	Site 4	Site 5
Individual 1	0.000	0.000	0.051	0.000	0.000
Individual 2	0.201	0.000	0.000	0.000	0.000
Individual 3	0.000	0.000	0.000	0.024	0.000

Eventually, the code simulates the number of captures of each animal at each site by each camera, which is in an $n_{indv} \times n_{site} \times n_{reps}$ array. The part of this array for the first three individuals at sites 1–5 by the first two cameras is shown below. The table only contains 0's, meaning that we fail to capture these individuals at these sites, likely due to the limited amount of time they spend there.

Camera 1	Site 1	Site 2	Site 3	Site 4	Site 5
Individual 1	0	0	0	0	0
Individual 2	0	0	0	0	0
Individual 3	0	0	0	0	0

Camera 2	Site 1	Site 2	Site 3	Site 4	Site 5
Individual 1	0	0	0	0	0
Individual 2	0	0	0	0	0
Individual 3	0	0	0	0	0

Next, let us develop a custom MCMC algorithm to fit the model. The interest is to estimate parameters ψ, κ, β, α, the centroids of territories (denoted ss) and N, given data y, x, w, o, and the coordinates of the site (denoted ll).

```
scr_mcmc <- function(y, x, w, o, ll, lon_min, lon_max, lat_min, lat_max,
nmcmc) {

 # Setup variables
 nindv <- dim(y)[1]
 nsite <- dim(y)[2]
 nreps <- dim(y)[3]
 ncovs <- dim(x)[2]
 npcvs <- dim(w)[3]
 ymax <- apply(y, 1:2, max)
 obs01 <- ifelse(apply(y, 1, sum) > 0, 1, 0) # indicate if an individual is
observed at all or not

 psi_save <- rep(0, nmcmc)
 kappa_save <- rep(0, nmcmc)
 beta_save <- matrix(0, ncovs+1, nmcmc)
 alpha_save <- matrix(0, npcvs, nmcmc)
 ss_save <- array(, dim=c(nindv, 2, 2000))
 N_save <- rep(0, 2000)

 # Priors
 log_kappa_mean <- 0
 log_kappa_sd <- 2
```

```
beta_mean <- 0
beta_sd <- 2
alpha_mean <- 0
alpha_sd <- 2

# Starting values
psi <- 0.5
kappa <- 1
beta <- rep(0, ncovs+1)
alpha <- rep(0, npcvs)
ss <- matrix(0, nindv, 2)
ss[,1] <- runif(nindv, lon_min, lon_max)
ss[,2] <- runif(nindv, lat_min, lat_max)
for (v in 1:nindv) {
 if (obs01[v] == 1) {
  ss[v,1] <- mean(ll[which(ymax[v,] > 0), 1])
  ss[v,2] <- mean(ll[which(ymax[v,] > 0), 2])
 }
} # v
d <- crossdist(ss, ll)
upsilon <- matrix(, nindv, nsite)
for (v in 1:nindv) {
 upsilon[v,] <- exp(-1 * kappa * d[v,] + cbind(1, x) %*% beta)
} # v
p <- matrix(1, nsite, nreps)
z <- obs01

# Tuning factors
kappa_tune <- 0.2
beta_tune <- 0.1
alpha_tune <- 0.05
ss_tune <- 1

# Begin MCMC loop
for (k in 1:nmcmc) {
 ### Sample psi
 psi <- rbeta(1, sum(z) + 0.001, sum(1-z) + 1)

 ### Sample kappa
 kappa_star <- exp(rnorm(1, log(kappa), kappa_tune))
 upsilon_star <- matrix(, nindv, nsite)
 for (v in 1:nindv) {
 upsilon_star[v,] <- exp(-1 * kappa_star * d[v,] + cbind(1, x) %*% beta)
 } # v
 prob1 <- prob2 <- rep(0, nindv)
```

```
  for (v in 1:nindv) {
   prob1[v] <- sum(dpois(y[v,,], upsilon_star[v,] * z[v] * p * o,
log=TRUE))
  prob2[v] <- sum(dpois(y[v,,], upsilon    [v,] * z[v] * p * o, log=TRUE))
  } # v
  mh1 <- sum(prob1) +
     dnorm(log(kappa_star), log_kappa_mean, log_kappa_sd, log=TRUE)
  mh2 <- sum(prob2) +
     dnorm(log(kappa    ), log_kappa_mean, log_kappa_sd, log=TRUE)
  mh <- exp(mh1 - mh2)
  if (mh > runif(1)) {
   kappa <- kappa_star
   upsilon <- upsilon_star
  }

  ### Sample beta
  beta_star <- rnorm(ncovs+1, beta, beta_tune)
  upsilon_star <- matrix(, nindv, nsite)
  for (v in 1:nindv) {
  upsilon_star[v,] <- exp(-1 * kappa * d[v,] + cbind(1, x) %*% beta_star)
  } # v
  prob1 <- prob2 <- rep(0, nindv)
  for (v in 1:nindv) {
   prob1[v] <- sum(dpois(y[v,,], upsilon_star[v,] * z[v] * p * o,
log=TRUE))
  prob2[v] <- sum(dpois(y[v,,], upsilon    [v,] * z[v] * p * o, log=TRUE))
  } # v
  mh1 <- sum(prob1) +
     sum(dnorm(beta_star, beta_mean, beta_sd, log=TRUE))
  mh2 <- sum(prob2) +
     sum(dnorm(beta    , beta_mean, beta_sd, log=TRUE))
  mh <- exp(mh1 - mh2)
  if (mh > runif(1)) {
   beta <- beta_star
   upsilon <- upsilon_star
  }

  ### Sample alpha
  alpha_star <- rnorm(npcvs, alpha, alpha_tune)
  p_star <- matrix(, nsite, nreps)
  for (i in 1:nsite) {
   p_star[i,] <- exp(w[i,,] %*% alpha_star)
  } # i
  prob1 <- prob2 <- matrix(0, nindv, nsite)
  prob1 <- prob2 <- rep(0, nindv)
```

```
  for (v in 1:nindv) {
   prob1[v] <- sum(dpois(y[v,,], upsilon[v,] * z[v] * p_star * o,
log=TRUE))
  prob2[v] <- sum(dpois(y[v,,], upsilon[v,] * z[v] * p   * o, log=TRUE))
  } # v
  mh1 <- sum(prob1) +
     sum(dnorm(alpha_star, alpha_mean, alpha_sd, log=TRUE))
  mh2 <- sum(prob2) +
     sum(dnorm(alpha   , alpha_mean, alpha_sd, log=TRUE))
  mh <- exp(mh1 - mh2)
  if (mh > runif(1)) {
   alpha <- alpha_star
   p <- p_star
  }

  ### Sample ss
  ss_star <- matrix(rnorm(nindv*2, ss, ss_tune), nindv, 2)
  d_star <- crossdist(ss_star, ll)
  upsilon_star <- matrix(, nindv, nsite)
  for (v in 1:nindv) {
  upsilon_star[v,] <- exp(-1 * kappa * d_star[v,] + cbind(1, x) %*% beta)
  } # v
  for (v in 1:nindv) {
   if (ss_star[v,1] > lon_min & ss_star[v,1] < lon_max &
     ss_star[v,2] > lat_min & ss_star[v,2] < lat_max) {
    mh1 <- sum(dpois(y[v,,], upsilon_star[v,] * z[v] * p * o, log=TRUE))
    mh2 <- sum(dpois(y[v,,], upsilon    [v,] * z[v] * p * o, log=TRUE))
    mh <- exp(mh1 - mh2)
    if (mh > runif(1)) {
     ss[v,] <- ss_star[v,]
     d[v,] <- d_star[v,]
     upsilon[v,] <- upsilon_star[v,]
    }
   }
  } # v

  ### Sample z
  for (v in 1:nindv) {
   if (obs01[v] == 0) {
    prod_temp <- psi * prod(exp(-1 * upsilon[v,] * p * o))
    psi_temp <- prod_temp / (prod_temp + 1 - psi)
    z[v] <- rbinom(1, 1, psi_temp)
   }
  } # v
```

```
  ### Save samples
  psi_save[k] <- psi
  kappa_save[k] <- kappa
  beta_save[,k] <- beta
  alpha_save[,k] <- alpha
  if (nmcmc > 2000) {
   if (k > nmcmc-2000) {
    ss_save[,,k-nmcmc+2000] <- ss
    N_save[k-nmcmc+2000] <- sum(z)
   }
  }
 } # k

 ### Write output
 list(psi_save=psi_save,
   kappa_save=kappa_save, beta_save=beta_save,
   alpha_save=alpha_save,
   ss_save=ss_save, N_save=N_save)

} # scr_mcmc
```

We then use the following code to implement the algorithm, which takes about 1.3 h to run. Then using code in Appendix 5.1, we can create traceplots for parameters ψ, κ, β, and α, which show good model convergence and consistency between parameter estimates and their true values used in data simulation (Fig. 5.1). However, it is worth noting that it is difficult to get excellent mixing for κ and β in this static and later dynamic SCR models, which warrants further investigation.

```
library(foreach)   # for parallel computing
library(doParallel) # for parallel computing

numCores <- round(detectCores() / 2) # only use half of the cores for
parallel computing
registerDoParallel(numCores) # setup parallel computing

nmcmc <- 50000 # number of iterations of MCMC computing
chain <- 3    # number of chains

start_time <- Sys.time() # start time of computing
out <- foreach (i=1:chain, .packages='boot') %dopar% {
 scr_mcmc(y=y, x=x, w=w, o=o, ll=ll,
     lon_min=lon_min, lon_max=lon_max, lat_min=lat_min,
```

```
lat_max=lat_max,
      nmcmc=nmcmc)
} # i
stopImplicitCluster() # clean up the cluster
end_time <- Sys.time() # end time of computing
time_taken <- end_time - start_time # computing time
```

We can also use the following code to plot the estimated centroids of territories and population size, in comparison to the true values (Fig. 5.2). The location of the estimated centroid is in general consistent with the simulated true location if an individual is captured, but the model has no capability of identifying the location of the centroid if an individual has never been captured. Nonetheless, population size can be correctly estimated.

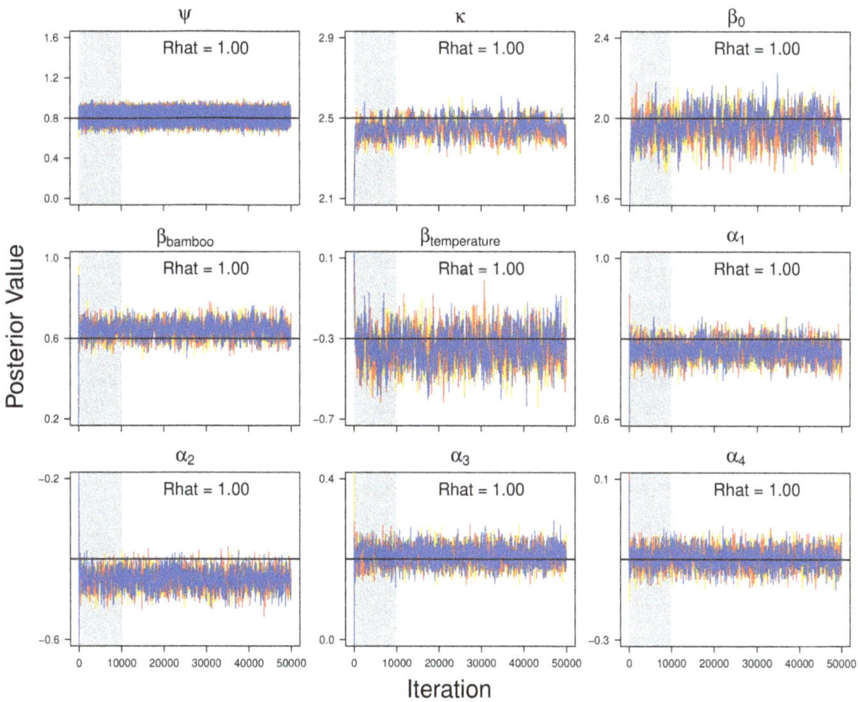

Fig. 5.1 Traceplots showing the posterior chains for ψ, β, κ, and α in the static SCR model. Rhat values are calculated after discarding the first 10,000 iterations. Horizontal lines represent the true values used in data simulation

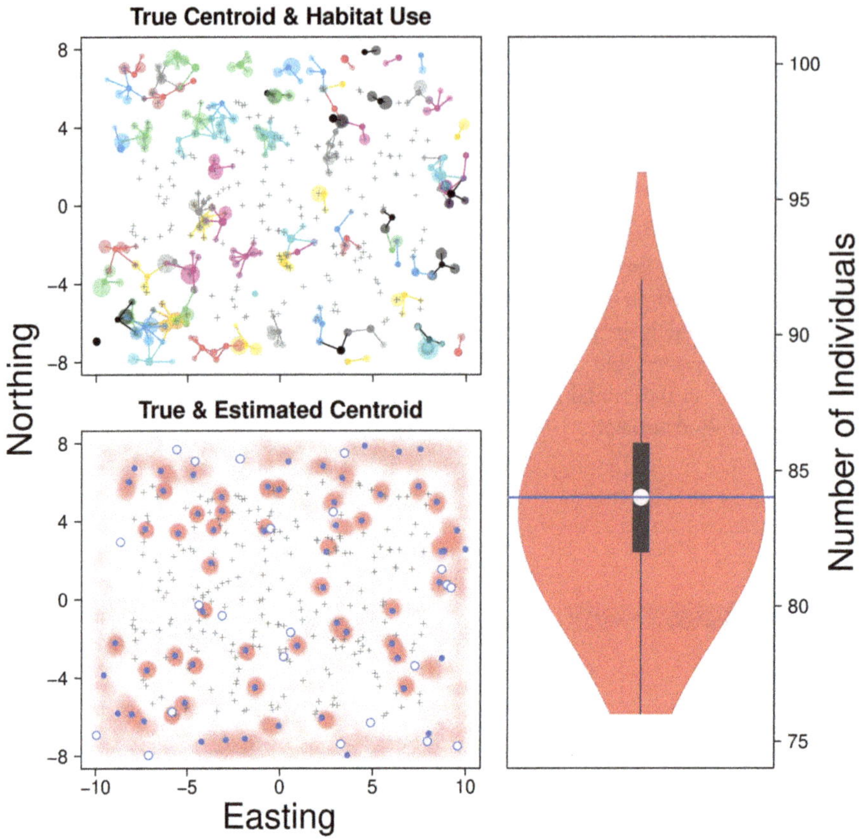

Fig. 5.2 The true centroids of territories (filled point) and amount of time of habitat use (transparent point; upper-left panel), the estimated centroids (pink clouds) in comparison to the true centroids (filled points for captured individuals, unfilled points for uncaptured individuals; lower-left panel), and estimated number of individuals (violin) in comparison to the true value (horizontal line; right panel). Grey plus signs represent the locations of camera traps

```
library(vioplot) # for making violin plots

ss_post <- array(, dim=c(nindv, 2, 6000))
ss_post[,,  1:2000] <- out[[1]]$ss
ss_post[,,2001:4000] <- out[[2]]$ss
ss_post[,,4001:6000] <- out[[3]]$ss
ss_sel <- ss_post[,,sample(1:6000, 100, replace=F)]
N_post <- c(out[[1]]$N, out[[2]]$N, out[[3]]$N)

layout(mat=matrix(c(rep(c(1,1,2,2),times=3),rep(3,8)), nrow=4,
ncol=5, byrow=FALSE))
```

```
par(mar=c(1,2,3,1))
par(oma=c(4,4,0,6))

plot(1, xlim=c(-10,10), ylim=c(-8,8), axes=F, xlab='', ylab='',
type='n')
for (v in 1:N) {
 col <- col2rgb(v)
 for (i in 1:nsite) {
  if (upsilon[v,i] > 0.5) {
   points(x=ll[i,1], y=ll[i,2], pch=16, cex=sqrt(upsilon[v,i])*1.4,
       col=rgb(col[1],col[2],col[3],alpha=120,maxColorValue=255))
   lines(x=c(ss[v,1],ll[i,1]), y=c(ss[v,2],ll[i,2]), lwd=1, col=v)
  }
 } # i
 points(ss[v,1], ss[v,2], cex=ifelse(sum(y[v,,])==0,1.6,1.2),
pch=16, col=v)
} # v
points(ll[which(o==1),], lwd=0.5, pch=3, col='grey50', cex=0.8)
axis(1, at=seq(-10,10,5), labels=rep('',5))
axis(2, at=seq(-08,08,4), cex.axis=1.6, las=2)
title(main='True Centroid & Habitat Use', cex.main=2)
box()

plot(1, xlim=c(-10,10), ylim=c(-8,8), axes=F, xlab='', ylab='',
type='n')
for (v in 1:N) {
 points(t(ss_sel[v,,]), pch=16, cex=3, col=rgb(240,128,128,alpha=6,
maxColorValue=255))
} # v
for (v in 1:N) {
 points(ss[v,1], ss[v,2], cex=ifelse(sum(y[v,,])==0,1.6,1.2),
pch=ifelse(sum(y[v,,])==0,21,16), col='royalblue', bg='white')
} # v
points(ll[which(o==1),], lwd=0.5, pch=3, col='grey50', cex=0.8)
axis(1, at=seq(-10,10,5), cex.axis=1.6)
axis(2, at=seq(-08,08,4), cex.axis=1.6, las=2)
title(main='True & Estimated Centroid', cex.main=2)
box()

axis(1, at=0, labels='Easting', tick=F, line=2.5, cex.axis=3)
title(ylab='Northing', cex.lab=3, line=1.5, outer=T)

par(mar=c(1,1,3,1))
plot(1, ylim=c(75, 100), type='n', axes=F, xlab='', ylab='')
vioplot(N_post, col='lightcoral', rectCol='grey36',
```

```
lineCol='grey36', border=NA, cex=3, add=T)
abline(h=N, col='royalblue', lwd=1.8)
axis(4, at=seq(75, 100, 5), cex.axis=1.6, las=2)
axis(4, at=87.5, label='Number of Individuals', line=4, cex.axis=3,
tick=F)
```

5.3 Dynamic Spatial Capture-Recapture Models

While static SCR models can provide information for habitat use, they do not allow changes in population size and the underlying BIDE processes over the years. Dynamic SCR models extend static SCR models to allow the number of animals in the study area to change across years due to mortality and recruitment (e.g., Gardner et al. 2010). Dynamic SCR models can also allow movement represented by the changes in the territory centroids (Ergon and Gardner 2014; Chandler and Clark 2014; Sutherland et al. 2015; Royle et al. 2016; Efford and Schofield 2020; Gowan et al. 2021). Therefore, dynamic SCR models are useful for understanding the BIDE processes and their linkage to habitat use. Next, we will introduce a series of dynamic SCR models. To provide a friendly learning curve, we will one by one add survival, recruitment, and movement into dynamic SCR models.

Dynamic SCR models describe the amount of time of habitat use by individual animals in a similar way as static SCR models, although now it is year specific:

$$\left.\upsilon_{v,i,t} = exp\left(-1 \times \kappa \times dist_{v,i,t} + \beta_0^{[\upsilon]} + \beta_1^{[\upsilon]} \times x_{i,t,1} + \ldots + \beta_{n_{covs}}^{[\upsilon]} \times x_{i,t,n_{covs}} \right)\right\}, \quad (5.4)$$

in which $\upsilon_{v,\,i,\,t}$ represents the true amount of time of individual v using site i in year t, $dist_{v,\,i,\,t}$ is the distance between the centroid of the individual v's territory and site i, κ is a decay parameter representing the effect of this distance on the logarithm of $\upsilon_{v,\,i,\,t}$, and β's are intercept and slope parameters representing the effects of environmental covariates x's on the logarithm of $\upsilon_{v,\,i,\,t}$.

Dynamic SCR models then describe the observation process as such:

$$\left.\begin{array}{r} y_{v,i,t,j} \sim Poisson\left(\upsilon_{v,i,t} \times z_{v,t} \times p_{i,t,j} \times o_i\right) \\ p_{i,t,j} = exp\left(\alpha_1 \times w_{i,t,j,1} + \ldots + \alpha_{n_{pcvs}} \times w_{i,t,j,n_{pcvs}}\right) \end{array}\right\}, \quad (5.5)$$

in which $y_{v,\,i,\,t,\,j}$ represents the number of times for which individual v is captured at site i in year t by the jth camera, $z_{v,\,t}$ is an indicator that equals 1 if individual v is alive in year t and 0 otherwise, $p_{i,\,t,\,j}$ is the detection rate and is a function of observation covariates w's with slope parameters α's a, and o_i is an indicator that equals 1 if camera is deployed at site i and 0 otherwise.

The models in Sects. 5.3.1, 5.3.2, and 5.3.3 differ in their ways of describing how $z_{v,\,t}$ changes over time. Also, $dist_{v,\,i,\,t}$ does not change over time (i.e., $dist_{v,\,i,\,t} = dist_{v,\,i}$

for any t) in Sects. 5.3.1 and 5.3.2 since movement (i.e., change in territory centroid) is not allowed in these sections; movement will be considered in Sect. 5.3.3.

5.3.1 With Survival

Next, we will introduce a dynamic SCR model that allows survival, but not recruitment or movement over the years. Without recruitment, the population will only become smaller and smaller, and without movement, the locations of territory centroids will not change over time. While this model surely is unrealistic, it can be a good start for us to learn to build dynamic SCR models. We will soon increase ecological realism by allowing recruitment and movement in Sects. 5.3.2 and 5.3.3.

The dynamic SCR model with survival describes the changes in $z_{v,\,t}$ due to local density- and environment-dependent survival processes (Milleret et al. 2023). Now, let us use our red panda as an example. The model assumes that the animal has a good chance to survive if its territory contains bamboo patches that are infrequently used by other individuals, which are large, or cold. Statistically speaking, we have

$$
\begin{aligned}
z_{v,t=1} &\sim Bernoulli(\psi)\\
z_{v,t\geq 2} &\sim Bernoulli\left(z_{v,t-1}\times \phi_{v,t-1}\right)\\
\phi_{v,t} &= logit^{-1}\left[\beta_0^{[\phi]} + \beta_{density}^{[\phi]} \times \frac{\sum_{i=1}^{n_{site}} \sum \left(\upsilon_{-v,i,t}\times z_{-v,t}\right)\times \upsilon_{v,i,t}}{\sum_{i=1}^{n_{site}} \upsilon_{v,i,t}}\right.\\
&\quad \left. +\beta_1^{[\phi]}\times \frac{\sum_{i=1}^{n_{site}} x_{i,t-1,1}\times \upsilon_{v,i,t}}{\sum_{i=1}^{n_{site}} \upsilon_{v,i,t}} + \ldots\right]\},
\end{aligned}
\tag{5.6}
$$

in which ψ is the probability of existence in the first year, $\phi_{v,\,t\,-\,1}$ is the survival probability of individual v in year $t-1$, and $\beta^{[\phi]}$'s are the intercept and slope parameters governing the effects of density and environmental covariates on survival. By using ψ instead of assuming all individuals exist in the first year, we have data augmentation here, which means some of the individuals will never exist throughout the study period.

Here, we calculate the population density in the territory of individual v as the weighted mean of the total amount of time of habitat use by any individuals other than v (i.e., $\sum(\upsilon_{-v,\,i,\,t}\times z_{-v,\,t})$), given that the individual is alive (i.e., $z_{-v,\,t}=1$), using the amount of time of habitat use by individual v, $\upsilon_{v,\,i,\,t}$, as the weight. In this way, frequent habitat use in the territory of individual v by other individuals can lead to a low survival probability of v through competition, assuming that $\beta_D^{[\phi]}$ is negative. Also, as mentioned above, $\upsilon_{-v,\,i,\,t}$ is a hypothesized amount of time of habitat use, which can be non-zero even when an individual does not exist. However, we make sure an individual can influence other individuals only if it exists by multiplying $z_{-v,\,t}$. Similarly, environmental conditions in the territory of individual v are calculated as the weighted means of these environmental covariates with $\upsilon_{v,\,i,\,t}$ as the weight, meaning that individuals that set their territories in suitable habitats (e.g., large bamboo patches) will have a high survival probability.

Next, we can use the following code to simulate data for the dynamic SCR model with survival. We assume a relatively strong decay parameter ($\kappa = 2.5$), a positive effect of bamboo patch size ($\beta^{[\upsilon]}_{bamboo} = 0.6$) and a negative effect of temperature ($\beta^{[\upsilon]}_{temperature} = -0.3$) on the true amount of time of habitat use on the log scale. We assume a negative density dependence ($\beta^{[\phi]}_{density} = -0.4$), a positive effect of bamboo patch size ($\beta^{[\phi]}_{bamboo} = 0.6$), and a negative effect of temperature ($\beta^{[\phi]}_{temperature} = -0.2$) on survival on the logit scale.

```
library(boot) # for using logit and inv.logit functions

set.seed(8)

# Basic values
nindv <- 100 # number of individuals (with data augmentation)
nsite <- 360 # number of sites
nyear <- 6  # number of years
nreps <- 5  # number of within-season replicates
ncovs <- 2  # number of environmental covariates
npcvs <- 3  # number of observational covariates

psi <- 0.8                # probability of first-year existence
kappa <- 2.5               # distance effect on habitat use
beta_upsilon <- c(2, 0.6, -0.3)  # intercept and slopes for habitat use
beta_phi <- c(1, -0.4, 0.6, -0.2) # intercept and slopes for apparent
survival
alpha <- c(-0.6, 0.4, -0.2)     # slopes for detection rate
lon_min <- -10              # lower bound of longitude or easting
lon_max <- +10              # upper bound of longitude or easting
lat_min <- - 8             # lower bound of latitude or northing
lat_max <- + 8             # upper bound of latitude or northing

# Define functions
crossdist <- function(ss, ll) { # function for calculating distance
 c1 <- complex(real=ss[,1], imaginary=ss[,2])
 c2 <- complex(real=ll[,1], imaginary=ll[,2])
 dist <- outer(c1, c2, function(z1, z2) Mod(z1 - z2))
} # crossdist

# Read in environmental covariates
load('data/covariates.RData')
ll <- cbind(cov$lon, cov$lat) # coordinates of sites

x1 <- (log(cov$size) - mean(log(cov$size))) / sd(log(cov$size)) #
standardized log habitat patch size
```

```
x2 <- (cov$temp - mean(cov$temp)) / sd(cov$temp) # standardized
temperature
x <- array(, dim=c(nsite, nyear, ncovs)) # environmental covariates
x[,,1] <- x1[,1:nyear]
x[,,2] <- x2[,1:nyear]

# Simulate data
### Process
ss <- matrix(, nindv, 2) # cooridnates of animal territory centroids
ss[,1] <- runif(nindv, lon_min, lon_max)
ss[,2] <- runif(nindv, lat_min, lat_max)

d <- crossdist(ss, ll) # distance between territory centroids and sites

upsilon <- array(0, dim=c(nindv, nsite, nyear)) # habitat use (i.e.,
time spent at a site)
for (v in 1:nindv) {
 for (t in 1:nyear) {
  upsilon[v,,t] <- exp(-1 * kappa * d[v,] + cbind(1, x[,t,]) %*%
beta_upsilon)
 } # t
} # v

z <- matrix(0, nindv, nyear) # indicator of individual existence
phi <- matrix(0, nindv, nyear-1) # apparent survival probability
z[,1] <- rbinom(nindv, 1, psi) # individual existence in the first year
for (v in 1:nindv) {
 for (t in 2:nyear) {
  phi[v,t-1] <- inv.logit(c(1, sum(upsilon[v,,t-1] * colSums(upsilon
[-v,,t-1] * z[-v,t-1])) / sum(upsilon[v,,t-1]), colSums(upsilon[v,,t-
1] * x[,t-1,]) / sum(upsilon[v,,t-1])) %*% beta_phi)
  z[v,t] <- rbinom(1, 1, z[v,t-1] * phi[v,t-1])
 } # t
} # v

### Spatial capture-recapture data
w <- array(rnorm(nsite * nyear * nreps * npcvs, 0, 1), dim=c(nsite,
nyear, nreps, npcvs)) # observational covariates

p <- array(0, dim=c(nsite, nyear, nreps)) # detection rate
for (i in 1:nsite) {
 for (t in 1:nyear) {
  p[i,t,] <- exp(w[i,t,,] %*% alpha)
 } # t
} # i
```

```
o <- ifelse(ll[,1] <= 8 & ll[,1] >= -8 & ll[,2] <= 6 & ll[,2] >= -6, 1, 0) #
indicator of sites with camera traps

y <- array(0, dim=c(nindv, nsite, nyear, nreps)) # frequency of capture
data
for (v in 1:nindv) {
 for (i in 1:nsite) {
  for (t in 1:nyear) {
   y[v,i,t,] <- rpois(nreps, upsilon[v,i,t] * z[v,t] * p[i,t,] * o[i])
  } # t
 } # i
} # v
```

The above code allows us to simulate the coordinates of territory centroids as an $n_{indv} \times 2$ matrix. The code also simulates the true amount of time of habitat use as an $n_{indv} \times n_{site} \times n_{year}$ array, the survival status as an $n_{indv} \times n_{year}$ matrix, and eventually the number of captures as an $n_{indv} \times n_{site} \times n_{year} \times n_{reps}$ array, all are similar to what we see in Sect. 5.2 but with an additional dimension of year.

Next, we will develop a custom MCMC algorithm to fit the model. The interest is to estimate parameters ψ, κ, $\beta^{[\upsilon]}$, $\beta^{[\phi]}$, α, and z given data y, x, w, o, and ll.

```
scr_survival_mcmc <- function(y, x, w, o, ll, lon_min, lon_max,
lat_min, lat_max, nmcmc) {

 # Setup variables
 nindv <- dim(y)[1]
 nsite <- dim(y)[2]
 nyear <- dim(y)[3]
 nreps <- dim(y)[4]
 ncovs <- dim(x)[3]
 npcvs <- dim(w)[4]
 ymax <- apply(y, 1:3, max)
 obs01 <- ifelse(apply(y, c(1,3), sum) > 0, 1, 0) # indicate if an
individual is observed at all or not

 psi_save <- rep(0, nmcmc)
 kappa_save <- rep(0, nmcmc)
 beta_upsilon_save <- matrix(0, ncovs+1, nmcmc)
 beta_phi_save <- matrix(0, ncovs+2, nmcmc)
 alpha_save <- matrix(0, npcvs, nmcmc)
 z_save <- array(, dim=c(nindv, nyear, 2000))

 # Priors
 log_kappa_mean <- 0
```

```
log_kappa_sd <- 2
beta_upsilon_mean <- 0
beta_upsilon_sd <- 2
beta_phi_mean <- 0
beta_phi_sd <- 2
alpha_mean <- 0
alpha_sd <- 2

# Starting values
psi <- 0.5
kappa <- 1
beta_upsilon <- rep(0, ncovs+1)
beta_phi <- rep(0, ncovs+2)
alpha <- rep(0, npcvs)
ss <- matrix(0, nindv, 2)
ss[,1] <- runif(nindv, lon_min, lon_max)
ss[,2] <- runif(nindv, lat_min, lat_max)
for (v in 1:nindv) {
 if (sum(obs01[v,]) > 0) {
  ss[v,1] <- mean(ll[which(rowSums(ymax[v,,]) > 0), 1])
  ss[v,2] <- mean(ll[which(rowSums(ymax[v,,]) > 0), 2])
 }
} # v
d <- crossdist(ss, ll)
upsilon <- array(0, dim=c(nindv, nsite, nyear))
for (v in 1:nindv) {
 for (t in 1:nyear) {
  upsilon[v,,t] <- exp(-1 * kappa * d[v,] + cbind(1, x[,t,]) %*%
beta_upsilon)
 } # t
} # v
phi <- matrix(0.5, nindv, nyear-1)
p <- array(0.5, dim=c(nsite, nyear, nreps))
zi <- obs01
for (v in 1:nindv) {
 if (sum(zi[v,]) > 0) {
  zi[v,1:max(which(zi[v,] == 1))] <- 1
 }
} # v
z <- zi

# Tuning factors
kappa_tune <- 0.15
beta_upsilon_tune <- 0.05
beta_phi_tune <- 0.25
```

```
 alpha_tune <- 0.05
 ss_tune <- 1

 # Begin MCMC loop
 for (k in 1:nmcmc) {
  ### Sample psi
  psi <- rbeta(1, sum(z[,1]) + 0.001, sum(1-z[,1]) + 1)

  ### Sample kappa
  kappa_star <- exp(rnorm(1, log(kappa), kappa_tune))
  upsilon_star <- array(0, dim=c(nindv, nsite, nyear))
  for (v in 1:nindv) {
   for (t in 1:nyear) {
    upsilon_star[v,,t] <- exp(-1 * kappa_star * d[v,] + cbind(1, x[,t,])
%*% beta_upsilon)
   } # t
  } # v
  prob1 <- prob2 <- matrix(0, nindv, nyear)
  for (v in 1:nindv) {
   for (t in 1:nyear) {
    prob1[v,t] <- sum(dpois(y[v,,t,], upsilon_star[v,,t] * z[v,t] * p[,
t,] * o, log=TRUE))
    prob2[v,t] <- sum(dpois(y[v,,t,], upsilon     [v,,t] * z[v,t] * p[,t,]
* o, log=TRUE))
   } # t
  } # v
  mh1 <- sum(prob1) +
     dnorm(log(kappa_star), log_kappa_mean, log_kappa_sd, log=TRUE)
  mh2 <- sum(prob2) +
     dnorm(log(kappa     ), log_kappa_mean, log_kappa_sd, log=TRUE)
  mh <- exp(mh1 - mh2)
  if (mh > runif(1)) {
   kappa <- kappa_star
   upsilon <- upsilon_star
  }

  ### Sample beta_upsilon
  beta_upsilon_star <- rnorm(ncovs+1, beta_upsilon,
beta_upsilon_tune)
  upsilon_star <- array(0, dim=c(nindv, nsite, nyear))
  for (v in 1:nindv) {
   for (t in 1:nyear) {
    upsilon_star[v,,t] <- exp(-1 * kappa * d[v,] + cbind(1, x[,t,]) %*%
beta_upsilon_star)
   } # t
```

```
  } # v
  prob1 <- prob2 <- matrix(0, nindv, nyear)
  for (v in 1:nindv) {
   for (t in 1:nyear) {
    prob1[v,t] <- sum(dpois(y[v,,t,], upsilon_star[v,,t] * z[v,t] * p[,
t,] * o, log=TRUE))
    prob2[v,t] <- sum(dpois(y[v,,t,], upsilon     [v,,t] * z[v,t] * p[,t,]
* o, log=TRUE))
   } # t
  } # v
  mh1 <- sum(prob1) +
      sum(dnorm(beta_upsilon_star, beta_upsilon_mean,
beta_upsilon_sd, log=TRUE))
  mh2 <- sum(prob2) +
      sum(dnorm(beta_upsilon     , beta_upsilon_mean, beta_upsilon_sd,
log=TRUE))
  mhB <- exp(mh1 - mh2)
  if (mhB > runif(1)) {
   beta_upsilon <- beta_upsilon_star
   upsilon <- upsilon_star
  }

  ### Sample beta_phi
  beta_phi_star <- rnorm(ncovs+2, beta_phi, beta_phi_tune)
  phi_star <- matrix(0, nindv, nyear-1)
  for (v in 1:nindv) {
   for (t in 2:nyear) {
    phi_star[v,t-1] <- inv.logit(c(1, sum(upsilon[v,,t-1] * colSums
(upsilon[-v,,t-1] * z[-v,t-1])) / sum(upsilon[v,,t-1]), colSums
(upsilon[v,,t-1] * x[,t-1,]) / sum(upsilon[v,,t-1])) %*%
beta_phi_star)
   } # t
  } # v
  mh1 <- sum(dbinom(z[,-1], 1, z[,-nyear] * phi_star, log=TRUE)) +
     sum(dnorm(beta_phi_star, beta_phi_mean, beta_phi_sd, log=TRUE))
  mh2 <- sum(dbinom(z[,-1], 1, z[,-nyear] * phi     , log=TRUE)) +
      sum(dnorm(beta_phi     , beta_phi_mean, beta_phi_sd, log=TRUE))
  mhP <- exp(mh1 - mh2)
  if (mhP > runif(1)) {
   beta_phi <- beta_phi_star
   phi <- phi_star
  }

  ### Sample alpha
  alpha_star <- rnorm(npcvs, alpha, alpha_tune)
```

```
  p_star <- array(, dim=c(nsite, nyear, nreps))
  for (i in 1:nsite) {
   for (t in 1:nyear) {
    p_star[i,t,] <- exp(w[i,t,,] %*% alpha_star)
   } # t
  } # i
  prob1 <- prob2 <- matrix(0, nindv, nyear)
  for (v in 1:nindv) {
   for (t in 1:nyear) {
    prob1[v,t] <- sum(dpois(y[v,,t,], upsilon[v,,t] * z[v,t] * p_star[,
t,] * o, log=TRUE))
    prob2[v,t] <- sum(dpois(y[v,,t,], upsilon[v,,t] * z[v,t] * p   [,t,]
* o, log=TRUE))
   } # t
  } # v
  mh1 <- sum(prob1) +
      sum(dnorm(alpha_star, alpha_mean, alpha_sd, log=TRUE))
  mh2 <- sum(prob2) +
      sum(dnorm(alpha   , alpha_mean, alpha_sd, log=TRUE))
  mhA <- exp(mh1 - mh2)
  if (mhA > runif(1)) {
   alpha <- alpha_star
   p <- p_star
  }

  ### Sample ss
  ss_star <- matrix(rnorm(nindv*2, ss, ss_tune), nindv, 2)
  d_star <- crossdist(ss_star, ll)
  upsilon_star <- array(0, dim=c(nindv, nsite, nyear))
  for (v in 1:nindv) {
   for (t in 1:nyear) {
    upsilon_star[v,,t] <- exp(-1 * kappa * d_star[v,] + cbind(1, x[,t,])
%*% beta_upsilon)
   } # t
  } # v
  for (v in 1:nindv) {
   if (ss_star[v,1] > lon_min & ss_star[v,1] < lon_max &
     ss_star[v,2] > lat_min & ss_star[v,2] < lat_max) {
    prob1 <- prob2 <- rep(0, nyear)
    for (t in 1:nyear) {
     prob1[t] <- sum(dpois(y[v,,t,], upsilon_star[v,,t] * z[v,t] * p[,
t,] * o, log=TRUE))
    prob2[t] <- sum(dpois(y[v,,t,], upsilon   [v,,t] * z[v,t] * p[,t,] *
o, log=TRUE))
    } # t
```

```
      mh1 <- sum(prob1)
      mh2 <- sum(prob2)
      mhS <- exp(mh1 - mh2)
      if (mhS > runif(1)) {
       ss[v,] <- ss_star[v,]
       d[v,] <- d_star[v,]
       upsilon[v,,] <- upsilon_star[v,,]
      }
     }
    } # v

    ### Sample z
    for (v in 1:nindv) {
     if (zi[v,1] == 0) {
      prod_temp <- psi * prod(exp(-1 * upsilon[v,,1] * p[,1,] * o))
      psi_temp <- prod_temp / (prod_temp + 1 - psi)
      z[v,1] <- rbinom(1, 1, psi_temp)
     }
     for (t in 2:nyear) {
      if (z[v,t-1] == 0 & zi[v,t] == 0) {
       z[v,t] <- 0
      } else if (z[v,t-1] == 1 & zi[v,t] == 0) {
      prod_temp <- phi[v,t-1] * prod(exp(-1 * upsilon[v,,t] * p[,t,] * o))
       psi_temp <- prod_temp / (prod_temp + 1 - phi[v,t-1])
       z[v,t] <- rbinom(1, 1, psi_temp)
      }
     } # t
    } # v

    ### Save samples
    psi_save[k] <- psi
    kappa_save[k] <- kappa
    beta_upsilon_save[,k] <- beta_upsilon
    beta_phi_save[,k] <- beta_phi
    alpha_save[,k] <- alpha
    if (nmcmc > 2000) {
     if (k > nmcmc-2000) {
      z_save[,,k-nmcmc+2000] <- z
     }
    }
   } # k
```

```
### Write output
list(psi_save=psi_save,
   kappa_save=kappa_save, beta_upsilon_save=beta_upsilon_save,
   beta_phi_save=beta_phi_save,
   alpha_save=alpha_save,
   z_save=z_save)

} # scr_survival_mcmc
```

We then use the following code to implement the algorithm, which takes about 11 h to run. Then using code in Appendix 5.2, we can create traceplots for the monitored parameters, which show good model convergence and consistency between parameter estimates of ψ, κ, $\beta^{[\upsilon]}$, $\beta^{[\phi]}$, and α and their true values used in data simulation (Fig. 5.3).

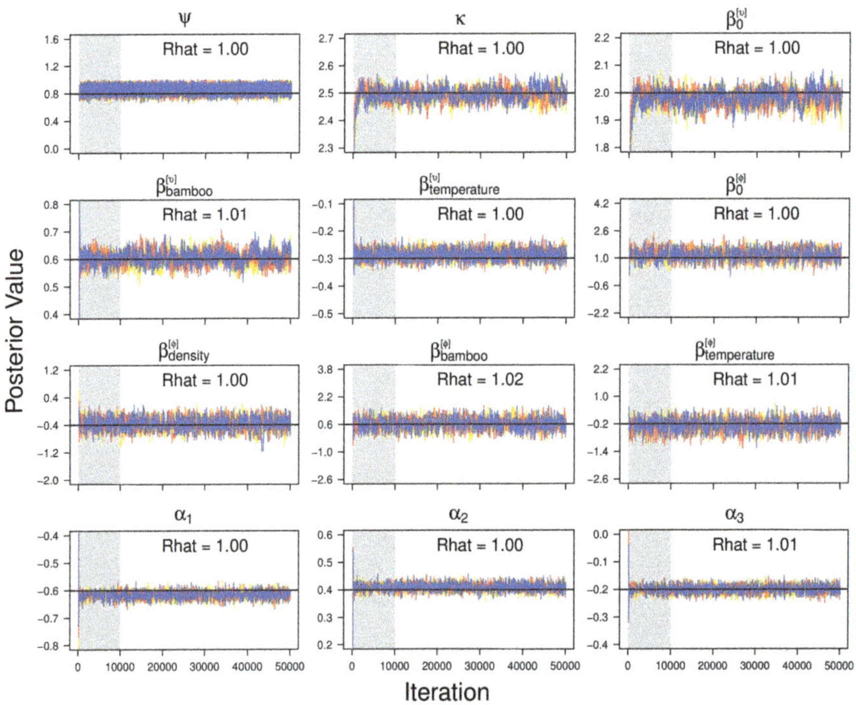

Fig. 5.3 Traceplots showing the posterior chains for ψ, κ, $\beta^{[\upsilon]}$, $\beta^{[\phi]}$, and α in the dynamic SCR model with survival. Rhat values are calculated after discarding the first 10,000 iterations. Horizontal lines represent the true values used in data simulation

```
library(foreach)   # for parallel computing
library(doParallel) # for parallel computing

numCores <- round(detectCores() / 2) # only use half of the cores for
parallel computing
registerDoParallel(numCores) # setup parallel computing

nmcmc <- 50000 # number of iterations of MCMC computing
chain <- 3   # number of chains

start_time <- Sys.time() # start time of computing
out <- foreach (i=1:chain, .packages='boot') %dopar% {
 scr_survival_mcmc(y=y, x=x, w=w, o=o, ll=ll,
         lon_min=lon_min, lon_max=lon_max, lat_min=lat_min,
lat_max=lat_max,
         nmcmc=nmcmc)
} # i
stopImplicitCluster() # clean up the cluster
end_time <- Sys.time() # end time of computing
time_taken <- end_time - start_time # computing time
```

Next, let us use the following code to plot the estimated survival history, which shows a satisfying agreement with the simulated true values if an individual is observed; by contrast, for unobserved individuals, including individuals that never exist, the model has a hard time to figure out which of them are actually alive (Fig. 5.4).

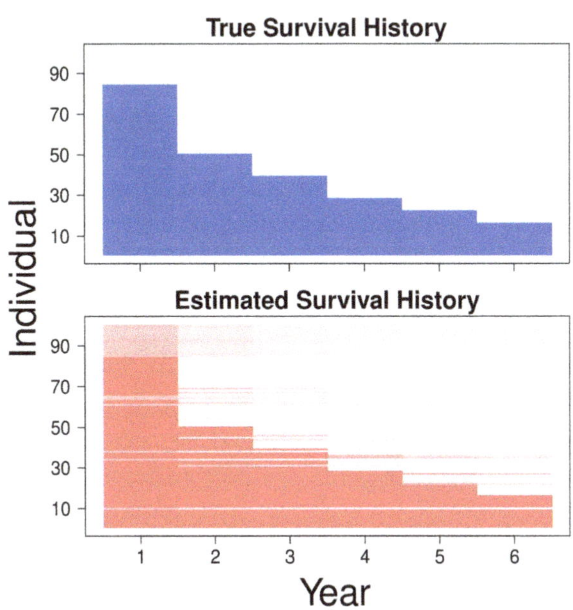

Fig. 5.4 The true (upper panel) and estimated (lower panel) survival history for each individual during the study period in the dynamic SCR model with survival. Individuals that never exist are put at the top of the panels

```
z_post <- array(, dim=c(nindv, nyear, 6000))
z_post[,,  1:2000] <- out[[1]]$z
z_post[,,2001:4000] <- out[[2]]$z
z_post[,,4001:6000] <- out[[3]]$z
z_mean <- apply(z_post, 1:2, mean)
nscale <- 100
z_scale <- round(z_mean * (nscale - 1)) + 1
col_scale <- colorRampPalette(c('white','lightcoral'))(nscale)
z_order <- z[rev(order(rowSums(z))),]
z_scale_order <- z_scale[rev(order(rowSums(z))),]

par(mfrow=c(2,1))
par(mar=c(1,0,2.5,1))
par(oma=c(4,5.5,0,0))

plot(1, xlim=c(.5,nyear+.5), ylim=c(.5,nindv+.5), type='n', axes=F)
for (i in 1:nindv) {
 for (t in 1:nyear) {
  rect(t-.5,i-.5,t+.5,i+.5, col=ifelse(z_order[i,t]==1,
'royalblue', 'white'), border=NA)
 } # t
} # i
axis(1, at=seq(1,nyear,1), labels=rep('',nyear))
axis(2, at=seq(10,nindv,20), cex.axis=1.5, las=2)
box()
title(main='True Survival History', line=.5, cex.main=2)

plot(1, xlim=c(.5,nyear+.5), ylim=c(.5,nindv+.5), type='n', axes=F)
for (i in 1:nindv) {
 for (t in 1:nyear) {
  rect(t-.5,i-.5,t+.5,i+.5, col=col_scale[z_scale_order[i,t]],
border=NA)
 } # t
} # i
axis(1, at=seq(1,nyear,1), cex.axis=1.5)
axis(2, at=seq(10,nindv,20), cex.axis=1.5, las=2)
box()
title(main='Estimated Survival History', line=.5, cex.main=2)

title(ylab='Individual', cex.lab=3, line=3.2, outer=T)
title(xlab='Year', cex.lab=3, line=2.8, outer=T)
```

5.3.2 With Survival and Recruitment

Next, we will allow both survival and recruitment in the dynamic SCR model. In this way, the population will have a chance to increase, making the model more realistic. We still do not allow movement, meaning that the locations of territory centroids will not change over time. The dynamic SCR model with survival and recruitment describes the changes in $z_{v,\ t}$ due to density- and environment-dependent survival and recruitment processes. That is to say, for our red panda, an animal is likely to survive if its territory is infrequently used by others, contains large bamboo patches, or is cold, given that it is already in the population, or is likely to be recruited (be born or move) to an area that is infrequently used by others, contains large bamboo patches, or is cold, given that it is not in the population yet. Using statistical language, we have:

$$\left.\begin{array}{c} z_{v,t=1} \sim Bernoulli(\psi) \\ z_{v,t\geq 2} \sim Bernoulli\left[z_{v,t-1} \times \phi_{v,t-1} + (1-e_{v,t-1}) \times \gamma_{v,t-1}\right] \\ \phi_{v,t} = logit^{-1}\left[\beta_0^{[\phi]} + \beta_{density}^{[\phi]} \times \frac{\sum_{i=1}^{n_{site}} \sum (\upsilon_{-v,i,t} \times z_{-v,t}) \times \upsilon_{v,i,t}}{\sum_{i=1}^{n_{site}} \upsilon_{v,i,t}} + \beta_1^{[\phi]} \times \frac{\sum_{i=1}^{n_{site}} x_{i,t-1,1} \times \upsilon_{v,i,t}}{\sum_{i=1}^{n_{site}} \upsilon_{v,i,t}} + \ldots\right] \\ \gamma_{v,t} = logit^{-1}\left[\beta_0^{[\gamma]} + \beta_{density}^{[\gamma]} \times \frac{\sum_{i=1}^{n_{site}} \sum (\upsilon_{-v,i,t} \times z_{-v,t}) \times \upsilon_{v,i,t}}{\sum_{i=1}^{n_{site}} \upsilon_{v,i,t}} + \beta_1^{[\gamma]} \times \frac{\sum_{i=1}^{n_{site}} x_{i,t-1,1} \times \upsilon_{v,i,t}}{\sum_{i=1}^{n_{site}} \upsilon_{v,i,t}} + \ldots\right] \end{array}\right\}, \tag{5.7}$$

in which ψ is the probability of existence in the first year; from the second year, each individual has a probability of $\phi_{v,\ t-1}$ to survival given it is alive in the previous year (i.e., $z_{v,\ t-1} = 1$), and a probability of $\gamma_{v,\ t-1}$ to be recruited to the population given it has not existed so far (i.e., $e_{v,\ t-1} = 0$); $e_{v,\ t}$ is an indicator that equals 1 if individual v has been alive in any of the years up until year t, and 0 otherwise; $\phi_{v,\ t-1}$ and $\gamma_{v,\ t-1}$ are functions of density and environmental covariates with intercept and slope parameters $\beta^{[\phi]}$'s and $\beta^{[\gamma]}$'s, respectively. The use of ψ means that we are using data augmentation here, but this time some non-existing individuals may be recruited to the population and thus start to exist. Nonetheless, some individuals will never exist throughout the study period.

The way we allow individuals to survive or be recruited is quite similar to that in the so-called superpopulation Jolly-Seber model (Jolly 1965; Seber 1965; Crosbie and Manly 1985; Schwarz and Arnason 1996), but we utilize the advantage of SCR to link these demographic rates to local density and environmental conditions (Milleret et al. 2023) rather than the overall conditions of the entire study area (e.g., Chandler and Clark 2014). An important choice is to allow any individuals, including non-existing individuals, to have a territory centroid and amount of time of habitat use. The amount of time of habitat use is then used to determine the survival probability for existing individuals or recruitment probability for non-existing individuals. However, only existing individuals can influence the survival and recruitment of other individuals.

Next, we can use the following code to simulate data for the dynamic SCR model with survival and recruitment. We assume a relatively strong decay parameter

($\kappa = 2.5$), a positive effect of bamboo patch size ($\beta^{[\upsilon]}_{bamboo} = 0.6$) and a negative effect of temperature ($\beta^{[\upsilon]}_{temperature} = -0.3$) on the true amount of time of habitat use on the log scale. We assume negative effects of density ($\beta^{[\phi]}_{density} = -0.4$, $\beta^{[\gamma]}_{density} = -0.6$), positive effects of bamboo patch size ($\beta^{[\phi]}_{bamboo} = 0.6$, $\beta^{[\gamma]}_{bamboo} = 0.8$), and negative effects of temperature ($\beta^{[\phi]}_{temperature} = -0.2$, $\beta^{[\gamma]}_{temperature} = -0.4$) on survival and recruitment, respectively, on the logit scale.

```
library(boot) # for using logit and inv.logit functions

set.seed(9)

# Basic values
nindv <- 100 # number of individuals (with data augmentation)
nsite <- 360 # number of sites
nyear <- 6  # number of years
nreps <- 5  # number of within-season replicates
ncovs <- 2  # number of environmental covariates
npcvs <- 2  # number of observational covariates

psi <- 0.4                     # probability of first-year existence
kappa <- 2.5                    # distance effect on habitat use
beta_upsilon <- c(2, 0.6, -0.3)      # intercept and slopes for habitat use
beta_phi  <- c( 1.0, -0.4, 0.6, -0.2) # intercept and slopes for apparent
survival
beta_gamma <- c(-1.0, -0.6, 0.8, -0.4) # intercept and slopes for
recruitment
alpha <- c(-0.6, 0.3)              # slopes for detection rate
lon_min <- -10                 # lower bound of longitude or easting
lon_max <- +10                 # upper bound of longitude or easting
lat_min <- - 8                # lower bound of latitude or northing
lat_max <- + 8                # upper bound of latitude or northing

# Define functions
crossdist <- function(ss, ll) { # function for calculating distance
 c1 <- complex(real=ss[,1], imaginary=ss[,2])
 c2 <- complex(real=ll[,1], imaginary=ll[,2])
 dist <- outer(c1, c2, function(z1, z2) Mod(z1 - z2))
} # crossdist

# Read in environmental covariates
load('data/covariates.RData')
ll <- cbind(cov$lon, cov$lat) # coordinates of sites with camera traps
```

```
x1 <- (log(cov$size) - mean(log(cov$size))) / sd(log(cov$size)) #
standardized log habitat patch size
x2 <- (cov$temp - mean(cov$temp)) / sd(cov$temp) # standardized
temperature
x <- array(, dim=c(nsite, nyear, ncovs)) # environmental covariates
x[,,1] <- x1[,1:nyear]
x[,,2] <- x2[,1:nyear]

# Simulate data
### Process
ss <- matrix(, nindv, 2) # cooridnates of animal territory centroids
ss[,1] <- runif(nindv, lon_min, lon_max)
ss[,2] <- runif(nindv, lat_min, lat_max)

d <- crossdist(ss, ll) # distance between territory centroids and sites

upsilon <- array(0, dim=c(nindv, nsite, nyear)) # habitat use (i.e.,
time spent at a site)
for (v in 1:nindv) {
 for (t in 1:nyear) {
  upsilon[v,,t] <- exp(-1 * kappa * d[v,] + cbind(1, x[,t,]) %*%
beta_upsilon)
 } # t
} # v

z <- matrix(0, nindv, nyear) # indicator of individual existence at a
given year
e <- matrix(0, nindv, nyear) # indicator of individual existence so far
phi  <- matrix(0, nindv, nyear-1) # apparent survival probability
gamma <- matrix(0, nindv, nyear-1) # recruitment probability
z[,1] <- rbinom(nindv, 1, psi) # individual existence in the first year
e[,1] <- z[,1]
for (v in 1:nindv) {
 for (t in 2:nyear) {
  phi [v,t-1] <- inv.logit(c(1, sum(upsilon[v,,t-1] * colSums(upsilon
[-v,,t-1] * z[-v,t-1])) / sum(upsilon[v,,t-1]), colSums(upsilon[v,,t-
1] * x[,t-1,]) / sum(upsilon[v,,t-1])) %*% beta_phi )
  gamma[v,t-1] <- inv.logit(c(1, sum(upsilon[v,,t-1] * colSums
(upsilon[-v,,t-1] * z[-v,t-1])) / sum(upsilon[v,,t-1]), colSums
(upsilon[v,,t-1] * x[,t-1,]) / sum(upsilon[v,,t-1])) %*% beta_gamma)
  z[v,t] <- rbinom(1, 1, z[v,t-1] * phi[v,t-1] + (1 - e[v,t-1]) * gamma[v,
t-1])
  e[v,t] <- ifelse(sum(z[v,1:t]) == 0, 0, 1)
 } # t
} # v
```

```
### Spatial capture-recapture data
w <- array(rnorm(nsite * nyear * nreps * npcvs, 0, 1), dim=c(nsite,
nyear, nreps, npcvs)) # observational covariates

p <- array(0, dim=c(nsite, nyear, nreps)) # detection rate
for (i in 1:nsite) {
 for (t in 1:nyear) {
  p[i,t,] <- exp(w[i,t,,] %*% alpha)
 } # t
} # i

o <- ifelse(ll[,1] <= 8 & ll[,1] >= -8 & ll[,2] <= 6 & ll[,2] >= -6, 1, 0) #
indicator of sites with camera traps

y <- array(0, dim=c(nindv, nsite, nyear, nreps)) # frequency of capture
data
for (v in 1:nindv) {
 for (i in 1:nsite) {
  for (t in 1:nyear) {
   y[v,i,t,] <- rpois(nreps, upsilon[v,i,t] * z[v,t] * p[i,t,] * o[i])
  } # t
 } # i
} # v
```

Next, let us develop a custom MCMC algorithm to fit the model. The interest is to estimate parameters ψ, κ, $\beta^{[\upsilon]}$, $\beta^{[\phi]}$, $\beta^{[\gamma]}$, α, and z given data y, x, w, o, and ll.

```
scr_survival_recruitment_mcmc <- function(y, x, w, o, ll, lon_min,
lon_max, lat_min, lat_max, nmcmc) {

 # Setup variables
 nindv <- dim(y)[1]
 nsite <- dim(y)[2]
 nyear <- dim(y)[3]
 nreps <- dim(y)[4]
 ncovs <- dim(x)[3]
 npcvs <- dim(w)[4]
 ymax <- apply(y, 1:3, max)
 obs01 <- ifelse(apply(y, c(1,3), sum) > 0, 1, 0)
 beta_phi_update <- rep(0, nmcmc)
 beta_gamma_update <- rep(0, nmcmc)

 psi_save <- rep(0, nmcmc)
 kappa_save <- rep(0, nmcmc)
```

```
beta_upsilon_save <- matrix(0, ncovs+1, nmcmc)
beta_phi_save <- matrix(0, ncovs+2, nmcmc)
beta_gamma_save <- matrix(0, ncovs+2, nmcmc)
alpha_save <- matrix(0, npcvs, nmcmc)
z_save <- array(, dim=c(nindv, nyear, 2000))

# Priors
log_kappa_mean <- 0
log_kappa_sd <- 2
beta_upsilon_mean <- 0
beta_upsilon_sd <- 2
beta_phi_mean <- 0
beta_phi_sd <- 2
beta_gamma_mean <- 0
beta_gamma_sd <- 2
alpha_mean <- 0
alpha_sd <- 2

# Starting values
psi <- 0.5
kappa <- 1
beta_upsilon <- rep(0, ncovs+1)
beta_phi  <- rep(0, ncovs+2)
beta_gamma <- rep(0, ncovs+2)
alpha <- rep(0, npcvs)
ss <- matrix(0, nindv, 2)
ss[,1] <- runif(nindv, lon_min, lon_max)
ss[,2] <- runif(nindv, lat_min, lat_max)
for (v in 1:nindv) {
 if (sum(obs01[v,]) > 0) {
  ss[v,1] <- mean(ll[which(rowSums(ymax[v,,]) > 0), 1])
  ss[v,2] <- mean(ll[which(rowSums(ymax[v,,]) > 0), 2])
 }
} # v
d <- crossdist(ss, ll)
upsilon <- array(0, dim=c(nindv, nsite, nyear))
for (v in 1:nindv) {
 for (t in 1:nyear) {
  upsilon[v,,t] <- exp(-1 * kappa * d[v,] + cbind(1, x[,t,]) %*%
beta_upsilon)
 } # t
} # v
phi  <- matrix(0.5, nindv, nyear-1)
gamma <- matrix(0.5, nindv, nyear-1)
p <- array(0.5, dim=c(nsite, nyear, nreps))
```

```
 zi <- obs01
 for (v in 1:nindv) {
  if (sum(zi[v,]) > 0) {
   zi[v,min(which(zi[v,] == 1)):max(which(zi[v,] == 1))] <- 1
  }
 } # v
 z <- zi
 e <- matrix(0, nindv, nyear)
 e[,1] <- z[,1]
 for (t in 2:nyear) {
  e[,t] <- ifelse(rowSums(z[,1:t])==0, 0, 1)
 } # t

 # Tuning factors
 kappa_tune <- 0.1
 beta_upsilon_tune <- c(0.03, 0.06, 0.06)
 beta_phi_tune <- 0.2
 beta_gamma_tune <- 0.4
 alpha_tune <- 0.05
 ss_tune <- 1

 # Begin MCMC loop
 for (k in 1:nmcmc) {
  ### Sample psi
  psi <- rbeta(1, sum(z[,1]) + 0.001, sum(1-z[,1]) + 1)

  ### Sample kappa
  kappa_star <- exp(rnorm(1, log(kappa), kappa_tune))
  upsilon_star <- array(0, dim=c(nindv, nsite, nyear))
  for (v in 1:nindv) {
   for (t in 1:nyear) {
    upsilon_star[v,,t] <- exp(-1 * kappa_star * d[v,] + cbind(1, x[,t,])
%*% beta_upsilon)
   } # t
  } # v
  prob1 <- prob2 <- matrix(0, nindv, nyear)
  for (v in 1:nindv) {
   for (t in 1:nyear) {
    prob1[v,t] <- sum(dpois(y[v,,t,], upsilon_star[v,,t] * z[v,t] * p[,
t,] * o, log=TRUE))
    prob2[v,t] <- sum(dpois(y[v,,t,], upsilon   [v,,t] * z[v,t] * p[,t,]
* o, log=TRUE))
   } # t
  } # v
  mh1 <- sum(prob1) +
```

```
    dnorm(log(kappa_star), log_kappa_mean, log_kappa_sd, log=TRUE)
  mh2 <- sum(prob2) +
    dnorm(log(kappa    ), log_kappa_mean, log_kappa_sd, log=TRUE)
  mhK <- exp(mh1 - mh2)
  if (mhK > runif(1)) {
   kappa <- kappa_star
   upsilon<- upsilon_star
  }

  ### Sample beta_upsilon
  beta_upsilon_star <- rnorm(ncovs+1, beta_upsilon,
beta_upsilon_tune)
  upsilon_star <- array(0, dim=c(nindv, nsite, nyear))
  for (v in 1:nindv) {
   for (t in 1:nyear) {
    upsilon_star[v,,t] <- exp(-1 * kappa * d[v,] + cbind(1, x[,t,]) %*%
beta_upsilon_star)
   } # t
  } # v
  prob1 <- prob2 <- matrix(0, nindv, nyear)
  for (v in 1:nindv) {
   for (t in 1:nyear) {
    prob1[v,t] <- sum(dpois(y[v,,t,], upsilon_star[v,,t] * z[v,t] * p[,
t,] * o, log=TRUE))
    prob2[v,t] <- sum(dpois(y[v,,t,], upsilon     [v,,t] * z[v,t] * p[,t,]
* o, log=TRUE))
   } # t
  } # v
  mh1 <- sum(prob1) +
     sum(dnorm(beta_upsilon_star, beta_upsilon_mean,
beta_upsilon_sd, log=TRUE))
  mh2 <- sum(prob2) +
     sum(dnorm(beta_upsilon     , beta_upsilon_mean, beta_upsilon_sd,
log=TRUE))
  mhB <- exp(mh1 - mh2)
  if (mhB > runif(1)) {
   beta_upsilon <- beta_upsilon_star
   upsilon <- upsilon_star
  }

  ### Sample beta_phi
  beta_phi_star <- rnorm(ncovs+2, beta_phi, beta_phi_tune)
  phi_star <- matrix(0, nindv, nyear-1)
  for (v in 1:nindv) {
   for (t in 2:nyear) {
```

```
    phi_star[v,t-1] <- inv.logit(c(1, sum(upsilon[v,,t-1] * colSums
(upsilon[-v,,t-1] * z[-v,t-1])) / sum(upsilon[v,,t-1]), colSums
(upsilon[v,,t-1] * x[,t-1,]) / sum(upsilon[v,,t-1])) %*%
beta_phi_star)
   } # t
  } # v
  prob1 <- prob2 <- rep(0, nyear-1)
  for (t in 1:(nyear-1)) {
   prob1[t] <- sum(dbinom(z[which(z[,t]==1),t+1], 1, phi_star[which
(z[,t]==1),t], log=TRUE))
   prob2[t] <- sum(dbinom(z[which(z[,t]==1),t+1], 1, phi    [which(z[,
t]==1),t], log=TRUE))
  } # t
  mh1 <- sum(prob1) +
     sum(dnorm(beta_phi_star, beta_phi_mean, beta_phi_sd, log=TRUE))
  mh2 <- sum(prob2) +
     sum(dnorm(beta_phi    , beta_phi_mean, beta_phi_sd, log=TRUE))
  mhP <- exp(mh1 - mh2)
  if (mhP > runif(1)) {
   beta_phi <- beta_phi_star
   phi <- phi_star
   beta_phi_update[k] <- 1
  }
  if (k > 1000) {
   if (sum(beta_phi_update[(k-9):k]) == 0) {
    beta_phi <- runif(ncovs+2, min=apply(beta_phi_save[,(k-99):k],1,
min), max=apply(beta_phi_save[,(k-99):k],1,max))
    phi <- matrix(0, nindv, nyear-1)
    for (v in 1:nindv) {
     for (t in 2:nyear) {
      phi[v,t-1] <- inv.logit(c(1, sum(upsilon[v,,t-1] * colSums
(upsilon[-v,,t-1] * z[-v,t-1])) / sum(upsilon[v,,t-1]), colSums
(upsilon[v,,t-1] * x[,t-1,]) / sum(upsilon[v,,t-1])) %*% beta_phi)
     } # t
    } # v
    beta_phi_update[k] <- 1
   }
  }

  ### Sample beta_gamma
  beta_gamma_star <- rnorm(ncovs+2, beta_gamma, beta_gamma_tune)
  gamma_star <- matrix(0, nindv, nyear-1)
  for (v in 1:nindv) {
   for (t in 2:nyear) {
    gamma_star[v,t-1] <- inv.logit(c(1, sum(upsilon[v,,t-1] * colSums
```

```
(upsilon[-v,,t-1] * z[-v,t-1])) / sum(upsilon[v,,t-1]), colSums
(upsilon[v,,t-1] * x[,t-1,]) / sum(upsilon[v,,t-1])) %*%
beta_gamma_star)
    } # t
   } # v
   prob1 <- prob2 <- rep(0, nyear-1)
   for (t in 1:(nyear-1)) {
    prob1[t] <- sum(dbinom(z[which(e[,t]==0),t+1], 1, gamma_star
[which(e[,t]==0),t], log=TRUE))
    prob2[t] <- sum(dbinom(z[which(e[,t]==0),t+1], 1, gamma   [which(e
[,t]==0),t], log=TRUE))
   } # t
   mh1 <- sum(prob1) +
       sum(dnorm(beta_gamma_star, beta_gamma_mean, beta_gamma_sd,
log=TRUE))
   mh2 <- sum(prob2) +
       sum(dnorm(beta_gamma    , beta_gamma_mean, beta_gamma_sd,
log=TRUE))
   mhP <- exp(mh1 - mh2)
   if (mhP > runif(1)) {
    beta_gamma <- beta_gamma_star
    gamma <- gamma_star
    beta_gamma_update[k] <- 1
   }
   if (k > 1000) {
    if (sum(beta_gamma_update[(k-9):k]) == 0) {
     beta_gamma <- runif(ncovs+2, min=apply(beta_gamma_save[,(k-99):
k],1,min), max=apply(beta_gamma_save[,(k-99):k],1,max))
     gamma <- matrix(0, nindv, nyear-1)
     for (v in 1:nindv) {
      for (t in 2:nyear) {
       gamma[v,t-1] <- inv.logit(c(1, sum(upsilon[v,,t-1] * colSums
(upsilon[-v,,t-1] * z[-v,t-1])) / sum(upsilon[v,,t-1]), colSums
(upsilon[v,,t-1] * x[,t-1,]) / sum(upsilon[v,,t-1])) %*% beta_gamma)
      } # t
     } # v
     beta_gamma_update[k] <- 1
    }
   }

   ### Sample alpha
   alpha_star <- rnorm(npcvs, alpha, alpha_tune)
   p_star <- array(, dim=c(nsite, nyear, nreps))
   for (i in 1:nsite) {
    for (t in 1:nyear) {
```

```
    p_star[i,t,] <- exp(w[i,t,,] %*% alpha_star)
   } # t
  } # i
  prob1 <- prob2 <- matrix(0, nindv, nyear)
  for (v in 1:nindv) {
   for (t in 1:nyear) {
    prob1[v,t] <- sum(dpois(y[v,,t,], upsilon[v,,t] * z[v,t] * p_star[,
t,] * o, log=TRUE))
    prob2[v,t] <- sum(dpois(y[v,,t,], upsilon[v,,t] * z[v,t] * p   [,t,]
* o, log=TRUE))
   } # t
  } # v
  mh1 <- sum(prob1) +
      sum(dnorm(alpha_star, alpha_mean, alpha_sd, log=TRUE))
  mh2 <- sum(prob2) +
      sum(dnorm(alpha   , alpha_mean, alpha_sd, log=TRUE))
  mhA <- exp(mh1 - mh2)
  if (mhA > runif(1)) {
   alpha <- alpha_star
   p <- p_star
  }

  ### Sample ss
  ss_star <- matrix(rnorm(nindv*2, ss, ss_tune), nindv, 2)
  d_star <- crossdist(ss_star, ll)
  upsilon_star <- array(0, dim=c(nindv, nsite, nyear))
  for (v in 1:nindv) {
   for (t in 1:nyear) {
    upsilon_star[v,,t] <- exp(-1 * kappa * d_star[v,] + cbind(1, x[,t,])
%*% beta_upsilon)
   } # t
  } # v
  for (v in 1:nindv) {
   if (ss_star[v,1] > lon_min & ss_star[v,1] < lon_max &
     ss_star[v,2] > lat_min & ss_star[v,2] < lat_max) {
    prob1 <- prob2 <- rep(0, nyear)
    for (t in 1:nyear) {
     prob1[t] <- sum(dpois(y[v,,t,], upsilon_star[v,,t] * z[v,t] * p[,
t,] * o, log=TRUE))
    prob2[t] <- sum(dpois(y[v,,t,], upsilon   [v,,t] * z[v,t] * p[,t,] *
o, log=TRUE))
    } # t
    mh1 <- sum(prob1)
    mh2 <- sum(prob2)
    mhS <- exp(mh1 - mh2)
```

```
    if (mhS > runif(1)) {
     ss[v,] <- ss_star[v,]
     d[v,] <- d_star[v,]
     upsilon[v,,] <- upsilon_star[v,,]
    }
   }
  } # v

  ### Sample z
  for (v in 1:nindv) {
   if (zi[v,1] == 0) {
    prod_temp <- psi * prod(exp(-1 * upsilon[v,,1] * p[,1,] * o))
    psi_temp <- prod_temp / (prod_temp + 1 - psi)
    z[v,1] <- rbinom(1, 1, psi_temp)
   }
   e[v,1] <- z[v,1]
   for (t in 2:nyear) {
    if (e[v,t-1] == 1 & z[v,t-1] == 0 & zi[v,t] == 0) {
     z[v,t] <- 0
    } else if (z[v,t-1] == 1 & zi[v,t] == 0) {
    prod_temp <- phi[v,t-1] * prod(exp(-1 * upsilon[v,,t] * p[,t,] * o))
     psi_temp <- prod_temp / (prod_temp + 1 - phi[v,t-1])
     z[v,t] <- rbinom(1, 1, psi_temp)
    } else if (e[v,t-1] == 0 & zi[v,t] == 0) {
     prod_temp <- gamma[v,t-1] * prod(exp(-1 * upsilon[v,,t] * p[,t,] *
o))
     psi_temp <- prod_temp / (prod_temp + 1 - gamma[v,t-1])
     z[v,t] <- rbinom(1, 1, psi_temp)
    }
    e[v,t] <- ifelse(sum(z[v,1:t]) == 0, 0, 1)
   } # t
  } # v

  ### Save samples
  psi_save[k] <- psi
  kappa_save[k] <- kappa
  beta_upsilon_save[,k] <- beta_upsilon
  beta_phi_save[,k] <- beta_phi
  beta_gamma_save[,k] <- beta_gamma
  alpha_save[,k] <- alpha
  if (nmcmc > 2000) {
   if (k > nmcmc-2000) {
    z_save[,,k-nmcmc+2000] <- z
   }
  }
 } # k
```

```
### Write output
list(psi_save=psi_save,
  kappa_save=kappa_save, beta_upsilon_save=beta_upsilon_save,
  beta_phi_save=beta_phi_save, beta_gamma_save=beta_gamma_save,
  alpha_save=alpha_save,
  z_save=z_save)

} # scr_survival_recruitment_mcmc
```

We then use the following code to implement the algorithm, which takes more than 14 h to run. Then using code in Appendix 5.3, we can create traceplots for the monitored parameters, which show good model convergence and consistency between parameter estimates of κ, $\beta^{[\upsilon]}$, $\beta^{[\phi]}$, $\beta^{[\gamma]}$, and α and their true values used in data simulation (Fig. 5.5).

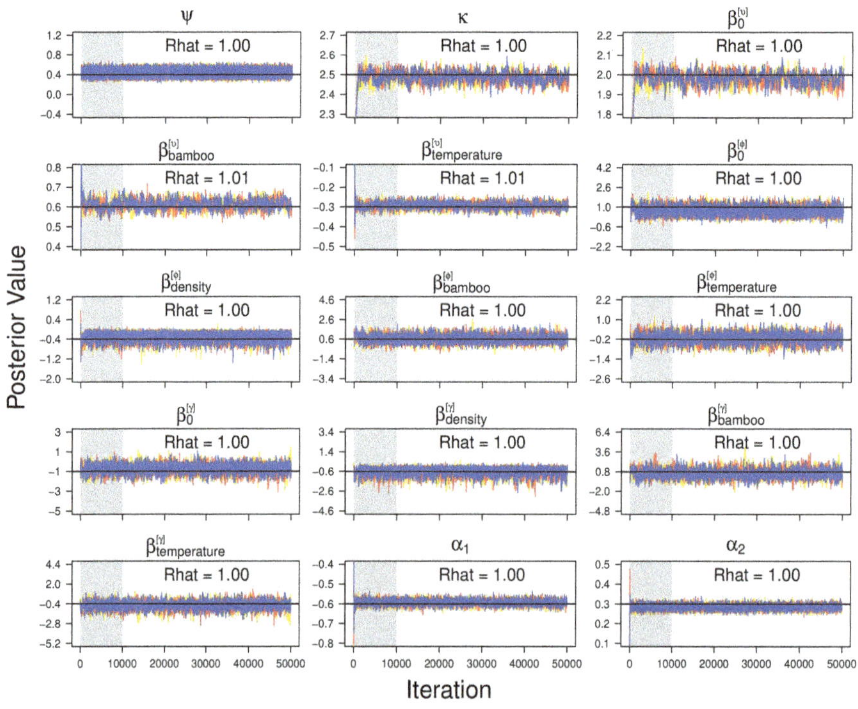

Fig. 5.5 Traceplots showing the posterior chains for κ, $\beta^{[\upsilon]}$, $\beta^{[\phi]}$, $\beta^{[\gamma]}$, and α in the dynamic SCR model with survival and recruitment. Rhat values are calculated after discarding the first 10,000 iterations. Horizontal lines represent the true values used in data simulation

```
library(foreach)  # for parallel computing
library(doParallel) # for parallel computing

numCores <- round(detectCores() / 2) # only use half of the cores for
parallel computing
registerDoParallel(numCores) # setup parallel computing

nmcmc <- 50000 # number of iterations of MCMC computing
chain <- 3     # number of chains

start_time <- Sys.time() # start time of computing
out <- foreach (i=1:chain, .packages='boot') %dopar% {
 scr_survival_recruitment_mcmc(
  y=y, x=x, w=w, o=o, ll=ll,
  lon_min=lon_min, lon_max=lon_max, lat_min=lat_min,
lat_max=lat_max,
  nmcmc=nmcmc)
} # i
stopImplicitCluster() # clean up the cluster
end_time <- Sys.time() # end time of computing
time_taken <- end_time - start_time # computing time
```

Next, let us use the following code to plot the estimated survival history, which shows a satisfying agreement with the simulated true values for captured individuals, but has a hard time identifying uncaptured individuals (Fig. 5.6).

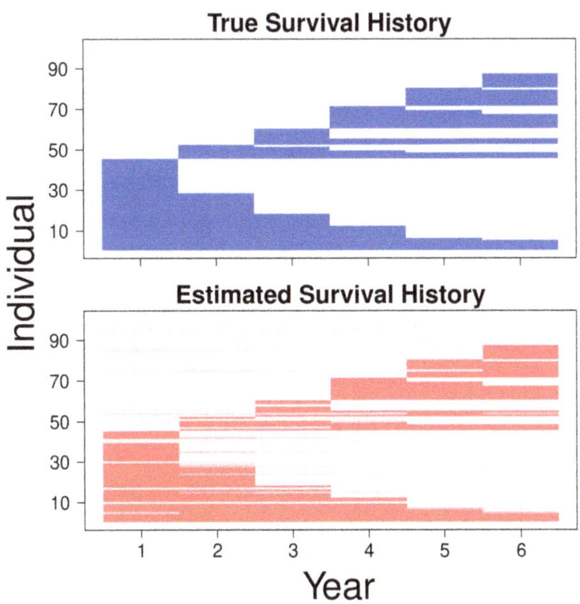

Fig. 5.6 The true (upper panel) and estimated (lower panel) survival history for each individual during the study period in the dynamic SCR model with survival and recruitment. Individuals that never exist are put at the top of the panels

```
z_post <- array(, dim=c(nindv, nyear, 6000))
z_post[,,  1:2000] <- out[[1]]$z
z_post[,,2001:4000] <- out[[2]]$z
z_post[,,4001:6000] <- out[[3]]$z
z_mean <- apply(z_post, 1:2, mean)
nscale <- 100
z_scale <- round(z_mean * (nscale - 1)) + 1
col_scale <- colorRampPalette(c('white','lightcoral'))(nscale)
f <- rep(0, nindv)
for (v in 1:nindv) {
 if (sum(z[v,]) > 0) {
  f[v] <- min(which(z[v,] == 1))
 }
} # v
tplot <- c(1:nyear, 0)
for (t in 1:(nyear+1)) {
 t1 <- z[which(f == tplot[t]),]
 t2 <- z_scale[which(f == tplot[t]),]
 if (length(which(f == tplot[t])) > 1) {
  t2 <- t2[rev(order(rowSums(t1))),]
  t1 <- t1[rev(order(rowSums(t1))),]
 }
 if (t == 1) {
  z_order <- t1
  z_scale_order <- t2
 } else {
  z_order <- rbind(z_order, t1)
  z_scale_order <- rbind(z_scale_order, t2)
 }
} # t

par(mfrow=c(2,1))
par(mar=c(1,0,2.5,1))
par(oma=c(4,5.5,0,0))

plot(1, xlim=c(.5,nyear+.5), ylim=c(.5,nindv+.5), type='n', axes=F)
for (i in 1:nindv) {
 for (t in 1:nyear) {
  rect(t-.5,i-.5,t+.5,i+.5, col=ifelse(z_order[i,t]==1,
'royalblue', 'white'), border=NA)
 } # t
} # i
axis(1, at=seq(1,nyear,1), labels=rep('',nyear))
axis(2, at=seq(10,nindv,20), cex.axis=1.5, las=2)
```

```
box()
title(main='True Survival History', line=.5, cex.main=2)

plot(1, xlim=c(.5,nyear+.5), ylim=c(.5,nindv+.5), type='n', axes=F)
for (i in 1:nindv) {
 for (t in 1:nyear) {
  rect(t-.5,i-.5,t+.5,i+.5, col=col_scale[z_scale_order[i,t]],
border=NA)
 } # t
} # i
axis(1, at=seq(1,nyear,1), cex.axis=1.5)
axis(2, at=seq(10,nindv,20), cex.axis=1.5, las=2)
box()
title(main='Estimated Survival History', line=.5, cex.main=2)

title(ylab='Individual', cex.lab=3, line=3.2, outer=T)
title(xlab='Year', cex.lab=3, line=2.8, outer=T)
```

5.3.3 With Survival, Recruitment, and Movement

Now, it is time to finally allow movement in the dynamic SCR model. This model still describes the changes in $z_{v,\ t}$ due to density- and environment-dependent survival and recruitment processes, as described in Sect. 5.3.2. However, this model is different from previous models by allowing a Gaussian movement between years (Chandler and Clark 2014; Royle et al. 2016). As such, the territory centroids of our red panda can change across years in the following way:

$$\begin{pmatrix} ss_{v,t}^{[x]} \\ ss_{v,t}^{[y]} \end{pmatrix} \sim Normal\left[\begin{pmatrix} ss_{v,t-1}^{[x]} \\ ss_{v,t-1}^{[y]} \end{pmatrix}, \begin{pmatrix} \sigma & 0 \\ 0 & \sigma \end{pmatrix}\right], \tag{5.8}$$

in which $ss_{v,t}^{[x]}$ is the horizontal coordinate or easting of the territory centroid of individual v in year t, $ss_{v,t}^{[y]}$ is the vertical coordinate or northing, and σ is the standard deviation of this Gaussian dispersal movement. Here, we assume the change in easting and northing are uncorrelated, and thus the 0's in the off-diagonals of the variance-covariance matrix. Readers can consider generalizing the variance-covariance matrix to allow correlated change in easting and northing, or other types of movement curves (Ergon and Gardner 2014; Sutherland et al. 2015). In this model, the distance between territory centroids and sites, $dist_{v,\ i,\ t}$, changes over time because it is calculated from ss_t and the coordinates of the sites, ll. Eventually, changes in $dist_{v,\ i,\ t}$ will lead to changes in the amount of time of habitat use, $\upsilon_{v,\ i,}$

$_t$ (Eq. 5.4), which in turn influences survival $\phi_{v,\ t}$ and recruitment $\gamma_{v,\ t}$ (Eqs. 5.6 and 5.7).

Next, we can use the following code to simulate data for the dynamic SCR model with survival, recruitment, and movement. We assume a relatively strong decay parameter ($\kappa = 2.5$), a positive effect of bamboo patch size ($\beta^{[\upsilon]}_{bamboo} = 0.6$) and a negative effect of temperature ($\beta^{[\upsilon]}_{temperature} = -0.3$) on the true amount of time of habitat use on the log scale. We assume negative effects of density ($\beta^{[\phi]}_{density} = -0.4$, $\beta^{[\gamma]}_{density} = -0.6$), positive effects of bamboo patch size ($\beta^{[\phi]}_{bamboo} = 0.6$, $\beta^{[\gamma]}_{bamboo} = 0.8$), and negative effects of temperature ($\beta^{[\phi]}_{temperature} = -0.2$, $\beta^{[\gamma]}_{temperature} = -0.4$) on survival and recruitment, respectively, on the logit scale. Eventually, we assume a moderate dispersal distance ($\sigma = 2.5$).

```
library(boot) # for using logit and inv.logit functions
library(truncnorm) # for using truncated normal distribution

set.seed(9)

# Basic values
nindv <- 100 # number of individuals (with data augmentation)
nsite <- 360 # number of sites
nyear <- 6  # number of years
nreps <- 5  # number of within-season replicates
ncovs <- 2  # number of environmental covariates
npcvs <- 4  # number of observational covariates

psi <- 0.4                  # probability of first-year existence
kappa <- 2.5                 # distance effect on habitat use
beta_upsilon <- c(2, 0.6, -0.3)     # intercept and slopes for habitat use
beta_phi  <- c( 1.0, -0.4, 0.6, -0.2) # intercept and slopes for apparent
survival
beta_gamma <- c(-1.0, -0.6, 0.8, -0.4) # intercept and slopes for
recruitment
sigma <- 2.5                 # standard deviation for Gaussian movement
alpha <- c(-0.8, 0.4, -0.2, 0)      # slopes for detection probability
lon_min <- -10                # lower bound of longitude or easting
lon_max <- +10                # upper bound of longitude or easting
lat_min <- - 8               # lower bound of latitude or northing
lat_max <- + 8               # upper bound of latitude or northing

# Define functions
crossdist <- function(ss, ll) { # function for calculating distance
 c1 <- complex(real=ss[,1], imaginary=ss[,2])
 c2 <- complex(real=ll[,1], imaginary=ll[,2])
```

```
 dist <- outer(c1, c2, function(z1, z2) Mod(z1 - z2))
} # crossdist

# Read in environmental covariates
load('data/covariates.RData')
ll <- cbind(cov$lon, cov$lat) # coordinates of sites with camera traps

x1 <- (log(cov$size) - mean(log(cov$size))) / sd(log(cov$size)) #
standardized log habitat patch size
x2 <- (cov$temp - mean(cov$temp)) / sd(cov$temp) # standardized
temperature
x <- array(, dim=c(nsite, nyear, ncovs)) # environmental covariates
x[,,1] <- x1[,1:nyear]
x[,,2] <- x2[,1:nyear]

# Simulate data
### Process
ss <- array(, dim=c(nindv, 2, nyear)) # cooridnates of animal territory
centroids
ss[,1,1] <- runif(nindv, lon_min, lon_max)
ss[,2,1] <- runif(nindv, lat_min, lat_max)
for (t in 2:nyear) {
 ss[,1,t] <- rtruncnorm(n=nindv, a=lon_min, b=lon_max, mean=ss[,1,t-
1], sd=sigma)
 ss[,2,t] <- rtruncnorm(n=nindv, a=lat_min, b=lat_max, mean=ss[,2,t-
1], sd=sigma)
} # t

d <- array(, dim=c(nindv, nsite, nyear)) # distance between territory
centroids and sites
for (t in 1:nyear) {
 d[,,t] <- crossdist(ss[,,t], ll)
} # t

upsilon <- array(0, dim=c(nindv, nsite, nyear)) # habitat use (i.e.,
time spent at a site)
for (v in 1:nindv) {
 for (t in 1:nyear) {
  upsilon[v,,t] <- exp(-1 * kappa * d[v,,t] + cbind(1, x[,t,]) %*%
beta_upsilon)
 } # t
} # v

z <- matrix(0, nindv, nyear) # indicator of individual existence at a
given year
```

```
e <- matrix(0, nindv, nyear) # indicator of individual existence so far
phi  <- matrix(0, nindv, nyear-1) # apparent survival probability
gamma <- matrix(0, nindv, nyear-1) # recruitment rate
z[,1] <- rbinom(nindv, 1, psi) # individual existence in the first year
e[,1] <- z[,1]
for (t in 2:nyear) {
 for (v in 1:nindv) {
  phi [v,t-1] <- inv.logit(c(1, sum(upsilon[v,,t-1] * colSums(upsilon
[-v,,t-1] * z[-v,t-1])) / sum(upsilon[v,,t-1]), colSums(upsilon[v,,t-
1] * x[,t-1,]) / sum(upsilon[v,,t-1])) %*% beta_phi )
  gamma[v,t-1] <- inv.logit(c(1, sum(upsilon[v,,t-1] * colSums
(upsilon[-v,,t-1] * z[-v,t-1])) / sum(upsilon[v,,t-1]), colSums
(upsilon[v,,t-1] * x[,t-1,]) / sum(upsilon[v,,t-1])) %*% beta_gamma)
  z[v,t] <- rbinom(1, 1, z[v,t-1] * phi[v,t-1] + (1 - e[v,t-1]) * gamma[v,
t-1])
  e[v,t] <- ifelse(sum(z[v,1:t]) == 0, 0, 1)
 } # v
} # t

### Spatial capture-recapture data
w <- array(rnorm(nsite * nyear * nreps * npcvs, 0, 1), dim=c(nsite,
nyear, nreps, npcvs)) # observational covariates

p <- array(0, dim=c(nsite, nyear, nreps)) # detection rate
for (i in 1:nsite) {
 for (t in 1:nyear) {
  p[i,t,] <- exp(w[i,t,,] %*% alpha)
 } # t
} # i

o <- ifelse(ll[,1] <= 8 & ll[,1] >= -8 & ll[,2] <= 6 & ll[,2] >= -6, 1, 0) #
indicator of sites with camera traps

y <- array(0, dim=c(nindv, nsite, nyear, nreps)) # frequency of capture
data
for (v in 1:nindv) {
 for (i in 1:nsite) {
  for (t in 1:nyear) {
   y[v,i,t,] <- rpois(nreps, upsilon[v,i,t] * z[v,t] * p[i,t,] * o[i])
  } # t
 } # i
} # v
```

Next, let us develop a custom MCMC algorithm to fit the model. The interest is to estimate parameters ψ, κ, $\beta^{[\upsilon]}$, $\beta^{[\phi]}$, $\beta^{[\gamma]}$, σ, α, and ss given data y, x, w, o, and ll.

```
scr_survival_recruitment_movement_mcmc <- function(y, x, w, o, ll,
lon_min, lon_max, lat_min, lat_max, nmcmc) {

 # Setup variables
 nindv <- dim(y) [1]
 nsite <- dim(y) [2]
 nyear <- dim(y) [3]
 nreps <- dim(y) [4]
 ncovs <- dim(x) [3]
 npcvs <- dim(w) [4]
 ymax <- apply(y, 1:3, max)
 obs01 <- ifelse(apply(y, c(1,3), sum) > 0, 1, 0)
 beta_phi_update <- rep(0, nmcmc)
 beta_gamma_update <- rep(0, nmcmc)

 psi_save <- rep(0, nmcmc)
 kappa_save <- rep(0, nmcmc)
 beta_upsilon_save <- matrix(0, ncovs+1, nmcmc)
 beta_phi_save <- matrix(0, ncovs+2, nmcmc)
 beta_gamma_save <- matrix(0, ncovs+2, nmcmc)
 sigma_save <- rep(0, nmcmc)
 alpha_save <- matrix(0, npcvs, nmcmc)
 ss_save <- array(, dim=c(nindv, 2, nyear, 2000))

 # Priors
 log_kappa_mean <- 0
 log_kappa_sd <- 2
 beta_upsilon_mean <- 0
 beta_upsilon_sd <- 2
 beta_phi_mean <- 0
 beta_phi_sd <- 2
 beta_gamma_mean <- 0
 beta_gamma_sd <- 2
 log_sigma_mean <- 0
 log_sigma_sd <- 2
 alpha_mean <- 0
 alpha_sd <- 2

 # Starting values
 psi <- 0.5
 kappa <- 1
 beta_upsilon <- rep(0, ncovs+1)
```

```
beta_phi  <- rep(0, ncovs+2)
beta_gamma <- rep(0, ncovs+2)
sigma <- 1
alpha <- rep(0, npcvs)
ss <- array(0, dim=c(nindv, 2, nyear))
ss[,1,] <- runif(nindv*nyear, lon_min, lon_max)
ss[,2,] <- runif(nindv*nyear, lat_min, lat_max)
for (v in 1:nindv) {
 for (t in 1:nyear) {
  if (obs01[v,t] > 0) {
   ss[v,1,t] <- mean(ll[which(ymax[v,,t] > 0), 1])
   ss[v,2,t] <- mean(ll[which(ymax[v,,t] > 0), 2])
  }
 } # t
} # v
d <- array(, dim=c(nindv, nsite, nyear))
for (t in 1:nyear) {
 d[,,t] <- crossdist(ss[,,t], ll)
} # t
upsilon <- array(0, dim=c(nindv, nsite, nyear))
for (v in 1:nindv) {
 for (t in 1:nyear) {
  upsilon[v,,t] <- exp(-1 * kappa * d[v,,t] + cbind(1, x[,t,]) %*%
beta_upsilon)
 } # t
} # v
phi  <- matrix(0.5, nindv, nyear-1)
gamma <- matrix(0.5, nindv, nyear-1)
p <- array(0.5, dim=c(nsite, nyear, nreps))
zi <- obs01
for (v in 1:nindv) {
 if (sum(zi[v,]) > 0) {
  zi[v,min(which(zi[v,] == 1)):max(which(zi[v,] == 1))] <- 1
 }
} # v
z <- zi
e <- matrix(0, nindv, nyear)
e[,1] <- z[,1]
for (t in 2:nyear) {
 e[,t] <- ifelse(rowSums(z[,1:t])==0, 0, 1)
} # t

# Tuning factors
kappa_tune <- 0.15
beta_upsilon_tune <- c(0.05, 0.06, 0.06)
```

```
 beta_phi_tune <- 0.2
 beta_gamma_tune <- 0.3
 sigma_tune <- 0.2
 alpha_tune <- 0.035
 ss_tune <- 1

 # Begin MCMC loop
 for (k in 1:nmcmc) {
  ### Sample psi (probability of existence)
  psi <- rbeta(1, sum(z[,1]) + 0.001, sum(1-z[,1]) + 1)

  ### Sample kappa
  kappa_star <- exp(rnorm(1, log(kappa), kappa_tune))
  upsilon_star <- array(0, dim=c(nindv, nsite, nyear))
  for (v in 1:nindv) {
   for (t in 1:nyear) {
    upsilon_star[v,,t] <- exp(-1 * kappa_star * d[v,,t] + cbind(1, x[,
t,]) %*% beta_upsilon)
   } # t
  } # v
  prob1 <- prob2 <- matrix(0, nindv, nyear)
  for (v in 1:nindv) {
   for (t in 1:nyear) {
   prob1[v,t] <- sum(dpois(y[v,,t,], upsilon_star[v,,t] * z[v,t] * p[,
t,] * o, log=TRUE))
   prob2[v,t] <- sum(dpois(y[v,,t,], upsilon   [v,,t] * z[v,t] * p[,t,]
* o, log=TRUE))
   } # t
  } # v
  mh1 <- sum(prob1) +
      dnorm(log(kappa_star), log_kappa_mean, log_kappa_sd, log=TRUE)
  mh2 <- sum(prob2) +
      dnorm(log(kappa   ), log_kappa_mean, log_kappa_sd, log=TRUE)
  mhK <- exp(mh1 - mh2)
  if (mhK > runif(1)) {
   kappa <- kappa_star
   upsilon <- upsilon_star
  }

  ### Sample beta_upsilon
  beta_upsilon_star <- rnorm(ncovs+1, beta_upsilon,
beta_upsilon_tune)
  upsilon_star <- array(0, dim=c(nindv, nsite, nyear))
  for (v in 1:nindv) {
   for (t in 1:nyear) {
```

```
    upsilon_star[v,,t] <- exp(-1 * kappa * d[v,,t] + cbind(1, x[,t,]) %*%
beta_upsilon_star)
    } # t
  } # v
  prob1 <- prob2 <- matrix(0, nindv, nyear)
  for (v in 1:nindv) {
   for (t in 1:nyear) {
    prob1[v,t] <- sum(dpois(y[v,,t,], upsilon_star[v,,t] * z[v,t] * p[,
t,] * o, log=TRUE))
    prob2[v,t] <- sum(dpois(y[v,,t,], upsilon   [v,,t] * z[v,t] * p[,t,]
* o, log=TRUE))
    } # t
  } # v
  mh1 <- sum(prob1) +
      sum(dnorm(beta_upsilon_star, beta_upsilon_mean,
beta_upsilon_sd, log=TRUE))
  mh2 <- sum(prob2) +
      sum(dnorm(beta_upsilon    , beta_upsilon_mean, beta_upsilon_sd,
log=TRUE))
  mhB <- exp(mh1 - mh2)
  if (mhB > runif(1)) {
   beta_upsilon <- beta_upsilon_star
   upsilon <- upsilon_star
  }

  ### Sample beta_phi
  beta_phi_star <- rnorm(ncovs+2, beta_phi, beta_phi_tune)
  phi_star <- matrix(0, nindv, nyear-1)
  for (v in 1:nindv) {
   for (t in 2:nyear) {
    phi_star[v,t-1] <- inv.logit(c(1, sum(upsilon[v,,t-1] * colSums
(upsilon[-v,,t-1] * z[-v,t-1])) / sum(upsilon[v,,t-1]), colSums
(upsilon[v,,t-1] * x[,t-1,]) / sum(upsilon[v,,t-1])) %*%
beta_phi_star)
    } # t
  } # v
  prob1 <- prob2 <- rep(0, nyear-1)
  for (t in 1:(nyear-1)) {
   prob1[t] <- sum(dbinom(z[which(z[,t]==1),t+1], 1, phi_star[which
(z[,t]==1),t], log=TRUE))
   prob2[t] <- sum(dbinom(z[which(z[,t]==1),t+1], 1, phi    [which(z[,
t]==1),t], log=TRUE))
  } # t
  mh1 <- sum(prob1) +
     sum(dnorm(beta_phi_star, beta_phi_mean, beta_phi_sd, log=TRUE))
```

```
  mh2 <- sum(prob2) +
      sum(dnorm(beta_phi   , beta_phi_mean, beta_phi_sd, log=TRUE))
  mhP <- exp(mh1 - mh2)
  if (mhP > runif(1)) {
   beta_phi <- beta_phi_star
   phi <- phi_star
   beta_phi_update[k] <- 1
  }
  if (k > 1000) {
   if (sum(beta_phi_update[(k-9):k]) == 0) {
    beta_phi <- runif(ncovs+2, min=apply(beta_phi_save[,(k-99):k],1,
min), max=apply(beta_phi_save[,(k-99):k],1,max))
    phi <- matrix(0, nindv, nyear-1)
    for (v in 1:nindv) {
     for (t in 2:nyear) {
      phi[v,t-1] <- inv.logit(c(1, sum(upsilon[v,,t-1] * colSums
(upsilon[-v,,t-1] * z[-v,t-1])) / sum(upsilon[v,,t-1]), colSums
(upsilon[v,,t-1] * x[,t-1,]) / sum(upsilon[v,,t-1])) %*% beta_phi)
     } # t
    } # v
    beta_phi_update[k] <- 1
   }
  }

  ### Sample beta_gamma
  beta_gamma_star <- rnorm(ncovs+2, beta_gamma, beta_gamma_tune)
  gamma_star <- matrix(0, nindv, nyear-1)
  for (v in 1:nindv) {
   for (t in 2:nyear) {
    gamma_star[v,t-1] <- inv.logit(c(1, sum(upsilon[v,,t-1] * colSums
(upsilon[-v,,t-1] * z[-v,t-1])) / sum(upsilon[v,,t-1]), colSums
(upsilon[v,,t-1] * x[,t-1,]) / sum(upsilon[v,,t-1])) %*%
beta_gamma_star)
   } # t
  } # v
  prob1 <- prob2 <- rep(0, nyear-1)
  for (t in 1:(nyear-1)) {
   prob1[t] <- sum(dbinom(z[which(e[,t]==0),t+1], 1, gamma_star
[which(e[,t]==0),t], log=TRUE))
   prob2[t] <- sum(dbinom(z[which(e[,t]==0),t+1], 1, gamma   [which(e
[,t]==0),t], log=TRUE))
  } # t
  mh1 <- sum(prob1) +
      sum(dnorm(beta_gamma_star, beta_gamma_mean, beta_gamma_sd,
log=TRUE))
```

```
  mh2 <- sum(prob2) +
      sum(dnorm(beta_gamma    , beta_gamma_mean, beta_gamma_sd,
log=TRUE))
  mhP <- exp(mh1 - mh2)
  if (mhP > runif(1)) {
   beta_gamma <- beta_gamma_star
   gamma <- gamma_star
   beta_gamma_update[k] <- 1
  }
  if (k > 1000) {
   if (sum(beta_gamma_update[(k-9):k]) == 0) {
    beta_gamma <- runif(ncovs+2, min=apply(beta_gamma_save[,(k-99):
k],1,min), max=apply(beta_gamma_save[,(k-99):k],1,max))
    gamma <- matrix(0, nindv, nyear-1)
    for (v in 1:nindv) {
     for (t in 2:nyear) {
      gamma[v,t-1] <- inv.logit(c(1, sum(upsilon[v,,t-1] * colSums
(upsilon[-v,,t-1] * z[-v,t-1])) / sum(upsilon[v,,t-1]), colSums
(upsilon[v,,t-1] * x[,t-1,]) / sum(upsilon[v,,t-1])) %*% beta_gamma)
     } # t
    } # v
    beta_gamma_update[k] <- 1
   }
  }

  ### Sample sigma
  sigma_star <- exp(rnorm(1, log(sigma), sigma_tune))
  prob1 <- prob2 <- matrix(0, nindv, nyear-1)
  for (v in 1:nindv) {
   for (t in 1:(nyear-1)) {
    if (obs01[v,t]==1 & obs01[v,t+1]==1) {
    prob1[v,t] <- sum(dnorm(ss[v,,t+1], ss[v,,t], sigma_star, log=T))
     prob2[v,t] <- sum(dnorm(ss[v,,t+1], ss[v,,t], sigma     , log=T))
    }
   } # t
  } # v
  mh1 <- sum(prob1) +
      dnorm(log(sigma_star), log_sigma_mean, log_sigma_sd, log=T)
  mh2 <- sum(prob2) +
      dnorm(log(sigma     ), log_sigma_mean, log_sigma_sd, log=T)
  mh <- exp(mh1 - mh2)
  if (mh > runif(1)) {
   sigma <- sigma_star
  }
```

```
  ### Sample alpha
  alpha_star <- rnorm(npcvs, alpha, alpha_tune)
  p_star <- array(, dim=c(nsite, nyear, nreps))
  for (i in 1:nsite) {
   for (t in 1:nyear) {
    p_star[i,t,] <- exp(w[i,t,,] %*% alpha_star)
   } # t
  } # i
  prob1 <- prob2 <- matrix(0, nindv, nyear)
  for (v in 1:nindv) {
   for (t in 1:nyear) {
    prob1[v,t] <- sum(dpois(y[v,,t,], upsilon[v,,t] * z[v,t] * p_star[,
t,] * o, log=TRUE))
    prob2[v,t] <- sum(dpois(y[v,,t,], upsilon[v,,t] * z[v,t] * p     [,t,]
* o, log=TRUE))
   } # t
  } # v
  mh1 <- sum(prob1) +
      sum(dnorm(alpha_star, alpha_mean, alpha_sd, log=TRUE))
  mh2 <- sum(prob2) +
      sum(dnorm(alpha     , alpha_mean, alpha_sd, log=TRUE))
  mhA <- exp(mh1 - mh2)
  if (mhA > runif(1)) {
   alpha <- alpha_star
   p <- p_star
  }

  ### Sample ss
  ss_star <- array(rnorm(nindv*2*nyear, ss, ss_tune), dim=c(nindv,
2, nyear))
  d_star <- array(, dim=c(nindv, nsite, nyear))
  for (t in 1:nyear) {
   d_star[,,t] <- crossdist(ss_star[,,t], ll)
  } # t
  upsilon_star <- array(0, dim=c(nindv, nsite, nyear))
  for (v in 1:nindv) {
   for (t in 1:nyear) {
    upsilon_star[v,,t] <- exp(-1 * kappa * d_star[v,,t] + cbind(1, x[,
t,]) %*% beta_upsilon)
   } # t
  } # v
  for (v in 1:nindv) {
   if (ss_star[v,1,1] > lon_min & ss_star[v,1,1] < lon_max &
     ss_star[v,2,1] > lat_min & ss_star[v,2,1] < lat_max) {
    mh1 <- sum(dpois(y[v,,1,], upsilon_star[v,,1] * z[v,1] * p[,1,] * o,
```

```
log=TRUE))
    mh2 <- sum(dpois(y[v,,1,], upsilon   [v,,1] * z[v,1] * p[,1,] * o,
log=TRUE))
    mhS <- exp(mh1 - mh2)
    if (mhS > runif(1)) {
     ss[v,,1] <- ss_star[v,,1]
     d[v,,1] <- d_star[v,,1]
     upsilon[v,,1] <- upsilon_star[v,,1]
    }
   }
   for (t in 2:nyear) {
    if (ss_star[v,1,t] > lon_min & ss_star[v,1,t] < lon_max &
      ss_star[v,2,t] > lat_min & ss_star[v,2,t] < lat_max) {
    mh1 <- sum(dpois(y[v,,t,], upsilon_star[v,,t] * z[v,t] * p[,t,] * o,
log=TRUE)) +
         sum(dnorm(ss_star[v,,t], ss[v,,t-1], sigma, log=T))
     mh2 <- sum(dpois(y[v,,t,], upsilon   [v,,t] * z[v,t] * p[,t,] * o,
log=TRUE)) +
         sum(dnorm(ss   [v,,t], ss[v,,t-1], sigma, log=T))
     mhS <- exp(mh1 - mh2)
     if (mhS > runif(1)) {
      ss[v,,t] <- ss_star[v,,t]
      d[v,,t] <- d_star[v,,t]
      upsilon[v,,t] <- upsilon_star[v,,t]
     }
    }
   } # t
  } # v

  ### Sample z
  for (v in 1:nindv) {
   if (zi[v,1] == 0) {
    prod_temp <- psi * prod(exp(-1 * upsilon[v,,1] * p[,1,] * o))
    psi_temp <- prod_temp / (prod_temp + 1 - psi)
    z[v,1] <- rbinom(1, 1, psi_temp)
   }
   e[v,1] <- z[v,1]
   for (t in 2:nyear) {
    if (e[v,t-1] == 1 & z[v,t-1] == 0 & zi[v,t] == 0) {
     z[v,t] <- 0
    } else if (z[v,t-1] == 1 & zi[v,t] == 0) {
    prod_temp <- phi[v,t-1] * prod(exp(-1 * upsilon[v,,t] * p[,t,] * o))
     psi_temp <- prod_temp / (prod_temp + 1 - phi[v,t-1])
```

```
      z[v,t] <- rbinom(1, 1, psi_temp)
     } else if (e[v,t-1] == 0 & zi[v,t] == 0) {
      prod_temp <- gamma[v,t-1] * prod(exp(-1 * upsilon[v,,t] * p[,t,] *
o))
      psi_temp <- prod_temp / (prod_temp + 1 - gamma[v,t-1])
      z[v,t] <- rbinom(1, 1, psi_temp)
     }
     e[v,t] <- ifelse(sum(z[v,1:t]) == 0, 0, 1)
    } # t
   } # v

   ### Save samples
   psi_save[k] <- psi
   kappa_save[k] <- kappa
   beta_upsilon_save[,k] <- beta_upsilon
   beta_phi_save[,k] <- beta_phi
   beta_gamma_save[,k] <- beta_gamma
   sigma_save[k] <- sigma
   alpha_save[,k] <- alpha
   if (nmcmc > 2000) {
    if (k > nmcmc-2000) {
     ss_save[,,,k-nmcmc+2000] <- ss
    }
   }
  } # k

  ### Write output
  list(psi_save=psi_save,
    kappa_save=kappa_save, beta_upsilon_save=beta_upsilon_save,
    beta_phi_save=beta_phi_save, beta_gamma_save=beta_gamma_save,
sigma_save=sigma_save,
    alpha_save=alpha_save,
    ss_save=ss_save)

} # scr_survival_recruitment_movement_mcmc
```

We then use the following code to implement the algorithm, which takes about 15 h to run. Then using code in Appendix 5.4, we can create traceplots for the monitored parameters, which show good model convergence and consistency between parameter estimates of ψ, κ, $\beta^{[\upsilon]}$, $\beta^{[\phi]}$, $\beta^{[\gamma]}$, σ, and α and their true values used in data simulation (Fig. 5.7).

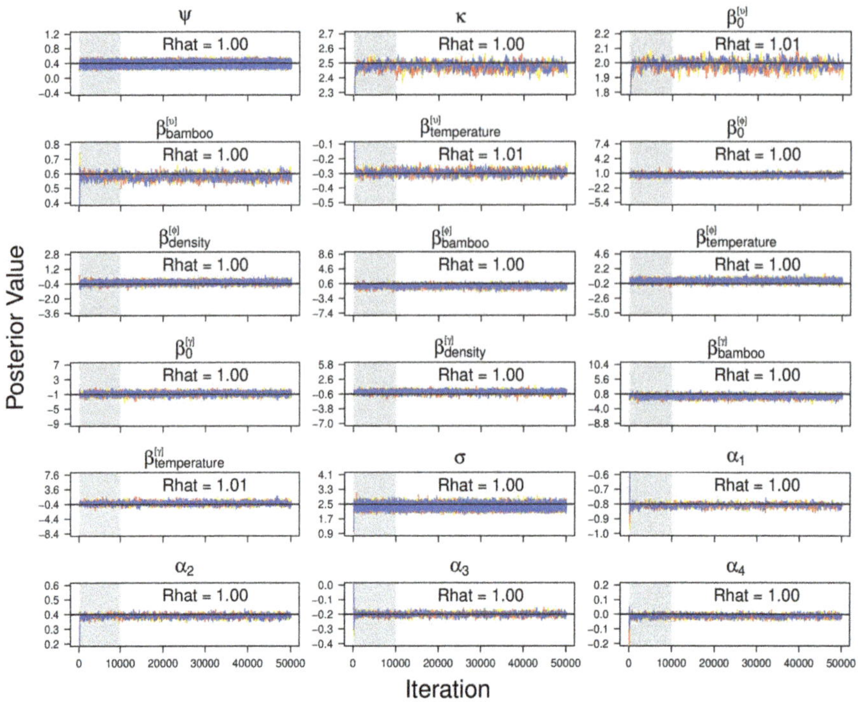

Fig. 5.7 Traceplots showing the posterior chains for ψ, κ, $\beta^{[\upsilon]}$, $\beta^{[\phi]}$, $\beta^{[\gamma]}$, σ, and α in the dynamic SCR model with survival, recruitment, and movement. Rhat values are calculated after discarding the first 10,000 iterations. Horizontal lines represent the true values used in data simulation

```
library(foreach)  # for parallel computing
library(doParallel) # for parallel computing

numCores <- round(detectCores() / 2) # only use half of the cores for
parallel computing
registerDoParallel(numCores) # setup parallel computing

nmcmc <- 50000 # number of iterations of MCMC computing
chain <- 3    # number of chains

start_time <- Sys.time() # start time of computing
out <- foreach (i=1:chain, .packages='boot') %dopar% {
 scr_survival_recruitment_movement_mcmc(
  y=y, x=x, w=w, o=o, ll=ll,
  lon_min=lon_min, lon_max=lon_max, lat_min=lat_min,
lat_max=lat_max,
  nmcmc=nmcmc)
```

```
} # i
stopImplicitCluster() # clean up the cluster
end_time <- Sys.time() # end time of computing
time_taken <- end_time - start_time # computing time
```

We can also use the following code to plot the estimated trajectories of movement (i.e., changes in territory centroids), which show satisfying agreement with the simulated true trajectories in years when an individual is captured but not so when an individual is not captured (Fig. 5.8). For example, the estimated trajectory of movement is quite consistent with the simulated true trajectory for individual 79, an individual that has been captured in every year of the study period; the estimated trajectories of movement are mostly consistent with the simulated true trajectories for individuals 19 and 47, except for the last year of the study period when these individuals are not captured; the estimated trajectory of movement cannot represent

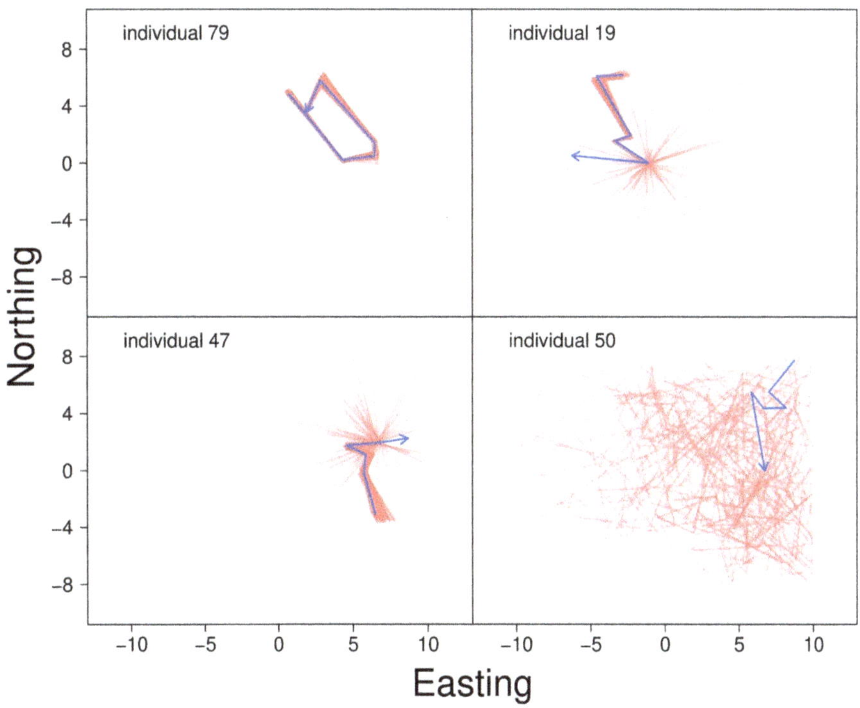

Fig. 5.8 The estimated trajectories of dispersal movement (i.e., changes in territory centroids; pink lines) in comparison to the simulated true trajectories (blue lines with arrows). A few individuals are selected, among which individual 79 is captured in every year of the study period, individuals 19 and 47 are captured in every but the last year of the study period, and individual 50 is not captured in any of the years of the study period

the simulated true trajectory at all for individual 50 as this individual has never been captured during the study period.

```
ss_post <- array(, dim=c(nindv, 2, nyear, 6000))
ss_post[,,,  1:2000] <- out[[1]]$ss
ss_post[,,,2001:4000] <- out[[1]]$ss
ss_post[,,,4001:6000] <- out[[1]]$ss
nplot <- 100
ss_plot <- ss_post[,,,sort(sample(1:6000,nplot,replace=F))]
id_plot <- c(79, 19, 47, 50)

par(mfrow=c(2,2))
par(mar=c(0,0,0,0))
par(oma=c(6,6,1,1))

for (v in 1:length(id_plot)) {
 id <- id_plot[v]
 plot(1, xlim=c(-12,12), ylim=c(-10,10), type='n', axes=F, xlab='',
ylab='')
 box()
 for (t in min(which(z[id,]==1)):(nyear-1)) {
  for (k in 1:nplot) {
   lines(x=c(ss_plot[id,1,t,k], ss_plot[id,1,t+1,k]), y=c(ss_plot
[id,2,t,k], ss_plot[id,2,t+1,k]),
      col=rgb(240,128,128,alpha=60,maxColorValue=255), lwd=2)
  } # k
 } # t
 for (t in min(which(z[id,]==1)):(nyear-2)) {
  lines(x=c(ss[id,1,t], ss[id,1,t+1]), y=c(ss[id,2,t], ss[id,2,t
+1]), col='royalblue', lwd=2)
 } # t
 arrows(ss[id,1,nyear-1], ss[id,2,nyear-1], ss[id,1,nyear], ss[id,2,
nyear], length=0.1, lwd=2, col='royalblue')
 if (v %in% c(1,3)) {
  axis(2, at=seq(-08,08,4), cex.axis=1.6, las=2)
 }
 if (v %in% c(3,4)) {
  axis(1, at=seq(-10,10,5), cex.axis=1.6)
 }
 text(x=-11, y=9, pos=4, labels=paste0('individual ', id), cex=1.6)
} # v
title(xlab='Easting' , cex.lab=3, line=4.0, outer=T)
title(ylab='Northing', cex.lab=3, line=3.5, outer=T)
```

5.4 Dynamic Spatial Capture-Recapture Models with Individual Genetic Information

In this chapter, we take advantage of the individual-based nature of SCR models to incorporate individual-level genetic information. Unlike the spatial integrated population models considered in Sect. 4.4.4 that group individuals into gene classes, this time we truly consider the uniqueness of each individual. Using our red panda as an example, we can assume that each individual has a unique genetic characteristic to determine its position along a continuous axis from r-selected tendency to K-selected tendency. Further, a more K-selected individual, in comparison to a more r-strategies one, tends to have higher survival and lower recruitment. We assume no dispersal movement for now, which surely can be extended. When surveying the population, this time researchers use fur traps instead of camera traps. The genetic materials obtained from these fur traps allow us to identify each individual, and obtain a continuous genetic score, as long as an individual is captured for at least once. The genetic score can tell us about the individual's genetic characteristics such that a higher genetic score indicates a more K-selected tendency, and vice versa.

Using statistical language, the model describes habitat use in the same way as in Sect. 5.3 such that:

$$\upsilon_{v,i,t} = exp\left(-1 \times \kappa \times dist_{v,i} + \beta_0^{[\upsilon]} + \beta_1^{[\upsilon]} \times x_{i,t,1} + \ldots + \beta_{n_{covs}}^{[\upsilon]} \times x_{i,t,n_{covs}} \right), \tag{5.9}$$

in which $\upsilon_{v,\ i,\ t}$ represents the true amount of time of individual v using site i in year t, $dist_{v,\ i}$ is the distance between the centroid of an individual v's territory and site i, κ is a decay parameter representing the effect of this distance on the logarithm of λ, and β's are intercept and slope parameters representing the effects of environmental covariates x's on the logarithm of $\upsilon_{v,\ i,\ t}$.The model then describes the observation process again in the same way as in Sect. 5.3 such that:

$$\left. \begin{array}{r} y_{v,i,t,j} \sim Poisson\left(\upsilon_{v,i,t} \times z_{v,t} \times p_{i,t,j} \times o_i\right) \\ p_{i,t,j} = exp\left(\alpha_1 \times w_{i,t,j,1} + \ldots + \alpha_{n_{pcvs}} \times w_{i,t,j,n_{pcvs}}\right) \end{array} \right\}, \tag{5.10}$$

in which $y_{v,\ i,\ t,\ j}$ represents the number of times for which individual v is captured at site i in year t by the jth trap, $z_{v,\ t}$ is an indicator that equals 1 if individual v is alive in year t and 0 otherwise, $p_{i,\ t,\ j}$ is the detection rate and is a function of observation covariates's with slope parameters α's, and o_i is an indicator that equals 1 if a trap is deployed at site i and 0 otherwise. In addition to SCR data, we also assume to obtain observed genetic score, q_v, which equals the true genetic score g_v as long as individual v is captured at least once during the study period, or is marked as a missing value if individual v has never been captured.

The key difference between this model and the model introduced in Sect. 5.3.2 is that now survival and recruitment are both influenced by the individual's genetic characteristics, in addition to local density and environmental conditions. More specifically, we have:

$$\left.\begin{array}{c} z_{v,t=1} \sim Bernoulli(\psi) \\ z_{v,t\geq 2} \sim Bernoulli\left[z_{v,t-1} \times \phi_{v,t-1} + (1-e_{v,t-1}) \times \gamma_{v,t-1}\right] \\ \phi_{v,t} = logit^{-1}\left[\beta_0^{[\phi]} + \beta_{density}^{[\phi]} \times \frac{\sum_{i=1}^{n_{site}} \sum (\upsilon_{-v,i,t} \times z_{-v,t}) \times \upsilon_{v,i,t}}{\sum_{i=1}^{n_{site}} \upsilon_{v,i,t}} + \beta_1^{[\phi]} \times \frac{\sum_{i=1}^{n_{site}} x_{i,t-1,1} \times \upsilon_{v,i,t}}{\sum_{i=1}^{n_{site}} \upsilon_{v,i,t}} + \ldots + \beta_{gene}^{[\phi]} \times g_v\right] \\ \gamma_{v,t} = logit^{-1}\left[\beta_0^{[\gamma]} + \beta_{density}^{[\gamma]} \times \frac{\sum_{i=1}^{n_{site}} \sum (\upsilon_{-v,i,t} \times z_{-v,t}) \times \upsilon_{v,i,t}}{\sum_{i=1}^{n_{site}} \upsilon_{v,i,t}} + \beta_1^{[\gamma]} \times \frac{\sum_{i=1}^{n_{site}} x_{i,t-1,1} \times \upsilon_{v,i,t}}{\sum_{i=1}^{n_{site}} \upsilon_{v,i,t}} + \ldots + \beta_{gene}^{[\gamma]} \times g_v\right] \end{array}\right\}, \tag{5.11}$$

in which ψ is the probability of existence in the first year, $z_{v,\ t}$ indicates if individual v is alive ($z_{v,\ t} = 1$) or not ($z_{v,\ t} = 0$) in year t, $e_{v,\ t}$ indicates if individual v has existed ($e_{v,\ t} = 1$) or not ($e_{v,\ t} = 0$) up until year t, $\phi_{v,\ t}$ is the survival probability and $\gamma_{v,\ t}$ is the recruitment probability, both of which are governed by density, environment, and gene (denoted g_v) with intercept and slope parameters $\beta^{[\phi]}$'s and $\beta^{[\gamma]}$'s, respectively. The genetic score g_v further follows a Normal distribution with a mean $\mu^{[gene]}$ and standard deviation $\sigma^{[gene]}$. In particular, we have $\beta_{gene}^{[\phi]}$ and $\beta_{gene}^{[\gamma]}$ in the model to represent the effect of genes on survival and recruitment, respectively.

As mentioned above, we do not consider dispersal movement in this model for simplicity, which can be further extended and evaluated. Also note that the individual characteristic considered here, gene, normally does not change over time. We can extend the model to allow other kinds of individual characteristics, such as body mass, that change temporally. It is also possible to extend the model to account for observation errors in the measures of individual characteristics.

Next, we can use the following code to simulate data for the dynamic SCR model with genetic information. We assume a relatively strong decay parameter ($\kappa = 2.5$), a positive effect of bamboo patch size ($\beta_{bamboo}^{[\upsilon]} = 0.6$) and a negative effect of temperature ($\beta_{temperature}^{[\upsilon]} = -0.3$) on the true amount of time of habitat use on the log scale. We assume negative effects of density ($\beta_{density}^{[\phi]} = -0.4$, $\beta_{density}^{[\gamma]} = -0.6$), positive effects of bamboo patch size ($\beta_{bamboo}^{[\phi]} = 0.6$, $\beta_{bamboo}^{[\gamma]} = 0.8$), and negative effects of temperature ($\beta_{temperature}^{[\phi]} = -0.2$, $\beta_{temperature}^{[\gamma]} = -0.4$) on survival and recruitment, respectively, on the logit scale. We assume a positive effect of genetic score on logit survival ($\beta_{gene}^{[\phi]} = 0.8$), and a negative effect of genetic score on logit recruitment ($\beta_{gene}^{[\phi]} = -0.2$).

```
library(boot) # for using logit and inv.logit functions

set.seed(10)

# Basic values
nindv <- 100 # number of individuals (with data augmentation)
nsite <- 360 # number of sites
nyear <- 6  # number of years
nreps <- 5  # number of within-season replicates
ncovs <- 2  # number of environmental covariates
npcvs <- 4  # number of observational covariates
```

```
psi <- 0.4                    # probability of first-year existence
kappa <- 2.5                  # distance effect on habitat use
beta_upsilon <- c(2, 0.6, -0.3)       # intercept and slopes for habitat
use
beta_phi  <- c( 1.0, -0.4, 0.6, -0.2,  0.8) # intercept and slopes for
apparent survival
beta_gamma <- c(-1.0, -0.6, 0.8, -0.4, -0.2) # intercept and slopes for
recruitment
mu_gene <- 2                  # mean of genetic score
sigma_gene <- 0.5               # standard deviation of genetic score
alpha <- c(-0.6, 0.4, -0.2, 0)      # intercept and slopes for detection
probability
lon_min <- -10                 # lower bound of longitude or easting
lon_max <- +10                 # upper bound of longitude or easting
lat_min <- - 8                # lower bound of latitude or northing
lat_max <- + 8                # upper bound of latitude or northing

# Define functions
crossdist <- function(ss, ll) { # function for calculating distance
 c1 <- complex(real=ss[,1], imaginary=ss[,2])
 c2 <- complex(real=ll[,1], imaginary=ll[,2])
 dist <- outer(c1, c2, function(z1, z2) Mod(z1 - z2))
} # crossdist

# Read in environmental covariates
load('data/covariates.RData')
ll <- cbind(cov$lon, cov$lat) # coordinates of sites

x1 <- (log(cov$size) - mean(log(cov$size))) / sd(log(cov$size)) #
standardized log habitat patch size
x2 <- (cov$temp - mean(cov$temp)) / sd(cov$temp) # standardized
temperature
x <- array(, dim=c(nsite, nyear, ncovs)) # environmental covariates
x[,,1] <- x1[,1:nyear]
x[,,2] <- x2[,1:nyear]

# Simulate data
### Process
ss <- matrix(, nindv, 2) # cooridnates of animal territory centroids
ss[,1] <- runif(nindv, lon_min, lon_max)
ss[,2] <- runif(nindv, lat_min, lat_max)

d <- crossdist(ss, ll) # distance between territory centroids and sites
```

```
upsilon <- array(0, dim=c(nindv, nsite, nyear)) # habitat use (i.e.,
time spent at a site)
for (v in 1:nindv) {
 for (t in 1:nyear) {
  upsilon[v,,t] <- exp(-1 * kappa * d[v,] + cbind(1, x[,t,]) %*%
beta_upsilon)
 } # t
} # v

z <- matrix(0, nindv, nyear) # indicator of individual existence at a
given year
e <- matrix(0, nindv, nyear) # indicator of individual existence so far
g <- rnorm(nindv, mu_gene, sigma_gene) # genetic score for each
individual
phi  <- matrix(0, nindv, nyear-1) # apparent survival probability
gamma <- matrix(0, nindv, nyear-1) # recruitment probability
z[,1] <- rbinom(nindv, 1, psi) # individual existence in the first year
e[,1] <- z[,1]
for (v in 1:nindv) {
 for (t in 2:nyear) {
  phi [v,t-1] <- inv.logit(c(1, sum(upsilon[v,,t-1] * colSums(upsilon
[-v,,t-1] * z[-v,t-1])) / sum(upsilon[v,,t-1]), colSums(upsilon[v,,t-
1] * x[,t-1,]) / sum(upsilon[v,,t-1]), (g[v] - mu_gene) / sigma_gene) %*
% beta_phi )
  gamma[v,t-1] <- inv.logit(c(1, sum(upsilon[v,,t-1] * colSums
(upsilon[-v,,t-1] * z[-v,t-1])) / sum(upsilon[v,,t-1]), colSums
(upsilon[v,,t-1] * x[,t-1,]) / sum(upsilon[v,,t-1]), (g[v] - mu_gene) /
sigma_gene) %*% beta_gamma)
  z[v,t] <- rbinom(1, 1, z[v,t-1] * phi[v,t-1] + (1 - e[v,t-1]) * gamma[v,
t-1])
  e[v,t] <- ifelse(sum(z[v,1:t]) == 0, 0, 1)
 } # t
} # v

### Spatial capture-recapture data
w <- array(rnorm(nsite * nyear * nreps * npcvs, 0, 1), dim=c(nsite,
nyear, nreps, npcvs)) # observational covariates

p <- array(0, dim=c(nsite, nyear, nreps)) # detection rate
for (i in 1:nsite) {
 for (t in 1:nyear) {
  p[i,t,] <- exp(w[i,t,,] %*% alpha)
 } # t
} # i
```

```
o <- ifelse(ll[,1] <= 8 & ll[,1] >= -8 & ll[,2] <= 6 & ll[,2] >= -6, 1, 0) #
indicator of sites with camera traps

y <- array(0, dim=c(nindv, nsite, nyear, nreps)) # frequency of capture
data
for (v in 1:nindv) {
 for (i in 1:nsite) {
  for (t in 1:nyear) {
   y[v,i,t,] <- rpois(nreps, upsilon[v,i,t] * z[v,t] * p[i,t,] * o[i])
  } # t
 } # i
} # v

q <- g # observed genetic score
q[which(apply(y,1,sum) == 0)] <- NA # genetic score is known only for
individuals that are captured for at least once
```

Next, we can develop a custom MCMC algorithm to fit the model. The interest is to estimate parameters ψ, κ, $\beta^{[\upsilon]}$, $\beta^{[\phi]}$, $\mu^{[gene]}$, $\sigma^{[gene]}$, α, g, and z given data y, q, x, w, o, and ll.

```
gene_scr_mcmc <- function(y, q, x, w, o, ll, lon_min, lon_max, lat_min,
lat_max, nmcmc) {

 # Setup variables
 nindv <- dim(y)[1]
 nsite <- dim(y)[2]
 nyear <- dim(y)[3]
 nreps <- dim(y)[4]
 ncovs <- dim(x)[3]
 npcvs <- dim(w)[4]
 ymax <- apply(y, 1:3, max)
 obs01 <- ifelse(apply(y, c(1,3), sum) > 0, 1, 0)
 beta_phi_update <- rep(0, nmcmc)
 beta_gamma_update <- rep(0, nmcmc)

 psi_save <- rep(0, nmcmc)
 kappa_save <- rep(0, nmcmc)
 beta_upsilon_save <- matrix(0, ncovs+1, nmcmc)
 beta_phi_save <- matrix(0, ncovs+3, nmcmc)
 beta_gamma_save <- matrix(0, ncovs+3, nmcmc)
 mu_gene_save <- rep(0, nmcmc)
 sigma_gene_save <- rep(0, nmcmc)
 alpha_save <- matrix(0, npcvs, nmcmc)
```

```
g_save <- matrix(0, nindv, 2000)
z_save <- array(, dim=c(nindv, nyear, 2000))

# Priors
log_kappa_mean <- 0
log_kappa_sd <- 2
beta_upsilon_mean <- 0
beta_upsilon_sd <- 2
beta_phi_mean <- 0
beta_phi_sd <- 2
beta_gamma_mean <- 0
beta_gamma_sd <- 2
mu_gene_mean <- 0
mu_gene_sd <- 2
sigma_gene_mean <- 2 # mean of sigma_gene square
sigma_gene_sd <- 10 # standard deviation of sigma_gene square
alpha_mean <- 0
alpha_sd <- 2

# Starting values
psi <- 0.5
kappa <- 1
beta_upsilon <- rep(0, ncovs+1)
beta_phi  <- rep(0, ncovs+3)
beta_gamma <- rep(0, ncovs+3)
mu_gene <- 0
sigma_gene <- 2
sigma_gene_a <- (sigma_gene_mean ^ 2) / (sigma_gene_sd ^ 2) + 2
sigma_gene_b <- 1 / (sigma_gene_mean * (sigma_gene_a - 1))
alpha <- rep(0, npcvs)
ss <- matrix(0, nindv, 2)
ss[,1] <- runif(nindv, lon_min, lon_max)
ss[,2] <- runif(nindv, lat_min, lat_max)
for (v in 1:nindv) {
 if (sum(obs01[v,]) > 0) {
  ss[v,1] <- mean(ll[which(rowSums(ymax[v,,]) > 0), 1])
  ss[v,2] <- mean(ll[which(rowSums(ymax[v,,]) > 0), 2])
 }
} # v
d <- crossdist(ss, ll)
upsilon <- array(0, dim=c(nindv, nsite, nyear))
for (v in 1:nindv) {
 for (t in 1:nyear) {
  upsilon[v,,t] <- exp(-1 * kappa * d[v,] + cbind(1, x[,t,]) %*%
beta_upsilon)
```

```
   } # t
  } # v
 phi  <- matrix(0.5, nindv, nyear-1)
 gamma <- matrix(0.5, nindv, nyear-1)
 g <- q
 g[which(is.na(g))] <- rnorm(length(which(is.na(g))), mean(g,na.
rm=T), sd(g,na.rm=T))
 p <- array(0.5, dim=c(nsite, nyear, nreps))
 zi <- obs01
 for (v in 1:nindv) {
  if (sum(zi[v,]) > 0) {
   zi[v,min(which(zi[v,] == 1)):max(which(zi[v,] == 1))] <- 1
  }
 } # v
 z <- zi
 e <- matrix(0, nindv, nyear)
 e[,1] <- z[,1]
 for (t in 2:nyear) {
  e[,t] <- ifelse(rowSums(z[,1:t])==0, 0, 1)
 } # t

 # Tuning factors
 kappa_tune <- 0.2
 beta_upsilon_tune <- 0.06
 beta_phi_tune <- 0.2
 beta_gamma_tune <- 0.4
 alpha_tune <- 0.035
 ss_tune <- 1
 g_tune <- 0.1

 # Begin MCMC loop
 for (k in 1:nmcmc) {
  ### Sample psi
  psi <- rbeta(1, sum(z[,1]) + 0.001, sum(1-z[,1]) + 1)

  ### Sample kappa
  kappa_star <- exp(rnorm(1, log(kappa), kappa_tune))
  upsilon_star <- array(0, dim=c(nindv, nsite, nyear))
  for (v in 1:nindv) {
   for (t in 1:nyear) {
    upsilon_star[v,,t] <- exp(-1 * kappa_star * d[v,] + cbind(1, x[,t,])
%*% beta_upsilon)
   } # t
  } # v
  prob1 <- prob2 <- matrix(0, nindv, nyear)
```

```
  for (v in 1:nindv) {
   for (t in 1:nyear) {
    prob1[v,t] <- sum(dpois(y[v,,t,], upsilon_star[v,,t] * z[v,t] * p[,
t,] * o, log=TRUE))
    prob2[v,t] <- sum(dpois(y[v,,t,], upsilon     [v,,t] * z[v,t] * p[,t,]
* o, log=TRUE))
   } # t
  } # v
  mh1 <- sum(prob1) +
      dnorm(log(kappa_star), log_kappa_mean, log_kappa_sd, log=TRUE)
  mh2 <- sum(prob2) +
      dnorm(log(kappa     ), log_kappa_mean, log_kappa_sd, log=TRUE)
  mhK <- exp(mh1 - mh2)
  if (mhK > runif(1)) {
   kappa <- kappa_star
   upsilon<- upsilon_star
  }

  ### Sample beta_upsilon
  beta_upsilon_star <- rnorm(ncovs+1, beta_upsilon,
beta_upsilon_tune)
  upsilon_star <- array(0, dim=c(nindv, nsite, nyear))
  for (v in 1:nindv) {
   for (t in 1:nyear) {
    upsilon_star[v,,t] <- exp(-1 * kappa * d[v,] + cbind(1, x[,t,]) %*%
beta_upsilon_star)
   } # t
  } # v
  prob1 <- prob2 <- matrix(0, nindv, nyear)
  for (v in 1:nindv) {
   for (t in 1:nyear) {
    prob1[v,t] <- sum(dpois(y[v,,t,], upsilon_star[v,,t] * z[v,t] * p[,
t,] * o, log=TRUE))
    prob2[v,t] <- sum(dpois(y[v,,t,], upsilon     [v,,t] * z[v,t] * p[,t,]
* o, log=TRUE))
   } # t
  } # v
  mh1 <- sum(prob1) +
      sum(dnorm(beta_upsilon_star, beta_upsilon_mean,
beta_upsilon_sd, log=TRUE))
  mh2 <- sum(prob2) +
      sum(dnorm(beta_upsilon     , beta_upsilon_mean, beta_upsilon_sd,
log=TRUE))
  mhB <- exp(mh1 - mh2)
  if (mhB > runif(1)) {
```

```
    beta_upsilon <- beta_upsilon_star
    upsilon <- upsilon_star
   }

  ### Sample beta_phi
  beta_phi_star <- rnorm(ncovs+3, beta_phi, beta_phi_tune)
  phi_star <- matrix(0, nindv, nyear-1)
  for (v in 1:nindv) {
   for (t in 2:nyear) {
    phi_star[v,t-1] <- inv.logit(c(1, sum(upsilon[v,,t-1] * colSums
(upsilon[-v,,t-1] * z[-v,t-1])) / sum(upsilon[v,,t-1]), colSums
(upsilon[v,,t-1] * x[,t-1,]) / sum(upsilon[v,,t-1]), (g[v] - mu_gene) /
sigma_gene) %*% beta_phi_star)
   } # t
  } # v
  prob1 <- prob2 <- rep(0, nyear-1)
  for (t in 1:(nyear-1)) {
   prob1[t] <- sum(dbinom(z[which(z[,t]==1),t+1], 1, phi_star[which
(z[,t]==1),t], log=TRUE))
   prob2[t] <- sum(dbinom(z[which(z[,t]==1),t+1], 1, phi    [which(z[,
t]==1),t], log=TRUE))
  } # t
  mh1 <- sum(prob1) +
     sum(dnorm(beta_phi_star, beta_phi_mean, beta_phi_sd, log=TRUE))
  mh2 <- sum(prob2) +
     sum(dnorm(beta_phi     , beta_phi_mean, beta_phi_sd, log=TRUE))
  mhP <- exp(mh1 - mh2)
  if (mhP > runif(1)) {
   beta_phi <- beta_phi_star
   phi <- phi_star
   beta_phi_update[k] <- 1
  }
  if (k > 1000) {
   if (sum(beta_phi_update[(k-9):k]) == 0) {
    beta_phi <- runif(ncovs+3, min=apply(beta_phi_save[,(k-99):k],1,
min), max=apply(beta_phi_save[,(k-99):k],1,max))
    phi <- matrix(0, nindv, nyear-1)
    for (v in 1:nindv) {
     for (t in 2:nyear) {
      phi[v,t-1] <- inv.logit(c(1, sum(upsilon[v,,t-1] * colSums
(upsilon[-v,,t-1] * z[-v,t-1])) / sum(upsilon[v,,t-1]), colSums
(upsilon[v,,t-1] * x[,t-1,]) / sum(upsilon[v,,t-1]), (g[v] - mu_gene) /
sigma_gene) %*% beta_phi)
     } # t
    } # v
```

```
    beta_phi_update[k] <- 1
   }
  }

  ### Sample beta_gamma
  beta_gamma_star <- rnorm(ncovs+3, beta_gamma, beta_gamma_tune)
  gamma_star <- matrix(0, nindv, nyear-1)
  for (v in 1:nindv) {
   for (t in 2:nyear) {
    gamma_star[v,t-1] <- inv.logit(c(1, sum(upsilon[v,,t-1] * colSums
(upsilon[-v,,t-1] * z[-v,t-1])) / sum(upsilon[v,,t-1]), colSums
(upsilon[v,,t-1] *x[,t-1,]) / sum(upsilon[v,,t-1]), (g[v] - mu_gene) /
sigma_gene) %*% beta_gamma_star)
   } # t
  } # v
  prob1 <- prob2 <- rep(0, nyear-1)
  for (t in 1:(nyear-1)) {
   prob1[t] <- sum(dbinom(z[which(e[,t]==0),t+1], 1, gamma_star
[which(e[,t]==0),t], log=TRUE))
   prob2[t] <- sum(dbinom(z[which(e[,t]==0),t+1], 1, gamma    [which(e
[,t]==0),t], log=TRUE))
  } # t
  mh1 <- sum(prob1) +
      sum(dnorm(beta_gamma_star, beta_gamma_mean, beta_gamma_sd,
log=TRUE))
  mh2 <- sum(prob2) +
      sum(dnorm(beta_gamma     , beta_gamma_mean, beta_gamma_sd,
log=TRUE))
  mhP <- exp(mh1 - mh2)
  if (mhP > runif(1)) {
   beta_gamma <- beta_gamma_star
   gamma <- gamma_star
   beta_gamma_update[k] <- 1
  }
  if (k > 1000) {
   if (sum(beta_gamma_update[(k-9):k]) == 0) {
    beta_gamma <- runif(ncovs+3, min=apply(beta_gamma_save[,(k-99):
k],1,min), max=apply(beta_gamma_save[,(k-99):k],1,max))
    gamma <- matrix(0, nindv, nyear-1)
    for (v in 1:nindv) {
     for (t in 2:nyear) {
      gamma[v,t-1] <- inv.logit(c(1, sum(upsilon[v,,t-1] * colSums
(upsilon[-v,,t-1] * z[-v,t-1])) / sum(upsilon[v,,t-1]), colSums
(upsilon[v,,t-1] *x[,t-1,]) / sum(upsilon[v,,t-1]), (g[v] - mu_gene) /
sigma_gene) %*% beta_gamma)
```

```
    } # t
   } # v
   beta_gamma_update[k] <- 1
  }
 }

 ### Sample sigma_gene
 temp_a <- 1 / (sum((q - mu_gene) ^ 2, na.rm=T) / 2 + 1 / sigma_gene_b)
 temp_b <- length(which(!(is.na(q)))) / 2 + sigma_gene_a
 sigma_gene <- sqrt(1 / rgamma(1, shape=temp_b, scale=temp_a))

 ### Sample mu_gene
 temp_sd <- sqrt(1 / (length(which(!(is.na(q)))) / (sigma_gene ^ 2) +
1 / 4))
 temp_mean <- (temp_sd ^ 2) * (sum(q, na.rm=T) / (sigma_gene ^ 2))
 mu_gene <- rnorm(1, temp_mean, temp_sd)

 ### Sample alpha
 alpha_star <- rnorm(npcvs, alpha, alpha_tune)
 p_star <- array(, dim=c(nsite, nyear, nreps))
 for (i in 1:nsite) {
  for (t in 1:nyear) {
   p_star[i,t,] <- exp(w[i,t,,] %*% alpha_star)
  } # t
 } # i
 prob1 <- prob2 <- matrix(0, nindv, nyear)
 for (v in 1:nindv) {
  for (t in 1:nyear) {
   prob1[v,t] <- sum(dpois(y[v,,t,], upsilon[v,,t] * z[v,t] * p_star[,
t,] * o, log=TRUE))
   prob2[v,t] <- sum(dpois(y[v,,t,], upsilon[v,,t] * z[v,t] * p    [,t,]
* o, log=TRUE))
  } # t
 } # v
 mh1 <- sum(prob1) +
     sum(dnorm(alpha_star, alpha_mean, alpha_sd, log=TRUE))
 mh2 <- sum(prob2) +
     sum(dnorm(alpha    , alpha_mean, alpha_sd, log=TRUE))
 mhA <- exp(mh1 - mh2)
 if (mhA > runif(1)) {
  alpha <- alpha_star
  p <- p_star
 }
```

```
  ### Sample ss
  ss_star <- matrix(rnorm(nindv*2, ss, ss_tune), nindv, 2)
  d_star <- crossdist(ss_star, ll)
  upsilon_star <- array(0, dim=c(nindv, nsite, nyear))
  for (v in 1:nindv) {
   for (t in 1:nyear) {
    upsilon_star[v,,t] <- exp(-1 * kappa * d_star[v,] + cbind(1, x[,t,])
%*% beta_upsilon)
   } # t
  } # v
  for (v in 1:nindv) {
   if (ss_star[v,1] > lon_min & ss_star[v,1] < lon_max &
     ss_star[v,2] > lat_min & ss_star[v,2] < lat_max) {
    prob1 <- prob2 <- rep(0, nyear)
    for (t in 1:nyear) {
     prob1[t] <- sum(dpois(y[v,,t,], upsilon_star[v,,t] * z[v,t] * p[,
t,] * o, log=TRUE))
    prob2[t] <- sum(dpois(y[v,,t,], upsilon [v,,t] * z[v,t] * p[,t,] *
o, log=TRUE))
    } # t
    mh1 <- sum(prob1)
    mh2 <- sum(prob2)
    mhS <- exp(mh1 - mh2)
    if (mhS > runif(1)) {
     ss[v,] <- ss_star[v,]
     d[v,] <- d_star[v,]
     upsilon[v,,] <- upsilon_star[v,,]
    }
   }
  } # v

  ### Sample g
  g_star <- q
  g_star[which(is.na(q))] <- rnorm(length(which(is.na(q))), g[which
(is.na(q))], g_tune)
  phi_star <- gamma_star <- matrix(0, nindv, nyear-1)
  for (v in 1:nindv) {
   if (is.na(q[v])) {
    prob1 <- prob2 <- rep(0, nyear-1)
    for (t in 2:nyear) {
     phi_star [v,t-1] <- inv.logit(c(1, sum(upsilon[v,,t-1] * colSums
(upsilon[-v,,t-1] * z[-v,t-1])) / sum(upsilon[v,,t-1]), colSums
(upsilon[v,,t-1] * x[,t-1,]) / sum(upsilon[v,,t-1]), (g_star[v] -
mu_gene) / sigma_gene) %*% beta_phi )
    gamma_star[v,t-1] <- inv.logit(c(1, sum(upsilon[v,,t-1] * colSums
```

```
(upsilon[-v,,t-1] * z[-v,t-1])) / sum(upsilon[v,,t-1]), colSums
(upsilon[v,,t-1] * x[,t-1,]) / sum(upsilon[v,,t-1]), (g_star[v] -
mu_gene) / sigma_gene) %*% beta_gamma)
     if (z[v,t-1] == 1) {
      prob1[t-1] <- dbinom(z[v,t], 1, phi_star[v,t-1], log=TRUE)
      prob2[t-1] <- dbinom(z[v,t], 1, phi    [v,t-1], log=TRUE)
     } else if (e[v,t-1] == 0) {
      prob1[t-1] <- dbinom(z[v,t], 1, gamma_star[v,t-1], log=TRUE)
      prob2[t-1] <- dbinom(z[v,t], 1, gamma    [v,t-1], log=TRUE)
     }
    } # t
    mh1 <- sum(prob1) +
       dnorm(g_star[v], mu_gene, sigma_gene, log=TRUE)
    mh2 <- sum(prob2) +
       dnorm(g   [v], mu_gene, sigma_gene, log=TRUE)
    mhg <- exp(mh1 - mh2)
    if (mhg > runif(1)) {
     g[v] <- g_star[v]
     phi[v,] <- phi_star[v,]
     gamma[v,] <- gamma_star[v,]
    }
   }
  } # v

  ### Sample z
  for (v in 1:nindv) {
   if (zi[v,1] == 0) {
    prod_temp <- psi * prod(exp(-1 * upsilon[v,,1] * p[,1,] * o))
    psi_temp <- prod_temp / (prod_temp + 1 - psi)
    psi_temp <- min(c(1, psi_temp))
    z[v,1] <- rbinom(1, 1, psi_temp)
   }
   e[v,1] <- z[v,1]
   for (t in 2:nyear) {
    if (e[v,t-1] == 1 & z[v,t-1] == 0 & zi[v,t] == 0) {
     z[v,t] <- 0
    } else if (z[v,t-1] == 1 & zi[v,t] == 0) {
    prod_temp <- phi[v,t-1] * prod(exp(-1 * upsilon[v,,t] * p[,t,] * o))
     psi_temp <- prod_temp / (prod_temp + 1 - phi[v,t-1])
     psi_temp <- min(c(1, psi_temp))
     z[v,t] <- rbinom(1, 1, psi_temp)
    } else if (e[v,t-1] == 0 & zi[v,t] == 0) {
     prod_temp <- gamma[v,t-1] * prod(exp(-1 * upsilon[v,,t] * p[,t,] *
o))
```

```
      psi_temp <- prod_temp / (prod_temp + 1 - gamma[v,t-1])
      psi_temp <- min(c(1, psi_temp))
      z[v,t] <- rbinom(1, 1, psi_temp)
    }
    e[v,t] <- ifelse(sum(z[v,1:t]) == 0, 0, 1)
  } # t
 } # v

 ### Save samples
 psi_save[k] <- psi
 kappa_save[k] <- kappa
 beta_upsilon_save[,k] <- beta_upsilon
 beta_phi_save[,k] <- beta_phi
 beta_gamma_save[,k] <- beta_gamma
 mu_gene_save[k] <- mu_gene
 sigma_gene_save[k] <- sigma_gene
 alpha_save[,k] <- alpha
 if (nmcmc > 2000) {
  if (k > nmcmc-2000) {
   g_save[,k-nmcmc+2000] <- g
   z_save[,,k-nmcmc+2000] <- z
  }
 }
} # k

### Write output
list(psi_save=psi_save,
   kappa_save=kappa_save, beta_upsilon_save=beta_upsilon_save,
   beta_phi_save=beta_phi_save, beta_gamma_save=beta_gamma_save,
   mu_gene_save=mu_gene_save, sigma_gene_save=sigma_gene_save,
   alpha_save=alpha_save,
   g_save=g_save, z_save=z_save)

} # gene_scr_mcmc
```

We then use the following code to implement the algorithm, which takes about 15 h to run. Then using code in Appendix 5.5, we can create traceplots for the monitored parameters, which show good model convergence and consistency between parameter estimates of ψ, κ, $\beta^{[\upsilon]}$, $\beta^{[\phi]}$, $\mu^{[gene]}$, $\sigma^{[gene]}$, and α and their true values used in data simulation (Fig. 5.9).

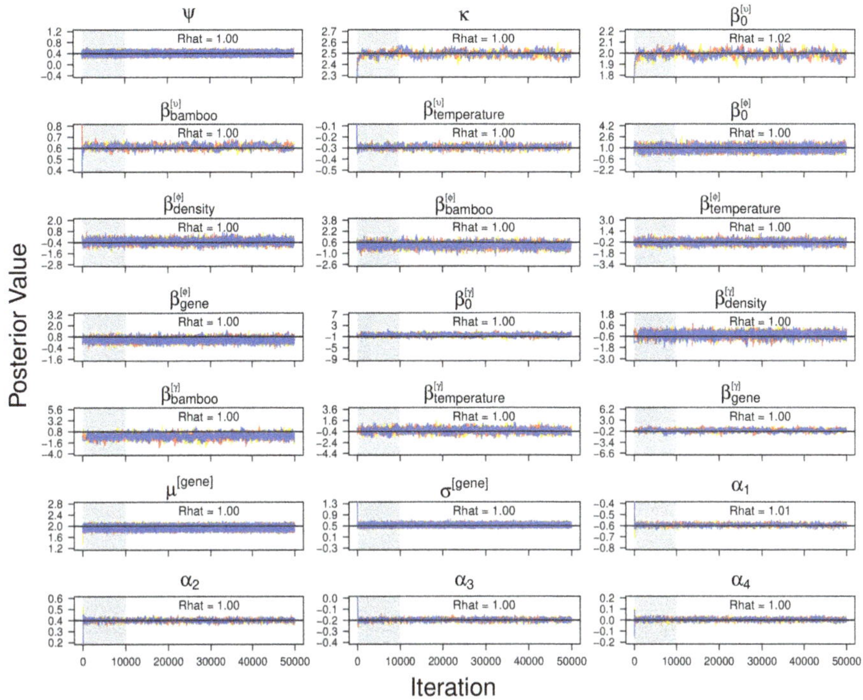

Fig. 5.9 Traceplots showing the posterior chains for ψ, κ, $\beta^{[\upsilon]}$, $\beta^{[\phi]}$, $\mu^{[gene]}$, $\sigma^{[gene]}$, and α in the dynamic SCR model with individual genetic information. Rhat values are calculated after discarding the first 10,000 iterations. Horizontal lines represent the true values used in data simulation

```
library(foreach)  # for parallel computing
library(doParallel) # for parallel computing

numCores <- round(detectCores() / 2) # only use half of the cores for
parallel computing
registerDoParallel(numCores) # setup parallel computing

nmcmc <- 50000 # number of iterations of MCMC computing
chain <- 3   # number of chains

start_time <- Sys.time() # start time of computing
out <- foreach (i=1:chain, .packages='boot') %dopar% {
 gene_scr_mcmc(y=y, q=q, x=x, w=w, o=o, ll=ll,
        lon_min=lon_min, lon_max=lon_max, lat_min=lat_min,
lat_max=lat_max,
        nmcmc=nmcmc)
} # i
stopImplicitCluster() # clean up the cluster
end_time <- Sys.time() # end time of computing
time_taken <- end_time - start_time # computing time
```

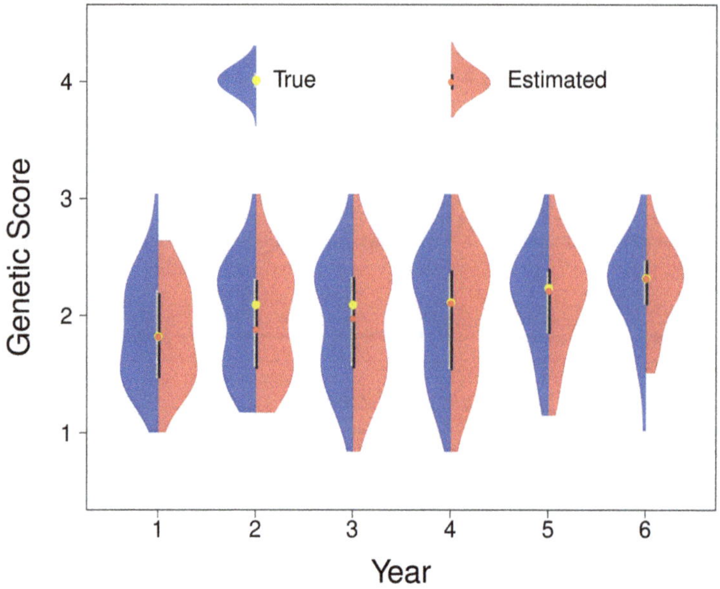

Fig. 5.10 Violin plots showing the true (blue) and estimated (red) distribution of genetic scores across the study population for each year. A higher genetic score indicates a more K-selected tendency, and vice versa

Next, we can use the following code to plot the distribution of genetic scores across the study population for each year, which shows great consistency between the estimates and the simulated true values (Fig. 5.10). In particular, the model can correctly represent the increase in average genetic score, which is likely due to the higher survival of more K-selected individuals who have higher genetic scores.

```
library(vioplot)

g_post <- cbind(out[[1]]$g_save, out[[2]]$g_save, out[[3]]$g_save)
g_med <- apply(g_post, 1, median)

par(mfrow=c(1,1))
par(mar=c(5.5,6,1,1))

plot(1, xlim=c(0.5,nyear+0.5), ylim=c(0.5,4.5), type='n', axes=F,
xlab='', ylab='')
for (t in 1:nyear) {
 z_post <- cbind(out[[1]]$z_save[,t,], out[[2]]$z_save[,t,], out
[[3]]$z_save[,t,])
 z_med <- apply(z_post, 1, median)
```

```
  vioplot(g[which(z[,t] == 1)], at=t, side='left', col='royalblue',
rectCol='grey80', lineCol=NA, colMed=NA, border=NA, add=T)
  vioplot(g_med[which(z_med == 1)], at=t, side='right',
col='lightcoral', rectCol='grey20', lineCol=NA, colMed=NA,
border=NA, add=T)
  points(median(g[which(z[,t] == 1)]) ~ t, pch=16, col='yellow',
cex=1.6)
  points(median(g_med[which(z_med == 1)]) ~ t, pch=16, col='tomato',
cex=1.2)
} # v

axis(1, at=seq(1,nyear,1), cex.axis=1.8)
axis(2, at=seq(0,6,1), cex.axis=1.8, las=2)
box()
title(xlab='Year', cex.lab=2.5, line=4)
title(ylab='Genetic Score', cex.lab=2.5, line=3.8)

vioplot(rnorm(1000,4,0.1), at=2, side='left', col='royalblue',
rectCol='grey80', lineCol=NA, colMed='yellow', pchMed=16,
border=NA, cex=1.6, add=T)
vioplot(rnorm(1000,4,0.1), at=4, side='right', col='lightcoral',
rectCol='grey20', lineCol=NA, colMed='tomato', pchMed=16,
border=NA, cex=1.2, add=T)
text(x=2.1, y=4, pos=4, label='True', cex=1.8)
text(x=4.5, y=4, pos=4, label='Estimated', cex=1.8)
```

5.5 Summary and Additional Reading

In this chapter, we talked about static, single-season SCR models and their dynamic, multi-season extensions, including an advanced dynamic SCR model that incorporates individual genetic information. While we have focused on the situation with perfect individual identification, SCR models have been developed for situations when only some of the individuals are identifiable (Chandler and Andrew Royle 2013; Augustine et al. 2018). The applications of SCR models are not limited to camera trap data but extend to other data types such as DNA data (Gardner et al. 2009; Augustine et al. 2020). Numerous studies have investigated sampling design for SCR models (Sollmann et al. 2012; Sun et al. 2014; Dupont et al. 2021). Lastly, recent studies have extended SCR models to integrate movement data (Chandler et al. 2022; Dupont et al. 2022; McClintock et al. 2022). Many other extensions of SCR models can be found in books and review papers (Royle et al. 2013b; Efford and Schofield 2022). Thus, here we conclude the chapter on SCR models. While we have introduced several families of models that are useful for characterizing spatially structured population dynamics, from relatively simple occupancy models all the way to advanced SCR models, we can still think about even more advanced models, which we will discuss in the next and the last chapter of the book.

References

Arnason AN (1972) Parameter estimates from mark-recapture experiments on two populations subject to migration and death. Res Popul Ecol (Kyoto) 13:97–113

Arnason AN (1973) The estimation of population size, migration rates and survival in a stratified population. Res Popul Ecol (Kyoto) 15:1

Augustine BC, Royle JA, Kelly MJ, Satter CB, Alonso RS, Boydston EE, Crooks KR (2018) Spatial capture–recapture with partial identity: an application to camera traps. Ann Appl Statistics 12:67–95

Augustine BC, Andrew Royle J, Linden DW, Fuller AK (2020) Spatial proximity moderates genotype uncertainty in genetic tagging studies. Proc Natl Acad Sci U S A 117:17903–17912

Borchers DL, Efford MG (2008) Spatially explicit maximum likelihood methods for capture-recapture studies. Biometrics 64:377–385

Chandler RB, Andrew Royle J (2013) Spatially explicit models for inference about density in unmarked or partially marked populations. Ann Appl Statistics 7:936–954

Chandler RB, Clark JD (2014) Spatially explicit integrated population models (E Cooch, Ed.). Methods Ecol Evol 5:1351–1360

Chandler RB, Crawford DA, Garrison EP, Miller KV, Cherry MJ (2022) Modeling abundance, distribution, movement and space use with camera and telemetry data. Ecology 103:e3583

Crosbie SF, Manly BFJ (1985) Parsimonious modelling of capture-mark-recapture studies. Biometrics 41:385–398

Dupont G, Royle JA, Nawaz MA, Sutherland C (2021) Optimal sampling design for spatial capture–recapture. Ecology 102:e03262

Dupont G, Linden DW, Sutherland C (2022) Improved inferences about landscape connectivity from spatial capture–recapture by integration of a movement model. Ecology 103:e3544

Efford MG (2004) Density estimation in live-trapping studies. Oikos 106:598–610

Efford MG, Schofield MR (2020) A spatial open-population capture-recapture model. Biometrics 76:392–402

Efford MG, Schofield MR (2022) A review of movement models in open population capture–recapture. Methods Ecol Evol 13:2106–2118

Ergon T, Gardner B (2014) Separating mortality and emigration: modelling space use, dispersal and survival with robust-design spatial capture-recapture data. Methods Ecol Evol 5:1327–1336

Gardner B, Royle JA, Wegan MT (2009) Hierarchical models for estimating density from DNA mark-recapture studies. Ecology 90:1106–1115

Gardner B, Reppucci J, Lucherini M, Royle JA (2010) Spatially explicit inference for open populations: estimating demographic parameters from camera-trap studies. Ecology 91:3376–3383

Gowan TA, Crum NJ, Roberts JJ (2021) An open spatial capture–recapture model for estimating density, movement, and population dynamics from line-transect surveys. Ecol Evol 11:7354–7365

Hooten MB, Schwob MR, Johnson DS, Ivan JS (2024) Geostatistical capture-recapture models. Spat Stat 59:100817

Jolly GM (1965) Explicit estimates from capture-recapture data with both death and immigration-stochastic model. Biometrika 52:225–248

Lewis WB, Cooper RJ, Chandler RB, Chitwood RW, Cline MH, Hallworth MT, Hatt JL, Hepinstall-Cymerman J, Kaiser SA, Rodenhouse NL, Sillett TS, Stodola KW, Webster MS,

Holmes RT (2023) Climate-mediated population dynamics of a migratory songbird differ between the trailing edge and range core. Ecol Monogr 93:e1559
McClintock BT, Abrahms B, Chandler RB, Conn PB, Converse SJ, Emmet RL, Gardner B, Hostetter NJ, Johnson DS (2022) An integrated path for spatial capture–recapture and animal movement modeling. Ecology 103:e3473
Milleret C, Dey S, Dupont P, Brøseth H, Turek D, de Valpine P, Bischof R (2023) Estimating spatially variable and density-dependent survival using open-population spatial capture–recapture models. Ecology:e3934
Royle JA, Young KV (2008) A hierarchical model for spatial capture recapture data. Ecology 89: 2281–2289
Royle JA, Dorazio RM, Link WA (2007) Analysis of multinomial models with unknown index using data augmentation. J Comput Graph Statistics 16:67–85
Royle JA, Chandler RB, Sun CC, Fuller AK (2013a) Integrating resource selection information with spatial capture-recapture. Methods Ecol Evol 4:520–530
Royle JA, Chandler RB, Sollmann R, Gardner B (2013b) Spatial capture-recapture. Academic Press
Royle JA, Fuller AK, Sutherland C (2016) Spatial capture–recapture models allowing Markovian transience or dispersal. Popul Ecol 58:53–62
Royle JA, Fuller AK, Sutherland C (2018) Unifying population and landscape ecology with spatial capture–recapture. Ecography 41:444–456
Schwarz CJ, Arnason AN (1996) A general methodology for the analysis of capture-recapture experiments in open populations. Biometrics 52:860–873
Seber GAF (1965) A note on the multiple-recapture census. Biometrika 52:249–260
Sollmann R, Gardner B, Belant JL (2012) How does spatial study design influence density estimates from spatial capture-recapture models? PLoS One (in press)
Sun CC, Fuller AK, Andrew Royle J (2014) Trap configuration and spacing influences parameter estimates in spatial capture-recapture models. PLoS One (in press)
Sutherland C, Fuller AK, Royle JA (2015) Modelling non-Euclidean movement and landscape connectivity in highly structured ecological networks. Methods Ecol Evol 6:169–177

Summary and Outlook

6

Here we are, in the last chapter of this book. Congratulations to you who have made it so far. Now, let us briefly review what we have talked about in this book and think about what we can do in the future.

6.1 Summary

It is always fascinating but meanwhile challenging to understand spatially structured population dynamics. Rooted from metapopulation models, modern Bayesian statistical models have been developed to understand more or less the BIDE processes while accounting for observation errors. In this book, we talked about four families of models from relatively simple occupancy models all the way to advanced spatial capture-recapture models. Occupancy models use detection/non-detection data to inform occupancy status; for dynamic occupancy models, two underlying processes are considered, persistence and colonization, which nonetheless are consequences of the BIDE processes. N-mixture models use count data to inform abundance; for dynamic N-mixture models, we could either separate movement (I and E) from local population growth (B − D), or gain (B + I) from loss (D + E), but it is still challenging to separate the entire BIDE processes with count data only. Spatial IPMs allow us to separate the entire BIDE processes by jointly analyzing count and capture-recapture data. We can also use spatial IPMs to understand how BIDE processes differ between groups of animals with different ages, body mass, or genetic characteristics, in a population. Spatial capture-recapture models push it even further to link individual habitat use with the BIDE processes by explicitly modeling uniqueness of and interactions between individuals. With these models, we should be well prepared to test some cool hypotheses in population ecology and make accurate predictions regarding the spatial structures and dynamics of natural populations.

Q. Zhao, *Bayesian Analysis of Spatially Structured Population Dynamics*, Analysis and Synthesis 253, https://doi.org/10.1007/978-3-031-64518-1_6

6.2 Outlook

There is still great room to further extend the models introduced in this book. Opportunities lie in the fact that there is a boom in data collection, by citizen scientists, by using advanced technologies such as unmanned aerial vehicles, generating not only big but also deep data. How to leverage the strengths while overcoming the weaknesses of different data sets remains a central question in statistical ecology. There is also a great need to consider uniqueness of and interactions between individuals in explaining and predicting spatial structures and dynamics of populations with analytical models.

6.2.1 Data Integration

From the models introduced in this book, we can see two things. On one hand, it has been really amazing to see how much information we can "squeeze" from relatively simple detection/non-detection or count data. For example, some of the advanced N-mixture models allow movement not only related to distance but also driven by density and environmental covariates. On the other hand, we can easily see, and we should never forget, the value of capture-recapture data: only by bringing in capture-recapture data in spatial IPMs could we separate the entire BIDE processes. Eventually, it is a good idea to combine data from unmarked populations covering large spatiotemporal extents with individual-level data containing strong demographic information.

In this book, we have illustrated data integration in integrated population models. The idea of data integration is very flexible and is way beyond what we have seen in this book (Pacifici et al. 2017; Miller et al. 2019; Zipkin et al. 2021). For example, people have combined detection/non-detection with count (Zipkin et al. 2017), presence-only with distance sampling (Farr et al. 2021), or telemetry with spatial capture-recapture data (Chandler et al. 2022; Dupont et al. 2022). Numerous studies have attempted to combine "big" data collected by citizen scientists with "deep" data collected by professional researchers (Robinson et al. 2018; Sun et al. 2019; Zhao et al. 2024). We can imagine, in near future, there will be models that combine data of detection/non-detection, count, capture-recapture, telemetry, and more, from both professional and citizen science.

6.2.2 Individual-Based Modeling

Traditionally, individual-based models are mostly simulation models that derive population-level patterns from individual-level information (Grimm and Railsback 2005). Recently, studies have started to use the idea of individual-based modeling in analytical models to estimate how individual-level processes affect demography (Plard et al. 2015; Gimenez et al. 2018; Fay et al. 2022; Schafer et al. 2022), and eventually population dynamics (Armstrong et al. 2021). Related to the above-mentioned point of data integration, analytical models that leverage bottom-up information from individual-level data and top-down information from population-level data are undoubtedly needed, but have not been well developed (Plard et al. 2019b). Budding directions include the combination of integral projection models (Easterling 1998; Easterling et al. 2000; Messerman et al. 2023) and integrated population models (Plard et al. 2019a), or spatial capture-recapture information with population survey data (Chandler and Clark 2014).

6.3 Final Remarks

Deal with space and time is fundamental in ecology. It will continue challenging ecologists and statisticians to model habitat connectivity, movement, or even more fundamentally, density dependence, with increasing need of improving computational efficiency. Hopefully, this book will serve as a valuable source of information for further development and application of Bayesian ecological modeling, particularly for spatially structured population dynamics.

References

Armstrong DP, Parlato EH, Frost PGH (2021) Incorporating individual variation in survival, reproduction and detection rates when projecting dynamics of small populations. Ecol Modell 455:109647

Chandler RB, Clark JD (2014) Spatially explicit integrated population models. Methods Ecol Evol 5:1351–1360

Chandler RB, Crawford DA, Garrison EP, Miller KV, Cherry MJ (2022) Modeling abundance, distribution, movement and space use with camera and telemetry data. Ecology 103:e3583

Dupont G, Linden DW, Sutherland C (2022) Improved inferences about landscape connectivity from spatial capture–recapture by integration of a movement model. Ecology 103:e3544

Easterling MR (1998) The integral projection model: theory, analysis and application. Chem Rev 96

Easterling MR, Ellner SP, Dixon PM (2000) Size-specific sensitivity: applying a new structured population model. Ecology 81:694–708

Farr MT, Green DS, Holekamp KE, Zipkin EF (2021) Integrating distance sampling and presence-only data to estimate species abundance. Ecology 102:e03204

Fay R, Authier M, Hamel S, Jenouvrier S, van de Pol M, Cam E, Gaillard JM, Yoccoz NG, Acker P, Allen A, Aubry LM, Bonenfant C, Caswell H, Coste CFD, Larue B, Le Coeur C, Gamelon M, Macdonald KR, Moiron M, Nicol-Harper A, Pelletier F, Rotella JJ, Teplitsky C, Touzot L, Wells CP, Sæther BE (2022) Quantifying fixed individual heterogeneity in demographic parameters: performance of correlated random effects for Bernoulli variables. Methods Ecol Evol 13:91–104

Gimenez O, Cam E, Gaillard JM (2018) Individual heterogeneity and capture–recapture models: what, why and how? Oikos 127:664–686

Grimm V, Railsback SF (2005) Individual-based modeling and ecology. Princeton University Press

Messerman AF, Clause AG, Gray LN, Krkošek M, Rollins HB, Trenham PC, Shaffer HB, Searcy CA (2023) Applying stochastic and Bayesian integral projection modeling to amphibian population viability analysis. Ecol Appl 33:e2783

Miller DAW, Pacifici K, Sanderlin JS, Reich BJ (2019) The recent past and promising future for data integration methods to estimate species' distributions. Methods Ecol Evol 10:22–37

Pacifici K, Reich BJ, Miller DAW, Gardner B, Stauffer G, Singh S, McKerrow A, Collazo JA (2017) Integrating multiple data sources in species distribution modeling: a framework for data fusion. Ecology 98:840–850

Plard F, Gaillard JM, Coulson T, Delorme D, Warnant C, Michallet J, Tuljapurkar S, Krishnakumar S, Bonenfant C (2015) Quantifying the influence of measured and unmeasured individual differences on demography. J Anim Ecol 84:1434–1445

Plard F, Turek D, Grüebler MU, Schaub M (2019a) IPM2: toward better understanding and forecasting of population dynamics. Ecol Monogr 89:e01364

Plard F, Fay R, Kéry M, Cohas A, Schaub M (2019b) Integrated population models: powerful methods to embed individual processes in population dynamics models. Ecology 100:e02715

Robinson OJ, Ruiz-Gutierrez V, Fink D, Meese RJ, Holyoak M, Cooch EG (2018) Using citizen science data in integrated population models to inform conservation. Biol Conserv 227:361–368

Schafer TLJ, Wikle CK, Hooten MB (2022) Bayesian inverse reinforcement learning for collective animal movement. Ann Appl Statistics 16:999–1013

Sun CC, Royle JA, Fuller AK (2019) Incorporating citizen science data in spatially explicit integrated population models. Ecology 100:e02777

Zhao Q, Latif QS, Nuse BL, Pavlacky Jr DC, Kilner CL, Ryder TB, Latimer CE (2024) Integrating counts from rigorous surveys and participatory science to better understand spatiotemporal variation in population processes. Methods Ecol Evol 15:1380–1393

Zipkin EF, Rossman S, Yackulic CB, Wiens JD, Thorson JT, Davis RJ, Grant EHC (2017) Integrating count and detection–nondetection data to model population dynamics. Ecology 98:1640–1650

Zipkin EF, Zylstra ER, Wright AD, Saunders SP, Finley AO, Dietze MC, Itter MS, Tingley MW (2021) Addressing data integration challenges to link ecological processes across scales. Front Ecol Environ 19:30–38

www.ingramcontent.com/pod-product-compliance
Ingram Content Group UK Ltd.
Pitfield, Milton Keynes, MK11 3LW, UK
UKHW022252091026
15624UKWH00015B/179
9783031645174